Introduction to Integration Suite Capabilities

Learn SAP API Management,
Open Connectors, Integration Advisor
and Trading Partner Management

Jaspreet Bagga

Apress®

Introduction to Integration Suite Capabilities: Learn SAP API Management, Open Connectors, Integration Advisor, and Trading Partner Management

Jaspreet Bagga
Austin, TX, USA

ISBN-13 (pbk): 978-1-4842-9629-5   ISBN-13 (electronic): 978-1-4842-9630-1
https://doi.org/10.1007/978-1-4842-9630-1

Managing Director, Apress Media LLC: Welmoed Spahr
Acquisitions Editor: Divya Modi
Development Editor: Laura Berendson
Coordinating Editor: Divya Modi
Copyeditor: Kim Burton

Cover designed by eStudioCalamar

Cover image by Bujin Zhao on Pixabay (www.pixabay.com)

Distributed to the book trade worldwide by Apress Media, LLC, 1 New York Plaza, New York, NY 10004, U.S.A. Phone 1-800-SPRINGER, fax (201) 348-4505, e-mail orders-ny@springer-sbm.com, or visit www.springeronline.com. Apress Media, LLC is a California LLC and the sole member (owner) is Springer Science + Business Media Finance Inc (SSBM Finance Inc). SSBM Finance Inc is a **Delaware** corporation.

For information on translations, please e-mail booktranslations@springernature.com; for reprint, paperback, or audio rights, please e-mail bookpermissions@springernature.com.

Apress titles may be purchased in bulk for academic, corporate, or promotional use. eBook versions and licenses are also available for most titles. For more information, reference our Print and eBook Bulk Sales web page at http://www.apress.com/bulk-sales.

Any source code or other supplementary material referenced by the author in this book is available to readers on GitHub (https://github.com/Apress). For more detailed information, please visit http://www.apress.com/source-code.

Paper in this product is recyclable

This book is for those who understand the significance of seamless connectivity and embrace the notion that integration is not merely a technical challenge but a strategic imperative. May the knowledge and insights shared within these pages empower you to leverage the SAP Integration Suite's capabilities beyond cloud integration, leading to enhanced operational efficiency, improved customer experiences, and sustainable growth of your integration platforms.

Together, let's explore the APIs, Integration Advisor, API Management, Trading Partner Management, Integration Assessment, Open Connectors, and Migration Assessment capabilities, transforming challenges into opportunities and laying the groundwork for a connected SAP Intelligent Enterprise.

May this book serve as a stepping stone, providing you with a solid foundation and a roadmap to implement tools and features of integration suites other than cloud integration, which address the complexities of enterprise integration under a single tool kit. Embrace the possibilities, think boldly, and harness the potential of the SAP Integration Suite capabilities to drive meaningful transformation.

I dedicate this book to all the businesses, organizations, and professionals who strive to optimize their SAP integration landscape and unlock the true potential of their digital ecosystems.

Table of Contents

About the Author

Jaspreet Bagga is an executive consultant with expertise in SAP and non-SAP integrations. He is a hands-on SAP architect who does pre-sales, solution architecture, and development work, leads the delivery of complex integration programs, manages global teams, and ensures that successful projects go live and meet goals. Jaspreet has made a lasting impact on more than 73+ global businesses, delivering more than 200 IT projects for Fortune 500 clients and businesses such as Walgreens, McKinsey & Company, Discovery Channel, Aflac, and Siemens, and government entities, such as the State of Nevada and the City of San Diego.

Jaspreet is the author of multiple books on SAP Integration Suite and certification exams. He was also awarded the top podium space for his published work in the *Journal of Research in Innovative Teaching & Learning* and for the Council of Supply Chain Management Professionals (CSCMP).

His latest interests are integrating SAP cloud applications, innovating on the SAP business technology platform, and mentoring SAP community newcomers on Integration topics.

Jaspreet is a thought leader in SAP Integration technologies and graduated from the State University of New York, Buffalo, Engineering School. He was also inducted as an official member of the Forbes Technology Council.

About the Technical Reviewer

Miguel Figueiredo is a passionate software professional with more than 30 years of experience in technical solution architecture. He has a degree in information systems, an MBA from Mackenzie University, and an international MBA in business administration from FIA Business School in partnership with Vanderbilt University.

Miguel gained experience delivering business intelligence solutions for Fortune 500 companies and multiple global corporations. As the SAP HANA Services Center of Excellence leader, he was responsible for evangelizing and best-practice adoption of data management and business intelligence in his region.

He advises companies to maximize value realization in their digital transformation journeys and move to cloud initiatives.

Miguel is dedicated to supporting his family and encouraging the development of good habits for the health of body and mind.

Acknowledgments

After penning *A Practical Guide to SAP Integration Suite: SAP's Cloud Middleware and Integration Solution*, writing a book of this magnitude would not have been possible without the support and contributions of many individuals.

First and foremost, I would like to express my sincere gratitude to my wife for her unwavering support, encouragement, and understanding. Your belief in me and my work has constantly motivated and inspired me.

I would like to extend my heartfelt appreciation to my associate, who has been an invaluable resource in the writing and research process.

Special thanks go to the Apress publishing team, reviewers, and technical experts who generously shared their time, expertise, and feedback to ensure the accuracy and relevance of the content. Your insights and suggestions have significantly enriched the book, and I am deeply grateful for your contributions.

Last, I want to express my deepest appreciation to the SAP Integration community, whose passion, knowledge-sharing, and collaborative spirit have been invaluable throughout my journey. Your expertise and dedication continue to inspire me.

Introduction

Welcome to the second book in this series, which dives deeper into the comprehensive capabilities of the SAP Integration Suite. This volume explores a wide range of functionalities beyond cloud integration, offering you a versatile Swiss knife to address your integration needs. Each chapter focuses on a specific aspect of the suite, providing insights into its unique features and demonstrating how it can be utilized to tackle various business scenarios.

This book, *Introduction to Integration Suite Capabilities: Learn SAP API Management, Open Connectors, Integration Advisor, and Trading Partner Management*, along with its related predecessor, *A Practical Guide to SAP Integration Suite: SAP's Cloud Middleware and Integration Solution*, form a complete learning set encompassing the entire SAP Integration Suite. Together, these two volumes serve as a practical guide to help you get hands-on knowledge of the suite's functionalities and leverage them effectively within your organization.

The first book comprehensively introduced cloud integration, covering theoretical and practical topics such as connectivity, message transformation, orchestration, and monitoring. It served as a foundation, equipping you with the necessary knowledge and skills to kickstart your integration journey.

This second book delves into the remaining capabilities of SAP Integration Suite, expanding your understanding of its vast potential. It begins by exploring APIs, their significance, and their role in modern integration landscapes. It then introduces the suite, giving you a holistic view of its architecture, components, and key features.

The subsequent chapters provide detailed explanations and practical examples of the various SAP Integration Suite components. From the Integration Advisor, which assists you in designing and implementing integrations efficiently, to SAP API Management, which empowers you to create, manage, and govern APIs effectively, each chapter sheds light on a specific functionality.

Additionally, you explore SAP Trading Partner Management, an essential tool for managing complex partner ecosystems, and the SAP Integration Assessment, which helps you evaluate your integration landscape and identify areas for improvement. The book also covers SAP Open Connectors, which simplify the integration with third-party applications, and Migration Assessment, a critical consideration when transitioning from legacy integration technologies.

This book aims to equip you with the tools and insights to make the most of the SAP Integration Suite by providing in-depth knowledge, practical examples, and best practices for each capability. Whether you are an integration consultant, a business owner, or an IT professional, this comprehensive learning set enables you to confidently navigate the Integration Suite's complexities and achieve seamless integration across your organization.

These two books form a cohesive resource, guiding you through every facet of the SAP Integration Suite. By studying the books in any order or independently and applying the principles and techniques covered, you will become proficient in harnessing the suite's capabilities and transforming your integration landscape into a strategic asset.

Embark on this journey of discovery, and let this complete learning set empower you to unlock the full potential of the SAP Integration Suite.

CHAPTER 1

■ ■ ■

Introduction to APIs

Welcome to the world of APIs, a fundamental building block for modern software development and integration. In today's digital era, where businesses rapidly adopt cloud technologies, APIs enable seamless communication and data exchange between different systems.

Application Programming Interfaces, commonly known as APIs, provide a standardized way for different applications to interact. They allow developers to expose certain functionalities of their applications and make them available to other applications or developers. This enables faster and more efficient business process development, integration, and automation.

APIs have become ubiquitous, powering everything from social media platforms to e-commerce websites, and have revolutionized how we interacts with the technology. They enable businesses to leverage the power of the cloud and connect their applications with third-party systems and services to create more comprehensive solutions.

This chapter explores the basics of APIs, their key features, and the various types of APIs available. It also delves into the importance of APIs in modern software development and integration. It provides some real-world examples of how APIs are being used to drive digital transformation across industries.

1.1 The Server

The discussion of APIs frequently centres on conceptual abstractions. Let's ground ourselves by starting with a physical object: the server. Simply, a server is a large computer. It is quicker and more powerful, yet it still contains all the same components as the laptop or desktop, you use for work. Servers typically lack a monitor, keyboard, and mouse, giving them an unwelcoming appearance. IT professionals link to them remotely to work on them (imagine a remote desktop-style connection).

There are several uses for servers, some deliver emails, while some store data. Web servers are the type that users interact with the most. When you visit a website, these servers provide a web page.

Websites are designed to cater to users' strengths. As humans, we possess an incredible ability to extract meaning from visual cues by connecting them with our past experiences and taking appropriate actions. Therefore, when you come across a small box on a web page labeled *First Name* it's instantly clear that it's prompting you to provide the term you use for legal identification.

What happens when you have a time-consuming task, such as copying 1,000 clients' contact details from one website to another? You would love giving this task to a machine to be completed fast and correctly. Sadly, the qualities that make websites ideal for people also make them challenging for computers to operate.

An API is an answer. An API is a programing layer that exposes a website's data into something a computer can understand. Similar to how a person may access and modify data by loading pages and submitting forms, a machine can view and edit data through it.

© Jaspreet Bagga 2023
J. Bagga, *Introduction to Integration Suite Capabilities*, https://doi.org/10.1007/978-1-4842-9630-1_1

1.1.1 What Is an API?

An Application Programming Interface (API) allows the two programs to interact with each other. An API specifies how a developer should ask an operating system or other software for services and how to provide data in various contexts and channels.

Using an API, any data can be shared. Function calls made up of verbs and nouns implement APIs. For example, on a real estate website, one API might list available houses according to location. At the same time, a second API offers the most recent interest rates, and a third API provides a mortgage calculator.

Companies can give external third-party developers, business associates, and internal divisions inside their organizations' access to the data and functionality of their applications by using an API. A specified interface enables goods and services to interact and use one other's data and capability. To interact with other goods and services, developers merely use the interface; they are not required to understand how an API is implemented. Over the past ten years, the use of APIs has increased so that many of the most well-liked web services today would only be viable.

Integrating multiple systems and services becomes remarkably simpler through the use of Application Programming Interfaces (APIs). They offer a convenient way to access data or functionalities from other programs, streamlining the process of system integration. Moreover, new apps that can integrate with existing programs or systems can be created via APIs.

Web developers frequently utilize APIs to gain access to services like social media platforms, payment gateways, and other web services. They are frequently used in mobile development to access functions like the GPS or camera on the device.

Due to their ability to connect various systems and services, APIs have become important for enterprises. Since APIs may be used to automate operations and increase productivity, this can be especially helpful for business that utilize a variety of applications or services.

1.1.2 How an API Is Used

An API is a collection of defined parameters that describe how computers or programs interact with one another. The web server and an application are separated by APIs, which serve as a middle layer to handle data transit between the two.

The following explains how an API functions.

1. To obtain information, a client application launches an API request (also known as an API call). Using a request verb, headers, and occasionally a request body, this request is processed by an application and sent to the web server via the API's Universal Resource Identifier (URI).

2. When the API receives a legitimate request, it calls the external application or web server.

3. The API receives a response from the server containing the data requested.

4. The API sends the data to the application that made the initial request.

APIs act as a bridge between the infrastructure that provides services and the applications that use them. Unlike user interfaces, which are created for human interaction, APIs are specially created for computer or application use. Due to this intermediary position, APIs automatically offer security benefits. To defend against server attacks, API endpoints isolate the application that consumes them from the underlying infrastructure. API calls also typically include authorization credentials, and access can be further restricted by an API gateway to reduce security risks. Additionally, several security precautions—such as cookies, HTTP headers, and query string parameters—are used during data transfer.

To understand the API, let's look at an example.

You go to a restaurant, and a waiter comes to you and asks for the order. You see the menu card and give the order to the waiter. The waiter then conveys your order to the chef. The chef starts preparing your food and performs various functions to make your food delicious. Finally, the chef prepared your food. The chef then gives your order to the waiter, and it finally comes to you, which you enjoy.

In this story, the waiter is the middleman or connection between you and the chef. Similarly, the APIs are the middleman between the user and the server, as depicted in Figure 1-1.

Figure 1-1. *API acting as the middleman between two parties* [1]

An API contains software-building tools and communication protocols that facilitate interaction between systems. An API may be for a database system, operating system, computer hardware, or a web-based system.

An API helps to communicate the browser with the database, as shown in Figure 1-2.

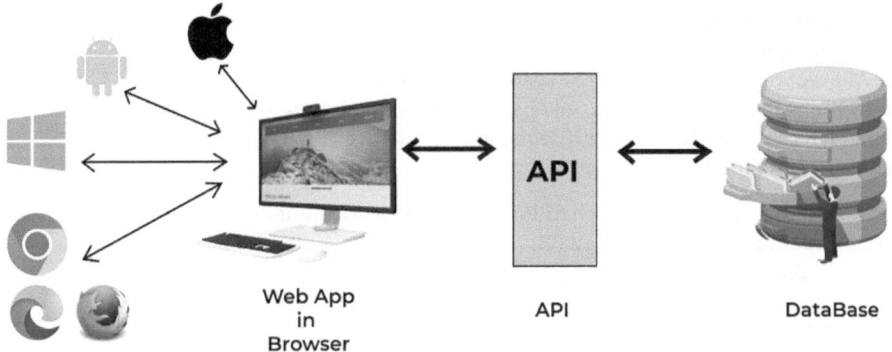

Figure 1-2. *API architecture connecting two systems*

---

[1] https://img.freepik.com/free-photo/male-chef-uniform-hat-apron-adding-spice-into-food-kitchen_176474-9019.jpg?w=1060&t=st=168542859 3~exp=1685429193~hmac=c2e404744d2335b8995f30d4a63de1161c15596d072b32 0798e1c10a14a37244

1.1.3 The Relevance of APIs

An API can help you manage current tools or create new ones while streamlining the process. The following are some of the principal advantages of APIs.

- **Enhanced cooperation**: Enhanced cooperation among various applications is made possible with APIs. A typical business uses more than 1,000 cloud applications, but many are not connected. APIs make it possible to integrate these platforms and applications seamlessly. This integration allows businesses to streamline processes and improve collaboration, enhancing productivity and performance. Without APIs, businesses may experience information silos and a lack of connectedness, impairing their overall performance.

- **Improved innovation**: APIs allow businesses to network with new business partners, offer fresh services to their existing clientele, and enter new markets. This flexibility can boost innovation, generate profits, and accelerate digital transformation. Stripe (Financial Service Company), for example, began as an API with just seven lines of code. Since then, the business has partnered with some of the biggest corporations in the world, grown to offer loans and business cards, and most recently achieved a valuation of $36 billion. This demonstrates how APIs can help company's growth and development in previously impractical ways.

- **Revenue from data**: APIs may make money by granting people access to priceless digital information. While some businesses might initially offer APIs for free to attract developers and possible business partners, they can eventually sell access to the API. For example, AccuWeather introduced a self-service developer portal to provide a selection of API packages. The business attracted 24,000 developers within ten months of its start, sold 11,000 API keys, and created a vibrant developer community. This example shows how APIs may generate valuable revenue for the businesses.

- **Security**: By placing an additional layer of security between the data and the server, APIs can improve the security of data transmission. Developers can further enhance API security by constructing API gateways to control and authenticate traffic and employing TLS encryption, tokens, signatures, and best practices for API management. As a result, the API is less vulnerable to assaults, and the data being communicated is secured against unauthorized access.

You have seen why APIs are needed, but which APIs can you use or encounter in your daily routine? The next section discusses some common examples of APIs.

1.1.4 Examples of APIs

APIs have developed into a valuable component of modern organizations because they enable organizations to expand access to their resources while preserving security and control. You might come across the following well-known instances of APIs.

- **Taking payments from outside parties**: For instance, an API powers the increasingly common Pay with PayPal feature on e-commerce websites. This enables customers to purchase online without disclosing personal information or allowing illegal access.

- **Comparing travel bookings**: Online travel agencies gather thousands of flights and present the most affordable options for each date and location. A program's users can access the most recent information about hotel and airline availability using APIs, which makes this service possible. APIs significantly minimize the time and effort required to search for available flights or lodging by enabling an independent interchange of data and requests.

- **All-purpose logins**: A well-known API example is a feature that allows users to connect to websites using their LinkedIn, Instagram, or Gmail profile login information. By utilizing an API from one of the more well-known services, any website can use this practical feature to instantly verify the user's identity, saving them the time and headache of creating a new profile for each website service or new membership.

- **Maps**: Google Maps is a popular example of a good GPS API. The app uses additional APIs, capabilities and its core APIs to show users directions or areas of interest on static or live maps. When planning routes or tracking moving objects, such as a transport vehicle, you can interface with the Maps API using geolocation and other data layers.

- **Twitter**: A writer, a special ID, a message, the timestamp of when it was sent, and geolocation metadata are just a few of the fundamental descriptive features present in every tweet. Developers can access public tweets and replies on Twitter, and they may post tweets using the company's API.

- **E-commerce**: This refers to the practice of carrying out commercial operations, such as buying and selling products online. One service that is virtually iconic of online shopping is PayPal. Moreover, e-commerce trademark marketplaces for both Amazon and Facebook exist.

1.1.5 Types of APIs

Most APIs today are web APIs that make data and functionality of an application available online. The four primary web API types are as follows.

- **Open APIs** are free and can be accessed through the HTTP protocol. They are also called *public APIs*, and have established API endpoints and request and response standards.

- **Internal APIs** are not visible to outside users. These internal development teams' efficiency and communication can be enhanced by using these secret APIs, which aren't accessible to people outside the business.

- **Partner APIs** are made available to or by strategic business partners. These APIs are often available to developers via a public API developer console in self-service mode. To use partner APIs, they must still undergo an onboarding procedure and obtain login information.

- **Composite APIs** combine several data or service APIs. These services enable developers to make a single call to many endpoints. In a microservices design, where information may be needed from multiple sources for a single activity, composite APIs are helpful.

1.1.6 Types of API Protocols

With the popularity of online APIs, many protocols have been created to give consumers a set of guidelines outlining acceptable data types and requests. Some API protocols make it easier to share standardized information.

1.1.6.1 REST APIs

There are no formal standards for (Representational State Transfer), a collection of web API architecture concepts. A REST API, commonly called a RESTful API, requires that its interface follow certain architectural guidelines. Although the two standards can be combined to create RESTful APIs, they are typically seen as rival ones.

The header can send audio or image files, but those files remain encoded as text and often in JSON or XML format. API developers can employ formatting options to make text transmissions useful in more complex ways. REST APIs are still utilized for many tasks; however, adhering to REST and HTTP's limitations requires ingenuity.

Furthermore, while the REST protocol offers many recommendations for how HTTP transfers should be written, there is no mechanism to enforce them. Because both the API provider and user must build their applications to withstand problematic queries and unexpected data payloads, this form of API can be less dependable in some circumstances. Web APIs can give data in a platform-agnostic fashion due to the lack of enforcement, giving API users more freedom in utilizing the data they get.

1.1.6.2 SOAP APIs

Users can transmit and receive data using SMTP and HTTP thanks to the API protocol known as SOAP (Simple Object Access Protocol), which was designed with XML. With SOAP APIs, it is simpler to transfer data across apps or software parts created in various languages or operating systems. The requirement of metadata files characterizing requests in SOAP makes exchanges more predictable, which is one of its main advantages. In contrast to REST, which is stateless, it also supports stateful requests. A more standardized protocol enables SOAP APIs to transmit data over more than simply HTTP and to successfully exchange more sophisticated data. For informed users, SOAP's usage of service interfaces rather than straightforward URL-based organization might also enhance discoverability. In general, SOAP works better for more complex systems where stability precedes performance or usability by a wide audience. It is commonly used in financial services and complex corporate software programs like Salesforce.

1.1.6.3 RPC APIs

Responses from RPC requests can be either XML or JSON files. In a few significant aspects, it varies from SOAP and REST APIs. Unlike other protocols, which call data resources, this one calls a method, as the name would imply. A RESTful API produces a document, but an RPC server responds with either an error message explaining why the function didn't work or a confirmation that it did. In other words, an RPC API handles actions, whereas a REST API handles resources.

Another significant distinction is that an RPC's URI identifies the server, whereas a REST API's routes display both the server and the query arguments. Since most businesses do not wish to make their RPC APIs accessible to the general public, they are rarely made public. RPC servers surpass the stateless/stateful gap between REST and SOAP since calling one modifies the server's state. RPC APIs are typically private because of the requirement for a high degree of privacy and trust between vendors and customers. Hence, compared to REST or SOAP APIs, search capabilities and predictability are less critical for RPC APIs, although reliability and performance are.

Distributed client-server applications are among the RPC API use cases that are most frequently encountered. Front-end developers can access server functions without worrying about specifics like opening and closing connections or processing inputs. Payloads are light and restricted to the parameters for the methods being called. Client applications can be hosted independently of the remote backend server that houses the functions and data since methods can be invoked from distant locations. A multi-threaded process can execute on the distant server without affecting the client application, which makes task threading simpler than calling methods locally.

1.1.6.4 GraphQL APIs

Even though it isn't truly a different protocol, GraphQL is a unique query language with recommended usage. Like a REST API, GraphQL transmits text data as part of each request's payload using HTTP, but it takes a different method.

Each endpoint in a REST API represents a separate data structure. You must match your objectives to the current schema and make the proper endpoint calls to obtain the desired information. GraphQL APIs normally only have one endpoint, which can accommodate an almost infinite number of data structures. Although the query can mix such fields in any sequence, the API user must know the accessible data fields. Data is returned in the form of the schema provided by the query, and queries are sent as the payload of HTTP POST requests.

In contrast to the rigid routing restrictions of a REST API, GraphQL offers customers a great deal of freedom inside a single query. Moreover, it can be difficult to cache data, putting the onus of keeping consistent query syntax on API consumers to obtain comparable data. Also, the user must know the available fields to create a query when using a GraphQL API. Compared to a comparable REST API, a GraphQL API requires more detailed bespoke documentation, and fewer methods are available to automate it.

1.1.7 Web Services, APIs, and Microservices

A web service is software that may be used by entering a web address. Web services are inherently dependent on a network. In actuality, any web service is an API because it exposes the data and functionality of an application. Yet not all APIs are web services.

A low-level programming language, such as JavaScript, may have been used to develop the API, traditionally used to describe an interface connected to an application. The modern API often follows REST principles, uses the JSON format, and is constructed for HTTP. As a result, it produces developer-friendly interfaces that are readily accessed and comprehended by applications created in Java, Ruby, Python, and many other languages.

With an API, users can expand on the features and information of another application. Given that they are present in applications as diverse as Spotify and Yahoo Finance, one might think of them as essential components that can be used to create nearly anything.

Thanks to API frameworks, developers can carry out operations that aren't dissimilar from routine activities. Consider the scenario where a server takes your order, puts it in, and then returns with the order once it is ready. This systematic procedure yields the desired result: a delicious supper. An online illustration might be someone registering for a new e-commerce site utilizing their Facebook account.

Service-oriented architecture (SOA) and *microservices architecture* are two popular architectural strategies for using APIs.

- **SOA** is a software design approach that separates functionalities into individual services across a network. The functional building blocks of SOA are typically accessible using industry-standard communication protocols through web services. Developers can build these services from scratch, but they are typically built by making legacy system functionality available as service interfaces.

- **Microservices architecture** is an alternative design that breaks a program into smaller, autonomous components is called microservices architecture. Testing, maintaining, and scaling the application is simplified by implementing it as a group of independent services. This approach has become more popular as the cloud computing era has progressed since it allows developers to work independently on one component from the others.

While SOA was an important milestone in the history of developing applications, microservices architecture is built for scalability, giving developers and businesses the flexibility and agility they need to build, update, test, and deploy applications at a detailed level with faster iteration cycles and more effective use of cloud computing resources.

1.1.8 Summary

This chapter overviewed APIs, including their significance in modern software development. The chapter examines the various API types that are accessible, as well as the various API protocols that programmers might employ to build them. Also covered are the requirement for APIs and several instances of their application. The chapter also emphasizes the distinctions between web services, APIs, and microservices. Readers unfamiliar with APIs or wanting to learn more about them and their advantages should start with this chapter.

The next chapter introduces SAP Integration Suite, and you learn how to establish a trial account to work on it and its capabilities.

CHAPTER 2

■ ■ ■

SAP Integration Suite

This chapter introduces SAP Integration Suite, beginning with a summary of the software's main features. The many features and advantages of SAP Integration Suite include Cloud Integration, API Management, Open Connectors, Integration Advisor, Trading Partner Management, and Integration Assessment and Migration Assessment.

Security is a top priority for cloud-based solutions, and with SAP Integration Suite, security is relevant. In this chapter, you learn more about the different security measures SAP has implemented to safeguard your data and systems. The chapter contains a step-by-step tutorial on how to set up a trial account and provision the capabilities you require if you are new to SAP Integration Suite. The chapter also goes over recent changes, how to access SAP Integration Suite, and typical installation difficulties.

2.1 Introduction to SAP Integration Suite

An organization may easily combine different applications, data sources, and business processes across its whole landscape with the help of SAP Integration Suite, a cloud-based integration platform offered by SAP. It offers a solitary platform for all integration kinds, including cloud-to-cloud, cloud-to-on-premises, and on-premise.

SAP Integration Suite provides a wide range of integration features, including API administration, data integration, process integration, event-driven integration, and more. It is built on top of the SAP cloud platform. Moreover, it supports several integration standards and paradigms, such as REST, SOAP, and OData.

Organizations may enhance operational efficiency, lower integration costs, and complexity, and optimize business processes and data flow with SAP Integration Suite. It lets businesses link their systems and data sources in real time, distributing accurate and current information throughout the business.

Furthermore, SAP Connectivity Suite offers prebuilt interfaces to several SAP and non-SAP systems, including Salesforce and Microsoft Dynamics. Additionally, it provides a low-code environment that enables developers to create integrations fast and deploy them using drag-and-drop tools and prebuilt connectors.

First, to better understand SAP Integration Suite, let's discuss the SAP Business Technology Platform.

2.1.1 SAP BTP

SAP Business Technology Platform (SAP BTP) is a cloud-based platform that enables businesses to create, expand, and connect SAP and non-SAP applications. It offers a variety of tools and services that simplify businesses to develop and deploy SAP applications on the cloud, and it is a crucial part of SAP's broader cloud strategy.

© Jaspreet Bagga 2023
J. Bagga, *Introduction to Integration Suite Capabilities*, https://doi.org/10.1007/978-1-4842-9630-1_2

The following are key features and capabilities of SAP BTP.

- A cloud-based development environment that dispenses with local infrastructure and enables businesses to create and test SAP applications in the cloud.

- A variety of integration tools and services make it simpler to interface SAP applications with other systems and technologies, such as connectors, data mapping tools, and API management tools.

- Support for well-known programming frameworks and languages, like SAPUI5, Node.js, and Java, makes it simpler for developers to create and implement SAP apps.

- Organizations can create more comprehensive and integrated cloud-based solutions by integrating other SAP cloud services, such as SAP Analytics Cloud and SAP Cloud Integration.

- SAP BTP is the platform for innovating and extending SAP solutions like S/4HANA, SuccessFactors, Ariba, SAP Analytics Cloud, and Concur.

- Built on the Cloud Foundry platform, SAP BTP offers a flexible and scalable environment for creating and deploying cloud-native applications.

- Include prebuilt AI models and interaction with well-known AI frameworks, this is a collection of tools and services for developing and deploying artificial intelligence (AI) and machine learning (ML) applications and models.

Overall, SAP BTP provides a powerful and flexible platform for organizations developing and deploying SAP applications in the cloud.

Companies today need access to real-time data to use cutting-edge technologies and industry best practices inside agile, integrated business processes. Integrating end-to-end processes, whether the solutions are from SAP, partners, or third parties, is a crucial component of SAP's strategy.

Lead-to-cash, source-to-pay, design-to-operate, and hire-to-retire are just a few business scenarios SAP actively promotes integration throughout.

For connected end-to-end business operations spanning SAP and external applications, SAP BTP offers integration options.

Figure 2-1 shows some of the major capabilities of SAP BTP, including automation, analytics, data, and integration.

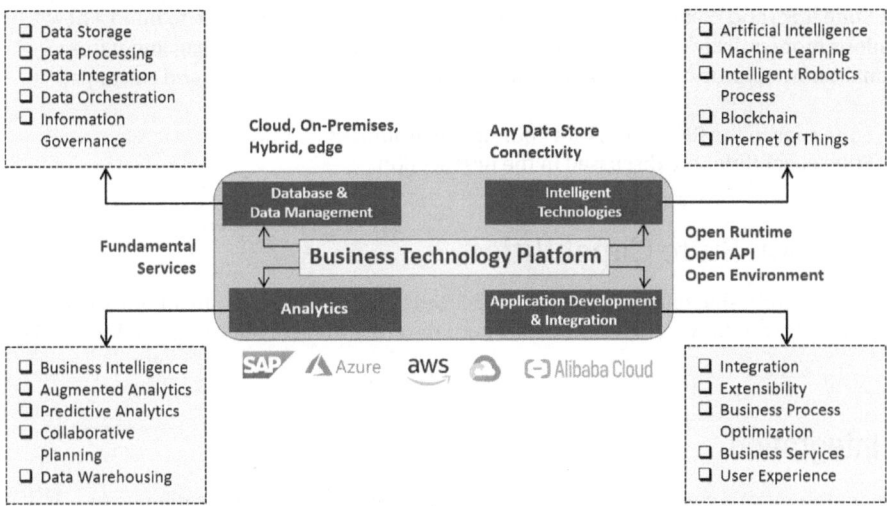

Figure 2-1. *SAP BTP capabilities*

You now know that SAP BTP is a cloud-based platform that enables businesses to create, expand, and connect SAP and non-SAP applications. Next, let's go into more detail about SAP Integration Suite and its capabilities.

2.1.2 SAP Integration Suite Overview

SAP Integration Suite is a collection of tools and services, as shown in Figure 2-2, that SAP offers businesses so they may interface their SAP systems with other enterprise software systems, including customer relationship management (CRM) and enterprise resource planning (ERP) systems. Because of this, companies may share data and procedures throughout their platforms, streamlining workflow and increasing productivity. SAP Integration Suite offers services for managing and monitoring the integration process and solutions for data, process, and message integration.

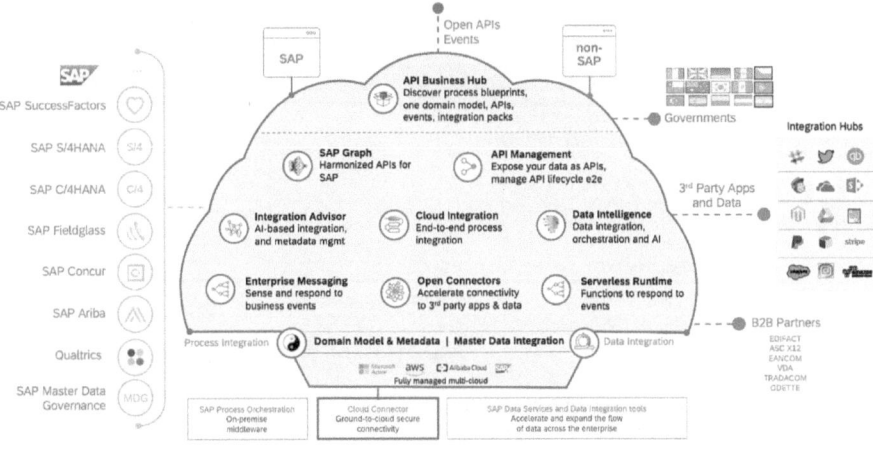

Figure 2-2. *SAP Integration Suite capabilities, source SAP SE*

SAP Integration Suite links and contextualizes processes and data, making it possible to build a new content-rich applications more quickly and with fewer dependencies on IT. With less engagement from integration professionals, new outputs can be produced using prebuilt integration packs and existing investments.

SAP Integration Suite runs in the SAP BTP Cloud Foundry environment.

SAP Integration Suite capabilities are discussed in the next section.

2.1.3 SAP Integration Suite Capabilities

SAP Integration Suite is a platform that offers a range of capabilities to help businesses streamline their operations and optimize their workflows. The following sections describe some of the key capabilities of SAP Integration Suite.

2.1.3.1 Cloud Integration

It is essential to consider the various components that must be harmonized as businesses migrate to the cloud. Businesses must integrate several systems and applications created and maintained using various technology stacks, according to various security standards, and having various business interface requirements. With the Cloud Integration feature of SAP Integration Suite, businesses can effortlessly and rapidly integrate these numerous applications, irrespective of their environment.

Process communications in real-time situations spanning several businesses, organizations, or departments within one organization by integrating SAP and non-SAP, cloud, and on-premises apps.

Figure 2-3 shows the Discover section of Cloud Integration in SAP Integration Suite. The list shown in the image is the prebuilt integration packages available by SAP. For organizations to rapidly and simply connect their SAP and non-SAP systems and applications, the prebuilt integration packages offered by SAP Cloud Integration are extremely important. These prebuilt packages include integration information and templates that have already been established; they can simply be changed to match particular business requirements, cutting down on the time and expense of integration projects.

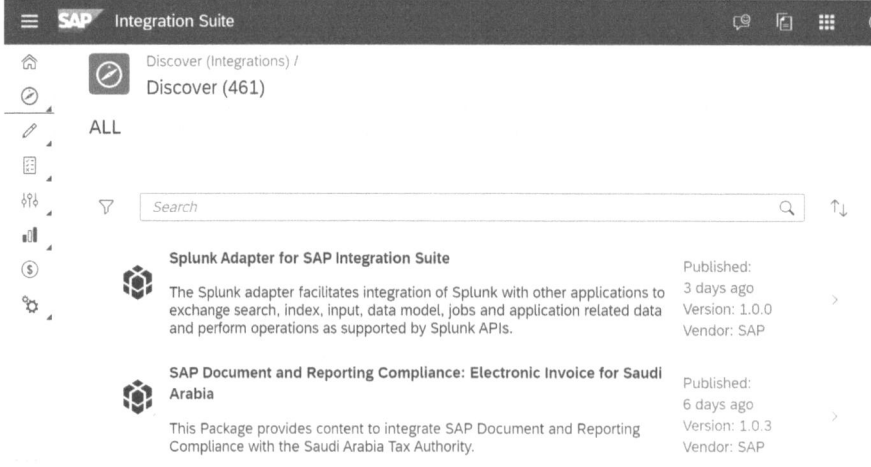

Figure 2-3. *Discover (Integrations) section of Cloud Integration*

2.1.3.2 API Management

The API Management feature of the suite offers a central layer for governing, managing, and metering APIs. It offers a web-based, code-free framework for creating new APIs, managing current APIs, enhancing APIs with security and access restrictions, creating logical groupings using product catalogs, and making APIs available to the community via a developer portal. To commercialize API access, pricing plans can also be connected to APIs. A plug point that can be utilized to enable the development of applications and extensions and the integration of businesses is the main goal of an API-based integration strategy.

Application programming interfaces (APIs) enable access to simple, scalable, and secure digital assets, which may be consumed.

Figure 2-4 shows the Discover API section of API Management in SAP Integration Suite. Discover APIs offers a wide selection of prebuilt API packages, SAP API management is crucial because it enables organizations to link their systems and applications quickly and effectively. These packages offer various integration scenarios, including integrating with cloud-based applications and services and linking SAP systems to non-SAP systems.

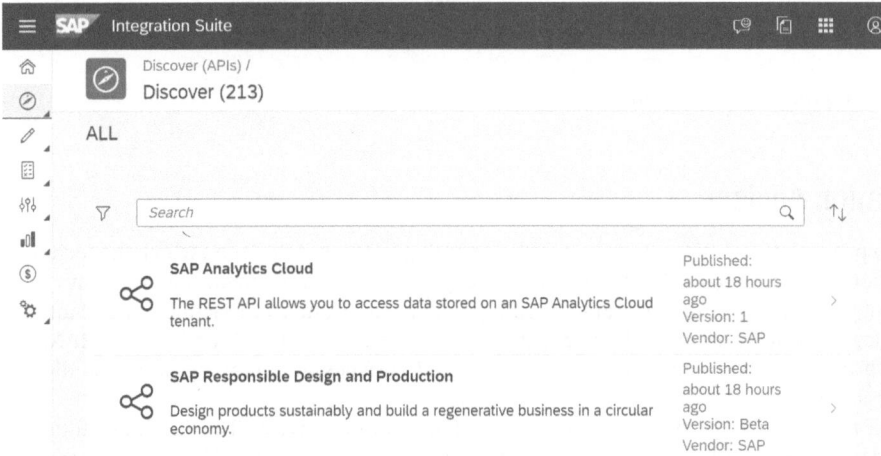

Figure 2-4. Discover section of SAP API Management

Businesses' efforts to expedite their digital transformation can be sped up by using Discover APIs Packages, which can considerably minimize the time and effort needed to design and deploy APIs. These easily adaptable packages allow organizations to customize them to meet their unique needs.

SAP API Management delivers tools and services that make it simple to manage and govern APIs throughout their life cycle in addition to prebuilt packages. This covers attributes that help guarantee APIs' dependability, scalability, and security, such as API documentation, testing, versioning, and security.

2.1.3.3 Open Connectors

With standardized authentication, error handling, and connectivity protocols, the Open Connectors feature of SAP Integration Suite offers third-party connectors. Rather than learning about the technology required to link with third-party systems, this enables developers to concentrate on creating business integrations.

Utilizing prebuilt connectors, create seamless interfaces with more than 160 non-SAP applications.

Figure 2-5 shows the SAP Open Connectors dashboard.

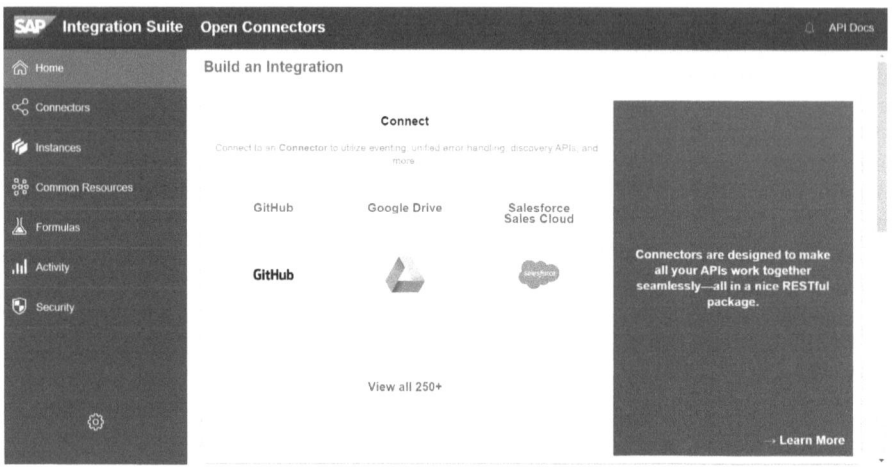

Figure 2-5. *SAP Open Connectors dashboard*

2.1.3.4 Integration Advisor

Any implementation project's integration process includes mapping business interfaces, which the technical developer and functional consultant do. But doing so can be expensive and time-consuming. You may hasten the creation of your business interfaces and mappings with the assistance of SAP Integration Suite's Integration Advisor feature. It features prebuilt information for EDI industry standards and SAP S/4HANA.

Create runtime artifacts quickly, construct business-oriented interfaces and mappings more rapidly, and put forth a lot less work.

Figure 2-6 shows the prebuilt type systems available in SAP Integration Advisor. The SAP Integration Advisor's prebuilt systems are important because they offer a quick and simple approach to developing linkages without major customization or configuration. These prebuilt systems are examples of regularly integrated technologies and systems businesses use, including SAP S/4HANA, Salesforce, and RESTful APIs.

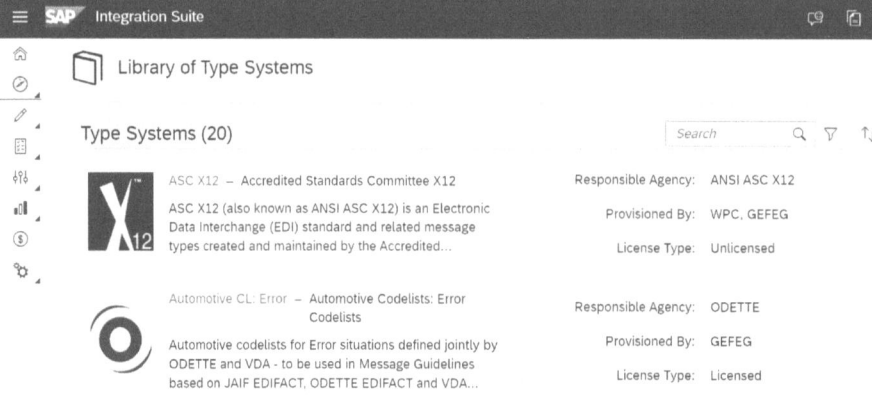

Figure 2-6. *SAP Integration Advisor dashboard*

2.1.3.5 Trading Partner Management

Trading Partner Management is a collection of products and services offered by SAP that lets organizations use SAP systems to manage their interactions with trading partners. This consists of tools for finding and integrating new trading partners, creating, and administering trading contracts, and keeping track of and controlling trading partners' performance. Businesses can enhance their supply chain processes and maximize their interactions with trading partners using Trading Partner Management. It is included in SAP Integration Suite, which offers several tools and services for linking SAP systems with other enterprise software systems.

Figure 2-7 shows the Trading Partner Management dashboard, where you can create a company profile, trading partners, agreement templates, and agreements. Each option has its own function, which is discussed in the upcoming chapters.

Figure 2-7. *SAP Trading Partner Management dashboard*

2.1.3.6 Integration Assessment

As an initial step, SAP defined Integration Assessment and templates. Integration Assessment comprises important ISA-M master data, such as defining integration domains, integration styles, use case patterns, and common key integration technology characteristics.

The Integration Assessment uses ISA-M (Integration Solution Advisory–Methodology) to assist enterprises in assessing and enhancing their integration capabilities. It is a systematic process with several steps to evaluate an organization's current integration landscape and offer suggestions for streamlining its integration procedures.

Figure 2-8 shows the Integration Assessment dashboard.

Figure 2-8. *SAP Integration Assessment dashboard*

2.1.3.7 Migration Assessment

A new feature called Migration Assessment has been added to the capabilities of SAP Integration Suite. This feature helps you estimate the technical work needed for the conversion process by evaluating how various integration scenarios can be moved. With this version, SAP is considering switching from its on-premises SAP Process Orchestration system to its SAP BTP-based Integration Suite to take advantage of the next generation of integration capabilities.

The Migration Assessment helps in evaluating the technical effort necessary for the migration process by analyzing potential migration paths for various integration scenarios.

Figure 2-9 shows the Request page in Migration Assessment, where two options are available: Data Extractions and Scenario Evaluations. Data Extractions is a procedure whereby the application collects data from a connected SAP Process Orchestration system, such as integration scenarios, mapping objects, communication channels, and other design time artifacts, and prepares the data for evaluation. While, Scenario Evaluations determine that after the information has been collected, the integration scenarios are assessed using the predetermined rules.

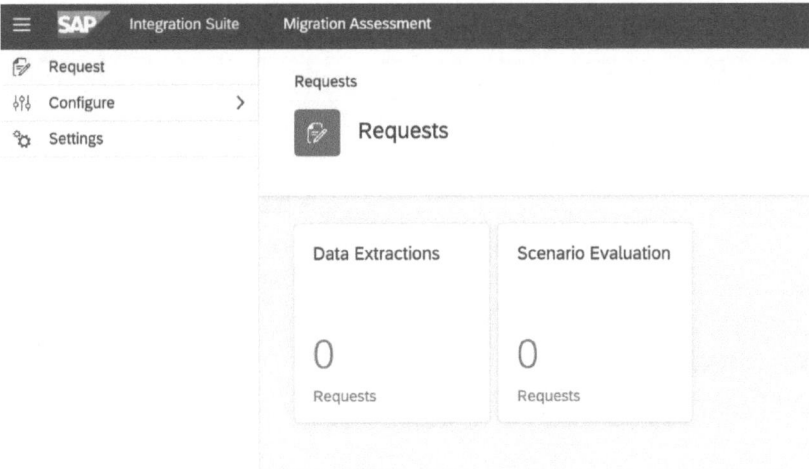

Figure 2-9. *Migration Assessment*

Having discussed the capabilities of SAP Integration Suite in the previous section, it's now time to take a closer look at some of the key features that make this cloud-based solution such a powerful tool for enterprise integration. From data mapping and transformation to real-time monitoring and analytics, SAP Integration Suite offers a range of features that enable organizations to optimize their business processes and achieve greater efficiency and agility. Let's dive in and explore some of these features in more detail.

2.1.4 Features of SAP Integration Suite

SAP Integration Suite includes various features that enable businesses to integrate their SAP systems with other enterprise software systems. The following are some of the key features of the SAP Integration Suite.

- **Build integration scenarios**: With Cloud Integration, you can explore, create, and run scenarios for process integration from beginning to end.

- **Manage APIs**: With API Management, you can find, create, and regulate APIs for API consumers.

- **Enable application connectivity**: To easily combine SAP and non-SAP applications, use Open Connectors to select prebuilt connectors from a catalog.

- **Put integration packs to use**: Using prebuilt integration packs for end-to-end scenarios like hire to retire, lead to cash, procure to pay, SuccessFactors, Jira, ServiceNow, and many more, you can accelerate the construction of integrations. For example, partner-developed integration packages are also available to copy/ configure. (Figure 2-10 shows this used in Cloud Integration.)

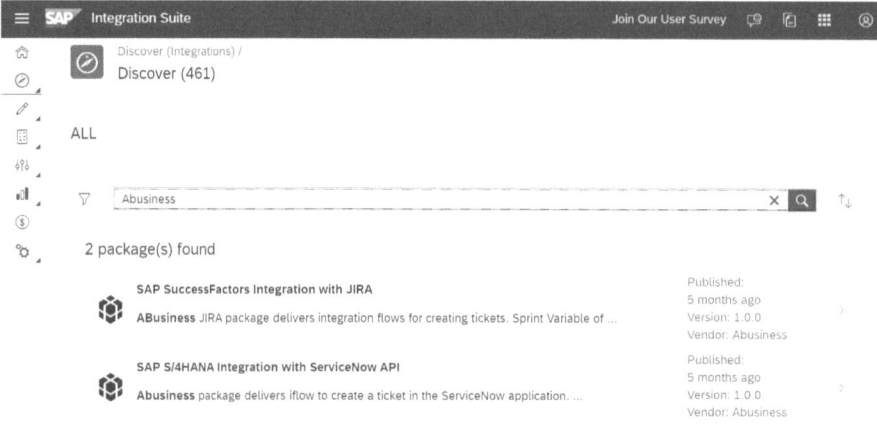

Figure 2-10. *Discover integration*

- **Implement mapping more simply**: Using Integration Advisor, create user interfaces and mappings for commercial applications using crowdsourcing and machine learning.

- **Organize trading partner integrations**: Using Trading Partner Management, construct, preserve, carry out, and keep track of business-to-business scenarios.

Now that you've explored some of the key features of SAP Integration Suite, it's clear that this cloud-based solution offers a wealth of tools for streamlining enterprise integration and improving business performance. But what are the benefits of these features? How can they help organizations achieve their goals and stay ahead of the competition? The next section looks at the benefits of SAP Integration Suite, from enhanced productivity and agility to improved data quality and customer satisfaction. Let's dive in and see how this robust solution can drive success for your organization.

2.1.5 Benefits of SAP Integration Suite

Using SAP Integration Suite to integrate SAP systems with other enterprise software systems has several advantages. The following are some of the main benefits.

- **Improved productivity:** By connecting SAP systems with other enterprise software systems, businesses may simplify operations and do away with the need for manual data input and process management.

- **Enhanced data accuracy:** Businesses may guarantee that their data is consistent and up-to-date by sharing data between SAP systems and other enterprise software systems. This increases the correctness of their information and decision-making.

- **Better customer service:** By connecting SAP systems with CRM systems, companies may give their clients more precise and timely information, enhancing the customer experience.

- **Increased flexibility:** By integrating SAP systems with other enterprise software systems, firms can more readily adjust to changing business needs and processes, enabling them to be more flexible and responsive to changing market conditions.

- **Reduced costs:** By automating procedures and removing the need for labor-intensive manual data entry and process management, firms can save money overall.

Having explored the benefits of SAP Integration Suite, it's important to take a step back and consider the bigger picture. To truly harness the power of this cloud-based solution, it's essential to understand the landscape in which it operates. This landscape encompasses a range of components and architecture that work together to enable seamless integration across your enterprise. From the Integration Suite cockpit to the Cloud Integration gateway, many moving parts must be understood and optimized to ensure the best possible outcomes.

The next section dives into SAP Integration Suite landscape and explores the key components that make it such a powerful tool for enterprise integration. Let's get started.

2.1.6 SAP BTP Integration Suite Landscape

SAP Integration Suite is a collection of tools and services that allow businesses to link their software with SAP BTP and other platforms and systems. To link SAP BTP with other platforms and systems, it offers a variety of connection methods, including messaging, APIs, data integration, and process integration.

SAP Integration Suite may be used to integrate various systems and platforms, including on-premises and cloud-based systems, and it is intended to be versatile and scalable. It is a crucial element of SAP BTP and was created to assist enterprises in streamlining their business operations and making the most of SAP BTP's capabilities.

Most of the customers use one non-production subaccount for DEV and QA. As a good practice or based on the industry-specific regulations, if you need to maintain different environments for data/business process separation then you can create N number of subaccounts for development, test, and production usage.

In Figure 2-11 and Figure 2-12, you can see the setup of the SAP BTP Integration Suite landscape for non-production and production environments.

- Integration Suite – Dev

- Integration Suite – Test

- Integration – Prod

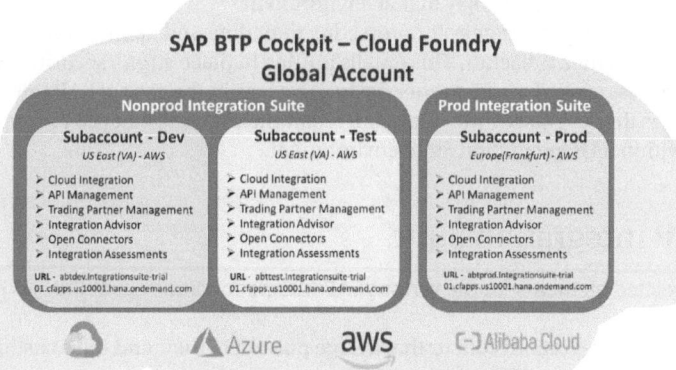

Figure 2-11. *SAP BTP Integration Suite subaccounts can be created in any of the four major cloud providers: AWS, GCP, Azure, and Alibaba Cloud*

> **Note**   SAP has discontinued issuing new Neo (SAP Data Center) Cloud Integration Enterprise licenses for new customers, and encourages existing customers to migrate to Cloud Foundry–based cloud integration. *Contact your SAP Account Executive for more details. Existing customers can continue to use their Neo-based licenses.
>
> For example, the API Management service is not available for new customers as a stand-alone BTP service anymore. It is only available in Cloud Foundry environment as part of the Integration Suite.

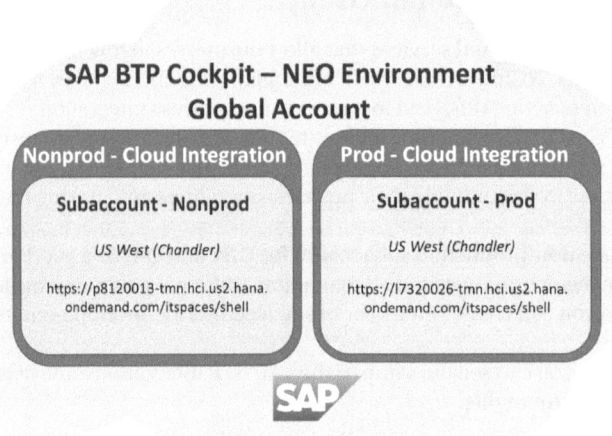

Figure 2-12.   *Subaccount in SAP Data Center—Neo*

The SAP Integration Suite landscape is a complicated and multifaceted environment, and it needs to be carefully managed to maintain optimal performance. But security is another important factor that should be at the forefront of your mind in today's increasingly digital environment.

Organizations must adopt a proactive strategy for protecting their data and applications considering the rising number of cyber threats and data breaches. This entails putting in place strong security measures that guard against unauthorized access, data theft, and other destructive acts in the case of SAP Integration Suite.

The next section examines the several security tools and recommendations that can assist protect your data's security and privacy within SAP Integration Suite environment.

2.1.7 Security: SAP Integration Suite

An overview of all security-related elements for the various SAP Integration Suite capabilities is provided in this security reference.

Customers using SAP Cloud Integration, concur that a large portion of their and their customers' sensitive data is processed and kept in a non-customer infrastructure.

An integration platform's primary function is to act as a hub for messages that might include sensitive client data. These messages must, first and foremost, be shielded from prying eyes and unlawful access.

Technically speaking, the integration platform is created as a cloud-based clustered and containerized integration platform. Different areas of the platform handle messages handled via integration flow from various clients (referred to as tenants).

In terms of CPU, data storage, and user access, tenants handling integration flows from various clients are carefully isolated from one another.

Figure 2-13 shows the high-level architecture of SAP Integration Suite key components.

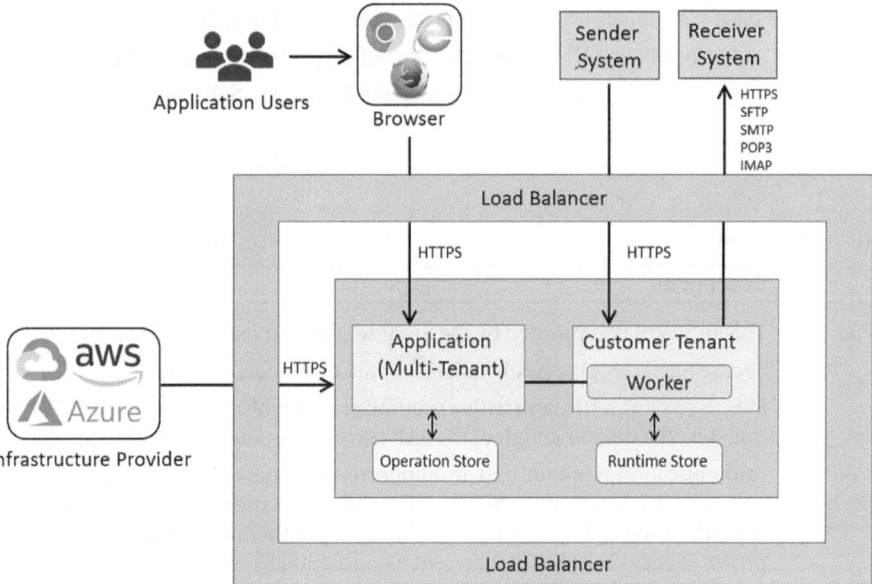

Figure 2-13. *High-level technical architecture of SAP Integration Suite*

2.1.7.1 Transport and Message-Level Security

The transfer of any data, whether between internal or distant components, may be secured using techniques like encryption.

Based on the specified transport protocol, the linked remote systems communicate during a scenario. These protocols provide a variety of choices to safeguard the sent data from unwanted access. The content of the exchanged messages can also be secured using digital encryption and signatures in addition to security at the transport level.

2.1.7.1.1 Transport Layer Security

Transport layer security in SAP Integration Suite refers to the steps taken to protect communication between two systems or applications at the transport layer. The third layer of the OSI (Open Systems Interconnection) paradigm is the transport layer, which oversees ensuring dependable end-to-end communication between two systems.

In SAP Integration Suite, transport-level security can be configured and enabled for various integration scenarios, such as web service calls, file transfers, and messaging. It is typically used to secure the communication between SAP systems and external systems, or between different SAP systems.

Depending on the underlying transport protocol, each adapter enables you to configure a certain security level.

Table 2-1 describes some of the adapter-specific transport-level security options that can be configured per your integration requirements.

■ **Note** SAP Integration Suite does not support unsecure communication at the transport level and follows the security best practices. For example, Inbound HTTP messages are not supported, and you must configure client/server SSL certificates for secure end-to-end communications. SAP Integration Suite supports only HTTPS communications.

Table 2-1. *Transport Layer Security*

Transport Protocol	Description
SFTP (Secure Shell File Transfer Protocol)	This protocol is supported by the SFTP sender and receiver adapters.
	Secure Shell (SSH) securely transmits data over open networks.
	SSH uses symmetric keys with a minimum length of 128 bits to secure FTP transfer. The default length of the SAP asymmetric key is 2048 bits.
	Authentication methods that are supported include: Username/password authentication, where the SFTP server utilizes the username and password to confirm the calling component. Public key authentication, where the SFTP server checks the caller component using a public key.
	When asymmetric key pairs are used, SFTP also ensures that the participants are utilizing only authorized public keys.
HTTP(S) (Hypertext Transfer Protocol Secure)	This protocol is supported by all adapters that allow communication over HTTPS, including the SOAP Adapter, IDoc Adapter, and the HTTP Adapter.
	Communication can be secured by using TLS. In this case, a symmetric key with a minimum length of 128 bits is employed (which is technically enforced). 2048 bits is the standard SAP asymmetric key length.
	Moreover, receiver adapters provide main propagation based on the SAP Cloud Connector.
	Simple authentication options include using client certificates, user credentials, or OAuth, depending on the sender or receiver adapter being used.
SMTP (Simple Mail Transfer Protocol)	Along with the Mail adapter, these protocols support email communication.
POP3 (Post Office Protocol)	Transport encryption is supported through the STARTTLS extended operation.
IMAP (Internet Message Access Protocol)	To authenticate against the email server, you can provide a username and password in plain text or encrypted form; the latter option is only available if the email server supports it.

2.1.7.1.2 Message-Level Security

Message-level security in SAP Integration Suite describes the steps taken to protect a message's payload (or message content) at the application layer. The OSI model's seventh layer, the application layer, oversees offering services that are particular to the program being utilized.

There are many approaches to establish message-level security, such employing encryption, digital signatures, or access control systems. These precautions guarantee that the message is only seen to the intended recipient and that the message's contents were not altered during transmission.

Message-level security can be set up and made available in SAP Integration Suite for a number of integration situations, including messaging, file transfers, and web service calls. Typically, it is used to protect sensitive or private data being communicated between SAP systems, either within themselves or with other systems.

In addition to the choices for transport-level security, you may also secure communication at the message level, where the contents of the sent and received messages can be secured using digital signatures and encryption. This may be accomplished using a variety of security standards, as listed in Table 2-2.

Utilizing certain integration flow security features, you may set message-level security choices.

Table 2-2 shows the supported standards and algorithms include.

Table 2-2. *Message-Level Security*

Standards	Security Features
PKCS#7/CMS Enveloped Data and Signed Data	Message content encryption and decryption
	Payload verification and signing
PKCS#7/CMS Enveloped and Signed Data	Payload encryption, decryption, signature, and verification
Pretty Good Privacy (PGP)	Message content encryption and decryption
	Encryption, decryption, message signing, and verification
XML Signature	Payload verification and signing
WS-Security	Signing and validating the SOAP body

2.1.7.2 Access SAP BTP Cockpit

As a license owner for enterprise accounts your S-user ID (SAP ID) can have access to multiple global accounts. This access is provided by SAP during license purchase to the account owners S-user ID, the account owner must create the access for administrators/developers to continue the further configuration by adding the right role collections/roles/authorizations as per the desired personas of BTP Integration Suite users. For trial BTP account users, you already have access to the global account.

2.1.7.2.1 Role Collection for Global Accounts

Two key predefined role collections are assigned to you when managing the SAP BTP cockpit. They are connected to account administrator personas and are useful for managing BTP global accounts.

Navigate to Security ➤ Role Collections in the SAP BTP cockpit. As shown in Figure 2-14, the role collections available are Global Account Administrator and Global Account Viewer.

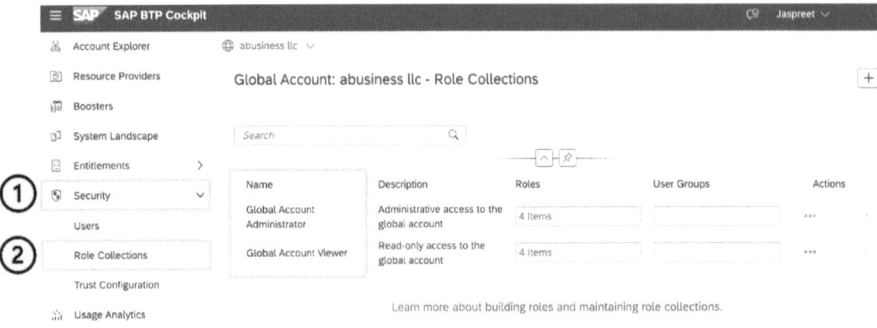

Figure 2-14. *Global Account Role Collections*

After you have been assigned a global BTP account administrator role, you should be able to see all the global accounts for which you have assigned access, as shown in Figure 2-15.

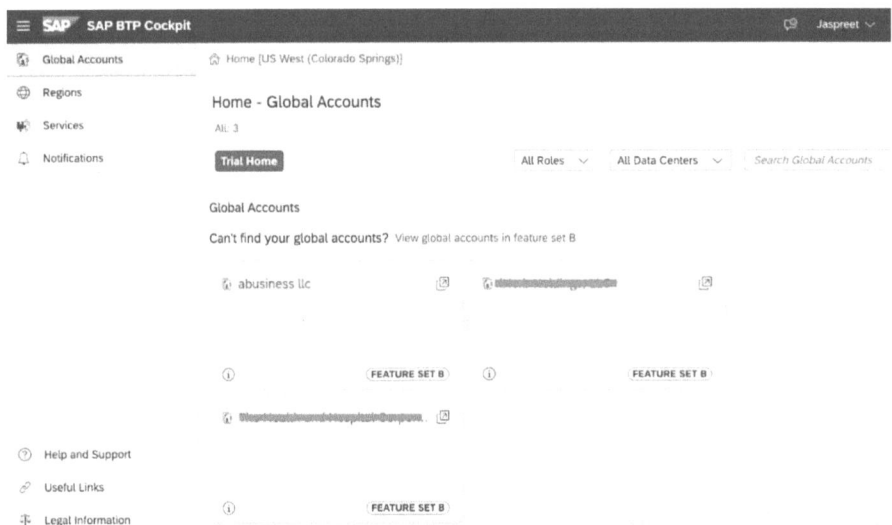

Figure 2-15. *SAP BTP landing page*

2.1.7.2.2 Role Collections for Subaccounts

There are three key predefined role collections that are assigned to you when managing SAP BTP subaccounts. These role collections are connected to subaccount administrator personas useful for managing BTP subaccount. Here are the role collections for subaccount administrator.

Navigate to Security ➤ Role Collections in the SAP BTP cockpit, as shown in Figure 2-16. The following are role collections associated with a subaccount.

- Subaccount Administrator

- Subaccount Service Administrator

- Subaccount Viewer

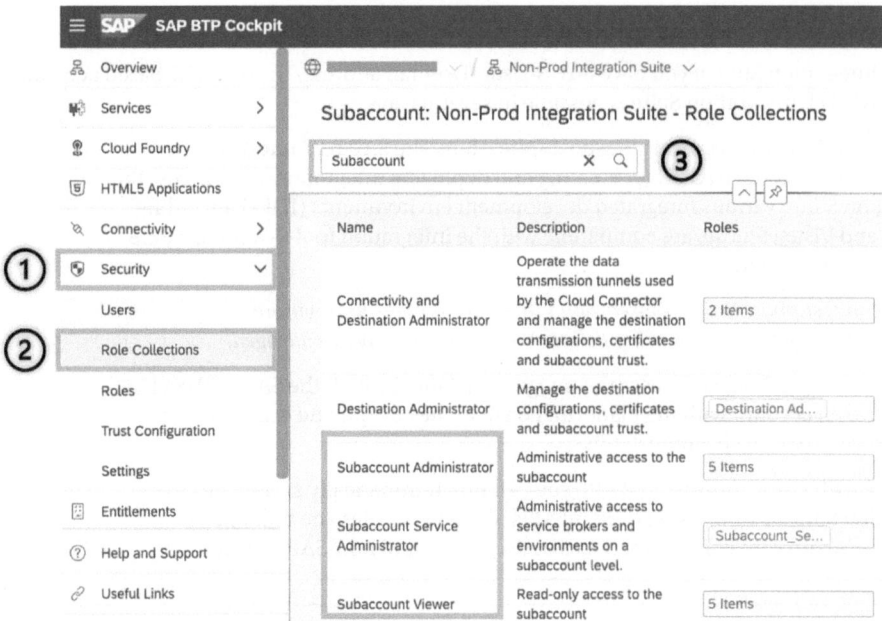

Figure 2-16. *Role collections for subaccounts*

For enterprise license users, once you have access to the global account and have assigned the subaccount administrator role to yourself/administrators, you can then create BTP Integration Suite-specific subaccounts.

Figure 2-17 shows two Neo subaccounts created inside the "abusiness llc" global account. This set up is mandatory and is the same for Cloud Foundry subaccounts as well.

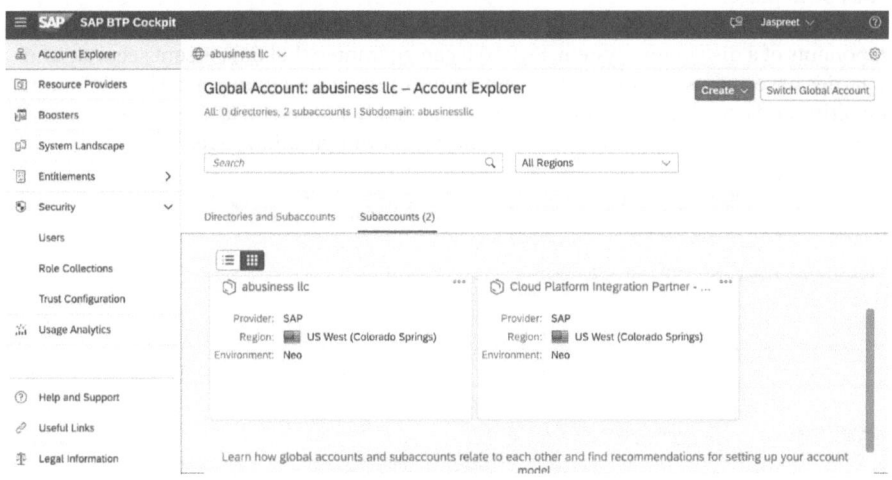

Figure 2-17. *Navigation from Global Account to Subaccount*

2.1.7.3 Access SAP BTP Integration Suite

You can access SAP Integration Suite if you have SAP S-User ID/email address/Universal ID based account. Accessing the SAP BTP Integration Suite is possible in several ways.

- **SAP Integration Suite browser-based application:** The SAP BTP integration tool, a graphical user interface (GUI) for creating and deploying integrations, can access the Integration Suite. Various integrated development environments (IDEs), including Eclipse and Visual Studio, are compatible with the integration tool as a stand-alone application or as a plug-in.

  SAP does not support Eclipse-based cloud integration plug-ins anymore. Developing integration flows using Eclipse Web-ID is strongly discouraged.

- **SAP BTP cockpit:** You can access the Integration Suite through the SAP BTP cockpit if you have an account with the platform. Log in to the cockpit and search for "integration" tools to accomplish this.

- **SAP BTP API:** Using the SAP BTP API, you can programmatically access the Integration Suite as well. Using RESTful APIs, this API enables you to monitor, automate, set up, and control interfaces between SAP and non-SAP services.

Note   Access to SAP BTP Integration Suite capabilities (Cloud Integration, Integration Advisor, Integration Assessment, etc.) can also be federated through customers own custom identity provider (Microsoft AD/Azure, Okta, etc.) for increased enterprise-level security by making use of SSO/MFA.

Establishing and monitoring trust relationships between various distributed system components is known as *trust configuration* in SAP BTP. To enable safe and dependable communication across these elements—which could include cloud-based applications, on-premises systems, and third-party services—trust configuration must be in place.

The Trust Configuration feature in SAP BTP is used to enable secure and dependable communication between various components of a distributed system. SAP BTP can guarantee that data is sent securely and that only authorized parties have access to it by creating and monitoring trust relationships. This maintains the confidentiality, integrity, and accessibility of data in the system and helps defend against cyberattacks, data breaches, and other security threats. Figure 2-18 shows the Trust Configuration page.

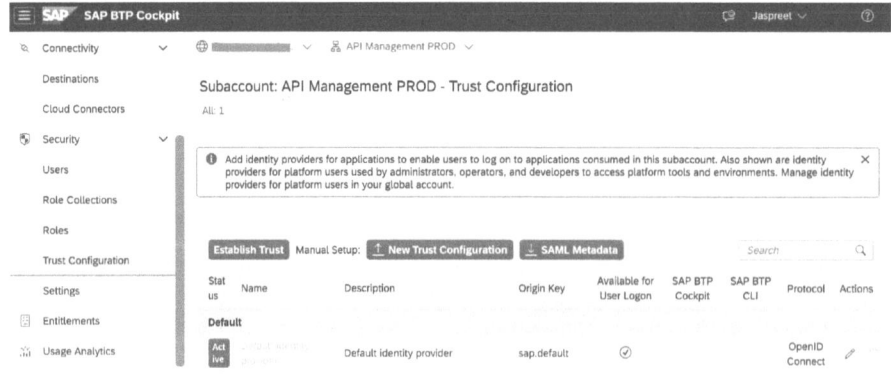

Figure 2-18. *SAP Integration Suite Trust Configuration*

If you have configured trust configuration security to use a custom identity provider instead of the default SAP identity provider, you can access Integration Suite capabilities. Log in using your enterprise credentials, as shown in Figure 2-19.

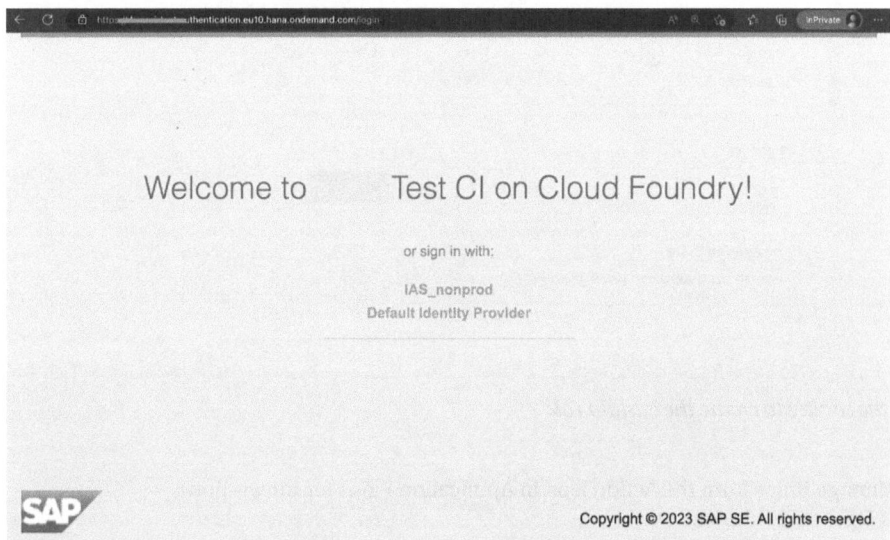

Figure 2-19. SAP Integration Suite login screen, custom identity provider enabled

2.1.7.4 Create Custom Roles

To give users specifically fine-grained access to different features and functionality within the platform, you create custom roles. For example, you should create a role that allows users to access certain API endpoints but not others.

The following goes through the procedure.

1. Go to the subaccount for your Cloud Foundry environment in the SAP BTP cockpit.

2. In the left-hand window, select Service Marketplace. To create a custom role, select Integration Suite, or you can also choose any other service, as shown in Figure 2-20.

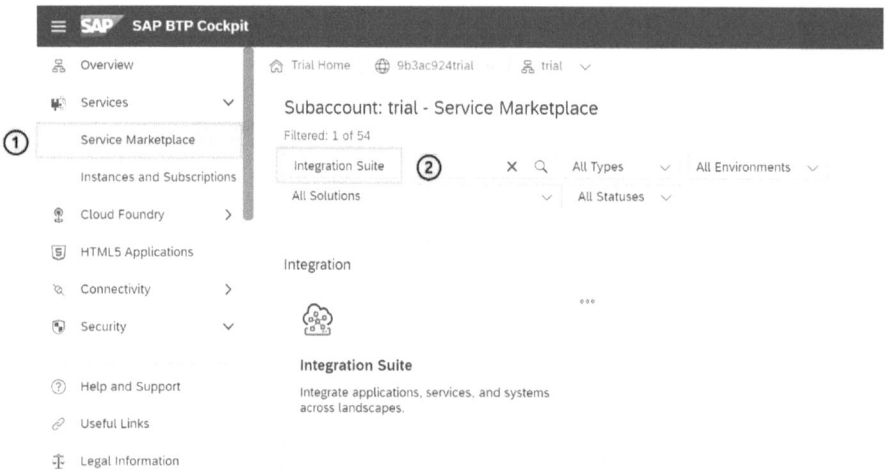

Figure 2-20. *Select the service to create the custom role*

3. Select **Manage Roles** from the Action icon in Application Plans for Integration Suite, as Figure 2-21 shows.

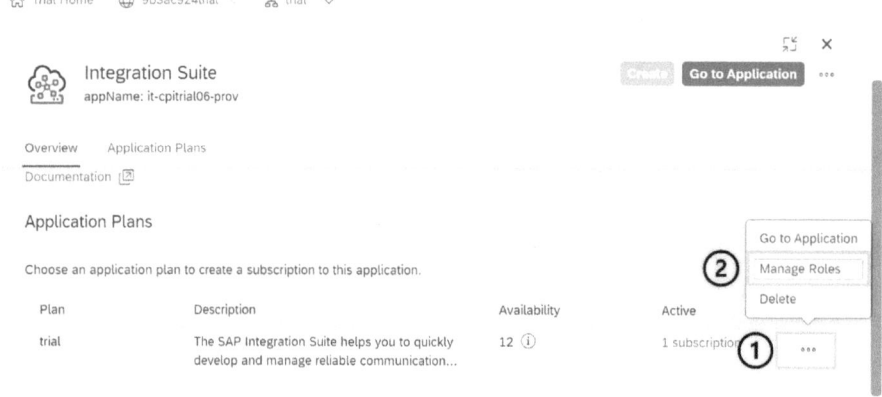

Figure 2-21. *Select Manage Roles*

4. Click + to add a new custom role, as shown in Figure 2-22.

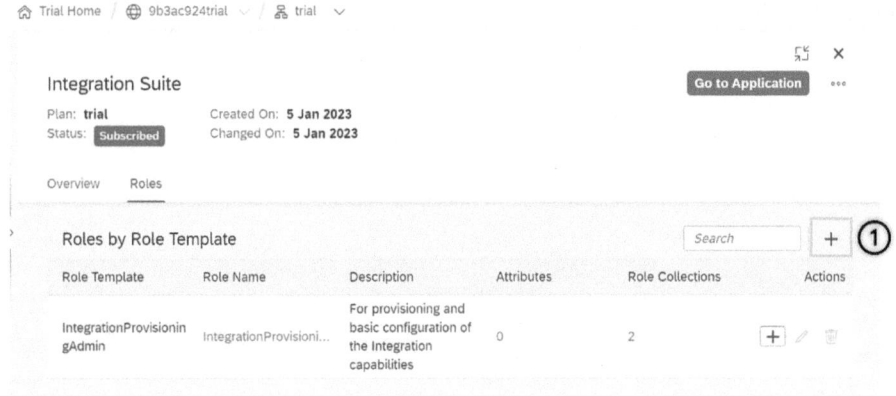

Figure 2-22. *Click + to add new role*

5. Enter information in the Create Role dialog box, as shown in Figure 2-23.

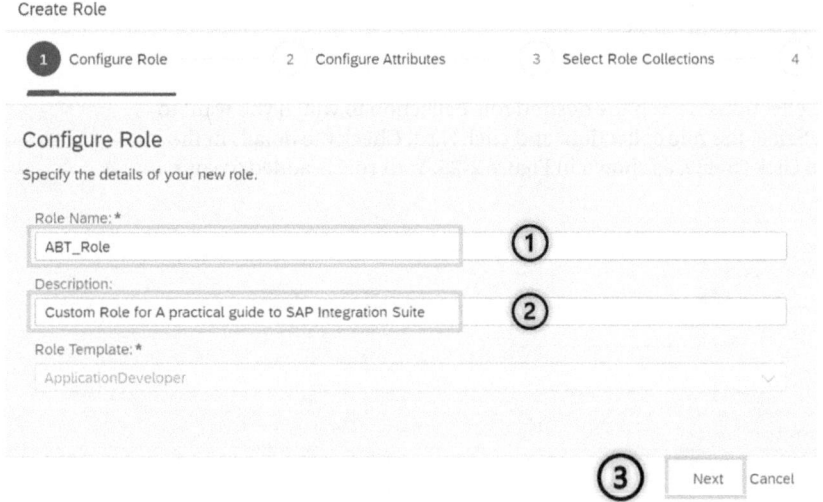

Figure 2-23. *Create Role (Configure Role)*

6. In Configure Attributes, keep the Source value Static for the custom role attributes. Give the values of the attributes in Values tab and press Enter, as shown in Figure 2-24.

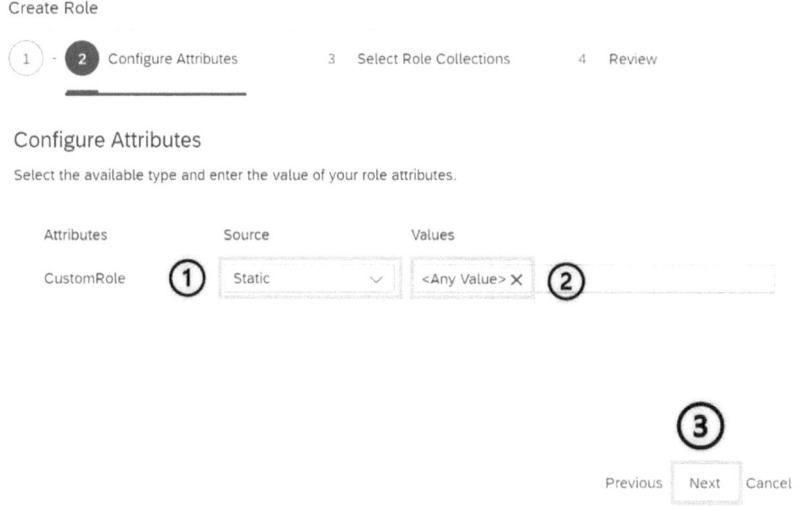

Figure 2-24. *Create Role (Configure Attributes)*

7. In Select Role Collections, search the desired role collection to which you want to assign the role. Select the role collection, and click Next. Check the details in the Review tab, and click Create, as shown in Figure 2-25. Your role is added to your role collection.

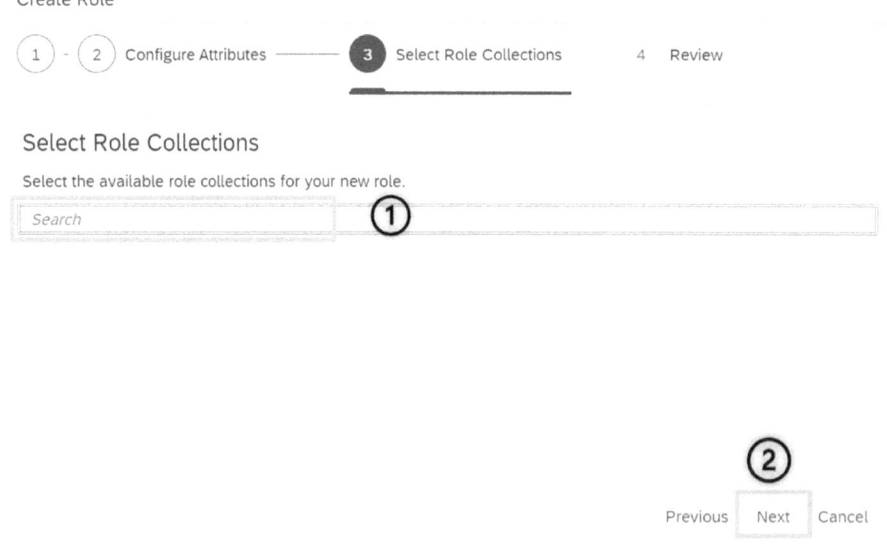

Figure 2-25. *Create Role (Select Role Collections)*

8. The custom role created in the list, as shown in Figure 2-26.

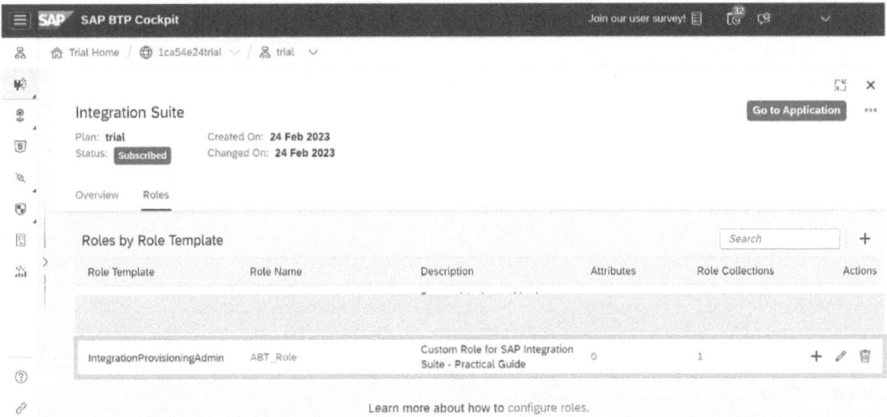

Figure 2-26. Custom role

9. Add the created role to the role collection, as shown in Figure 2-27.

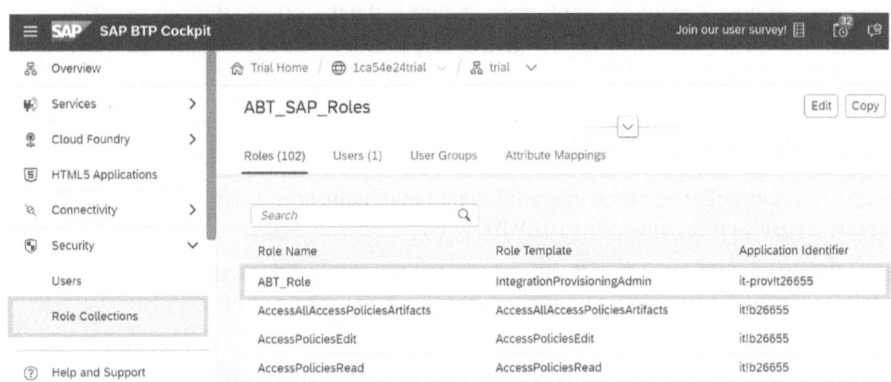

Figure 2-27. Role added to role collection

10. Assign role collection to the user, as Figure 2-28 shows.

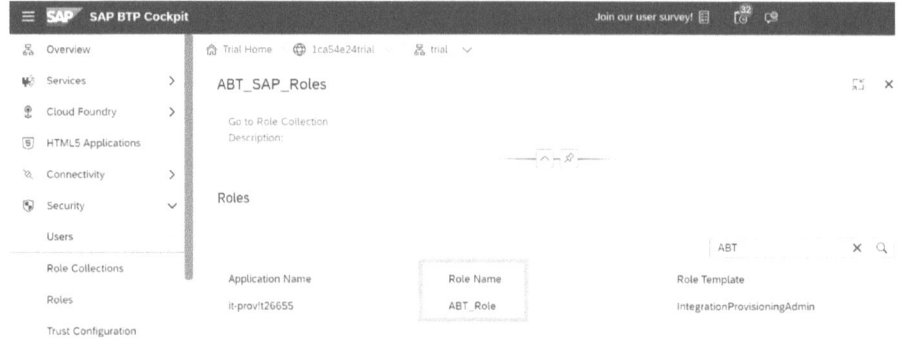

Figure 2-28. *User role collection*

After exploring the process of developing custom roles in SAP Integration Suite, let's now focus on role-based access control, a related but no less significant security feature. Organizations may make sure that only users with the proper degree of permission have access to critical data and apps by establishing particular roles and assigning them to users inside SAP Integration Suite environment. To properly enhance security, however, it's also vital to organize these roles into collections that correspond to the particular needs and demands of your business. Creating bespoke roles is simply the first step in this process. The process of developing custom role collections in SAP Integration Suite is covered in next section, along with how important it is to protect user privacy and security.

2.1.7.5 Create Custom Role Collection

In SAP BTP, creating a custom role collection is an essential step in managing access to resources and data. The following steps create a custom role collection in SAP BTP.

1. Every capability has a different role collection that must be assigned. You can create a role collection and assign it to the user after giving the correct user and roles.

2. Go to **Security ➤ Roles Collection**, and click the + on the right panel, as shown in Figure 2-29.

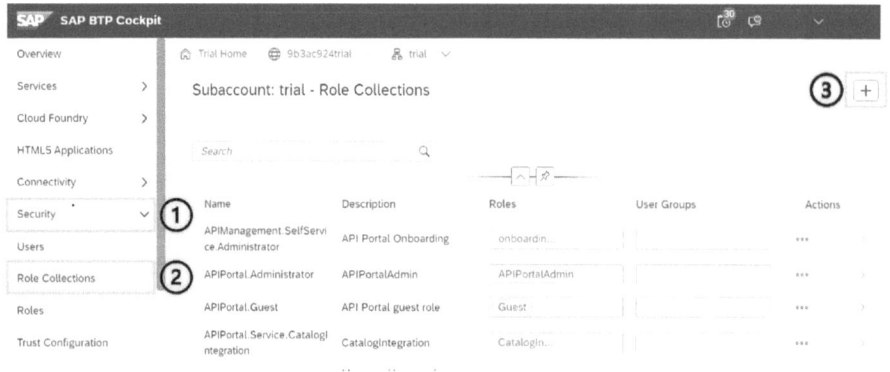

Figure 2-29. *Create custom role collection*

3. To create a new role collection, enter a name and description (optional), and click **Create**, as shown in Figure 2-30.

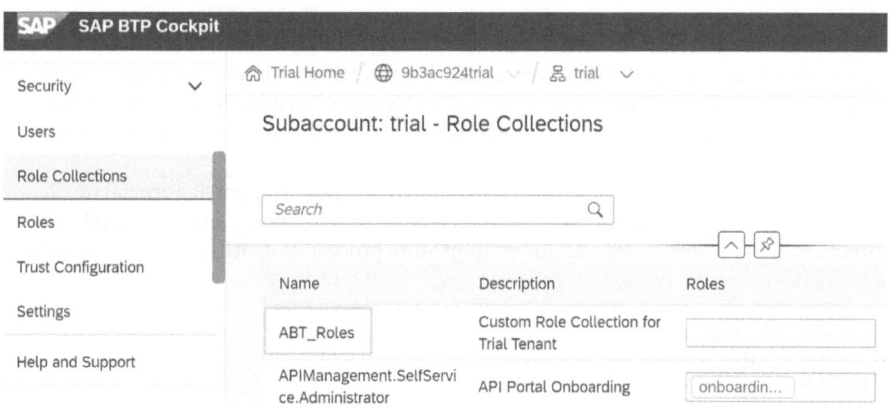

Create Role Collection

Name:* ABT_Roles

Description: Custom Role Collection for Trial Tenant

Create   Cancel

Figure 2-30. *Basic information for new role collection*

4. The custom role collection created is added to the list of role collections, as shown in Figure 2-31.

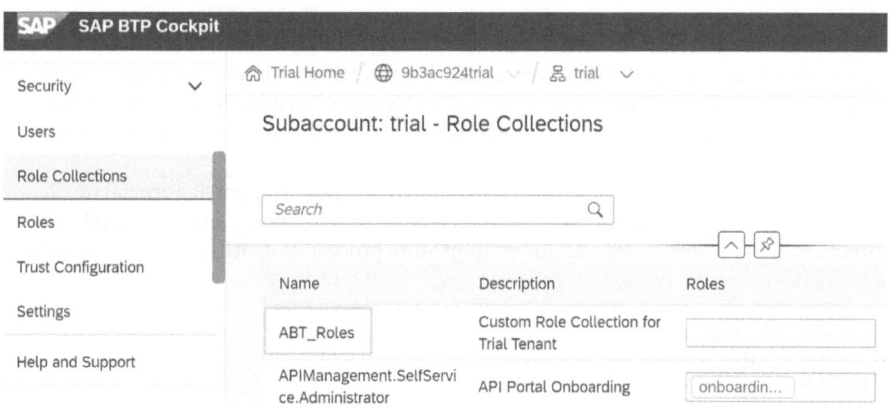

Figure 2-31. *ABT_Roles added to list of role collections*

5. To add the roles to the custom role collection, click the roles collection you have created. Open it in Edit mode. For the user, search your username and assign it to the role collection. For the roles, search the roles which you want to assign. Click **Save** after assigning the roles, as shown in Figure 2-32.

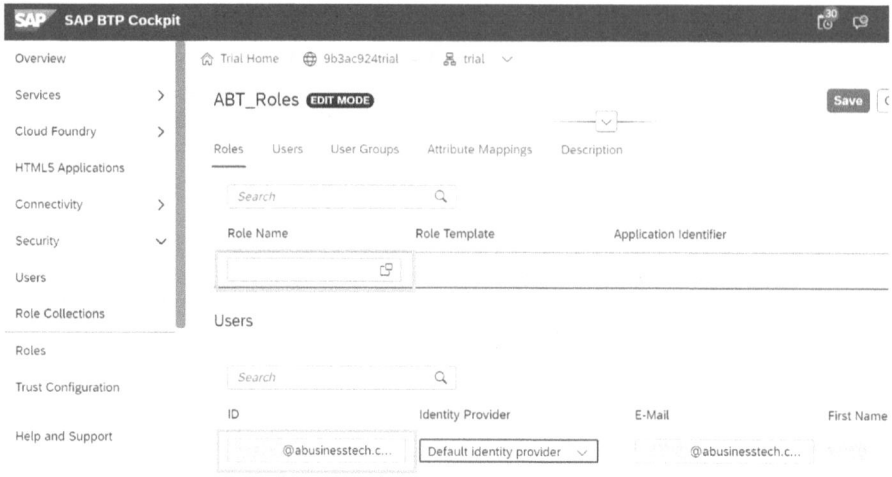

Figure 2-32. *Assign role collections to user*

6.   Assign the different role collections according to your requirement to the user.

Assigning these roles to certain users is the next crucial step in protecting your SAP Integration Suite environment. This section looked at the process of generating SAP custom role collections. A key component of efficient security management is role-based access control, which enables businesses to specify and enforce granular permissions across all their apps and systems.

However, to properly utilize this strategy inside SAP Integration Suite environment, it's crucial to carefully monitor user access and make sure that each user is given the right roles depending on their job duties and level of authority. Let's delve into the SAP Integration Suite process of distributing role collections to users.

2.1.7.6 Assign Role Collections to Users

To access Integration Suite, you must have the appropriate roles and role collection assigned.

You can organize the roles you create into collections using role collections. Users logged into the SAP ID service can be assigned the role collections you have defined.

The following goes through the procedure.

1.   Open the SAP BTP cockpit in your web browser and select the appropriate subaccount.

2.   Select **Security ➤ Role Collections** from the left-hand window.

3.   Select the role collection you wish to add users to.

4.   Select **Edit** from the **Users** section.

5.   The user ID of the user you want to add to the role collection should be entered. If the user is only present in one of the associated identity providers, you must select the provider and enter the user's email address.

6.   Click + to add additional users (Add a user).

7.   Save your changes, as shown in Figure 2-33.

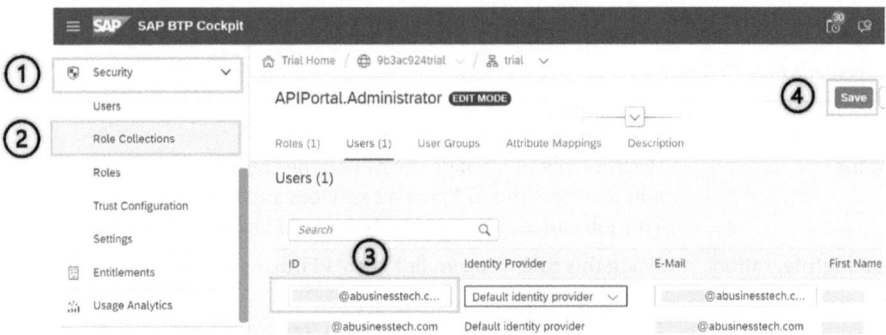

Figure 2-33. *Assign role collection to user*

2.1.7.7 Access Management for API Management

In the Cloud Foundry environment, you can maintain the enterprise roles and role collections for the API portal and API Business Hub that can be used for user management. One or more roles may be found in a role collection. Role collections that you create are then used to allocate roles to people. The SAP BTP cockpit allows you to view details regarding the role collections that have been preserved and the roles that are present inside a role collection.

The API portal comes with the predefined roles shown in Figure 2-34. All accounts that have subscribed to the API portal by default have access to, can see, and share these roles.

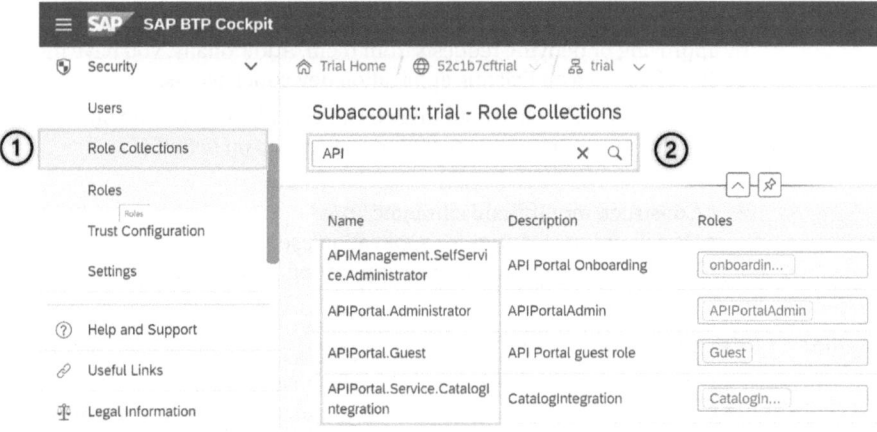

Figure 2-34. *API Management Role Collections*

Table 2-3 lists role collections and their API Management descriptions.

Table 2-3. *API Management Role Collections*

Role Collection	Description
APIPortal.Administrator	Use this role to manage the API proxies by adding extra policies and to access the API portal's services and user interface (UI). This job can also be used to manage APIs via the API Designer.
APIPortal.Service.CatalogIntegration	Using this role, you can link the API business hub enterprise to the API portal.
APIManagement.Selfservice. Administrator	Use this role to gain access to the API portal's Settings page and throughout the onboarding process.
APIPortal.Guest	Use this role to gain read-only access to the API portal. All policies, API providers, and metrics are available for viewing; they cannot be changed.

The API business hub enterprise preconfigured roles are described in Table 2-4.

Table 2-4. *API Business Hub Role Collections*

Role Collection	Description
AuthGroup.SelfService.Admin	To gain access to the API business hub enterprise, use this role during onboarding.
AuthGroup.API.Admin	Use this role to control an application developer's access to the portal by approving or denying requests from them. Additionally, you have the option of denying an Existing application developer access. • Manage a user's roles by adding new roles and removing old ones. • Admin can also carry out the following actions on behalf of an application developer. ○ Construct, modify, and eliminate apps. ○ Design unique features for application use. ○ When creating or updating an application, provide the app key and secret.
AuthGroup.ContentAuthor	Use this role to • Publish material to the enterprise's API business hub. • Create a link between the API business hub enterprise and the API portal.
AuthGroup.API.Application Developer	Access the API business hub enterprise with this role. Construct, modify, and eliminate apps. View analytics data on application performance, usage, and error rate. View and download bills for subscribed applications.
AuthGroup.Content.Admin	Create and update categories using this role.
AuthGroup.Site.Admin	Use this role for updating configuration. Make portal adjustments, including logo uploads, name and description updates, and footer link changes.

See the SAP Help Portal[1] for more information on access management for API Management.

2.1.7.8 Access Management for Integration Assessment

There are several predefined Integration Assessment role collections that you may assign to account users when managing users using the SAP BTP cockpit. These roles are connected to certain personas useful for integration projects based on the primary duties. Figure 2-35 shows the role collection associated with Integration Assessment.

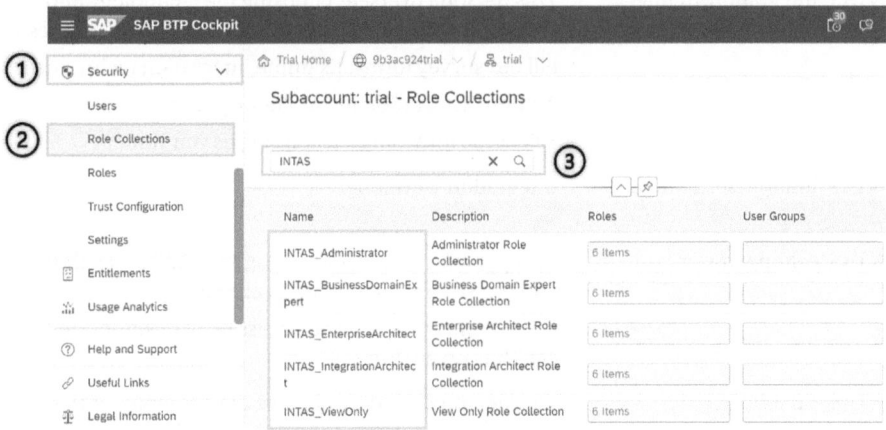

Figure 2-35. *Integration Assessment Role Collections*

The Integration Advisor comes with the predefined roles listed in Table 2-5. All users that need access to Integration Advisor based on their job responsibilities must be assigned one or more role/role collections.

[1] https://help.sap.com/docs/integration-suite/sap-integration-suite/identity-and-access-management-for-api-management

Table 2-5. *Role Collection for Integration Assessment*

Persona	Role Collection	Description
Business Domain Expert	INTAS_BusinessDomainExpert	This persona formulates inquiries and offers preliminary background.
		For the project managers and business users, create business solution requests. Fill out survey forms.
		Make requests for the interface.
Integration Architect	INTAS_IntegrationArchitect	This persona oversees choosing the technology and completing the technical questions contained request.
		Fill out survey forms. Examine integration techniques.
		Analyze integration technology options.
		Select the integration technologies you want.
Enterprise Architect	INTAS_EnterpriseArchitect	Integration Developer
Administrator	INTAS_Administrator	This persona oversees ensuring that data is accurate and consistent and offered to all users.
		An administrator is given access to all positions and can thus do all duties.
View Only	INTAS_ViewOnly	This persona can only observe the outcomes of ISA-M data, questionnaires, applications, technologies, and business solution requests.

The role templates needed to complete the tasks associated with Integration Assessment are summarized in Table 2-6.

Table 2-6. *Task and Permission for Integration Assessment*

Area	Task	Roles (Scopes)	Role Template	Persona
Configure	View applications	IntasApplication.Read	IntasApplicationRead	Business Domain Expert
	Edit/delete applications	IntasApplication.Read IntasApplication.Write IntasApplication.Delete	IntasApplicationWrite	Enterprise Architect Integration Architect
	View Technologies	IntasTechnology.Read	IntasTechnologyRead	Business Domain Expert
	Edit/delete technologies-general sections	IntasTechnology.Read IntasTechnology.Write	IntasTechnology GeneralWrite	Integration Architect
	Edit/delete technologies— all sections	IntasTechnology.Read IntasTechnology.Write IntasTechnology.Delete IntasTechnologyIsa.Write	IntasTechnologyWrite	Enterprise Architect

(*continued*)

Table 2-6. (*continued*)

Area	Task	Roles (Scopes)	Role Template	Persona
Request	View business solution request	IntasRequest.Read	IntasRequestRead	All persons are allowed.
	Edit business solution request	IntasRequest.Read IntasRequest.Write IntasRequest.Delete	IntasRequestWrite	Enterprise Architect Integration Architect Business Domain Expert
	View interface assessment	IntasAssessmentRequirements.Read IntasAssessmentSelection.Read IntasAssessment.Read	IntasAssessmentRead	Business Domain Expert
	Edit interface assessment	IntasAssessmentRequirements.Read IntasAssessmentRequirements.Write IntasAssessment-Selection.Read IntasAssessment-Selection.Write IntasAssessment.Read IntasAssessment.Write IntasAssessment.Delete	IntasAssessmentWrite	Enterprise Architect Integration Architect
Settings	View ISA-M data	IntasIsa.Read	IntasIsaRead	Integration Architect Business Domain Expert
	Edit/delete ISA-M data	IntasIsa.Read IntasIsa.Write IntasIsa.Delete	IntasIsaWrite	Enterprise Architect
	View questionnaire	IntasQuestionnaire.Read	IntasQuestionnaireRead	Integration Architect Business Domain Expert
	Edit/delete Questionnaire	IntasQuestionnaire.Read IntasQuestionnaire.Write IntasQuestionnaire.Delete	IntasQuestionnaireWrite	Enterprise Architect

Visit the SAP Help Portal[2] for more information on access management for Integration Assessment. The next section discusses access management for Integration Advisor.

---

[2] https://help.sap.com/docs/integration-suite/sap-integration-suite/9a09cdd9131f41 0db0c391d1a51c564d.html

2.1.7.9 Access Management for Integration Advisor

There are predefined Integration Advisor role collections that you may assign to account users when managing users using the SAP BTP cockpit. These roles are connected to certain personas useful for integration projects based on the primary duties, as shown in Figure 2-36.

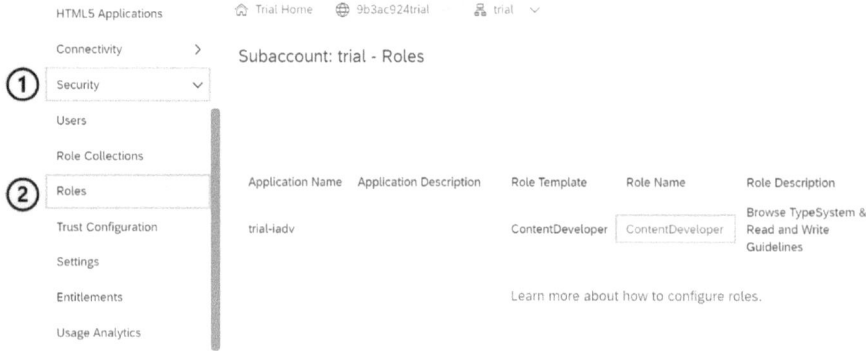

Figure 2-36. *Role name in Integration Advisor*

In SAP Integration Advisor, a role collection named trial-content-developer contains a ContentDeveloper role. The main purpose of the role collection ID is to browse the type systems and read and write the guidelines, as shown in Figure 2-37.

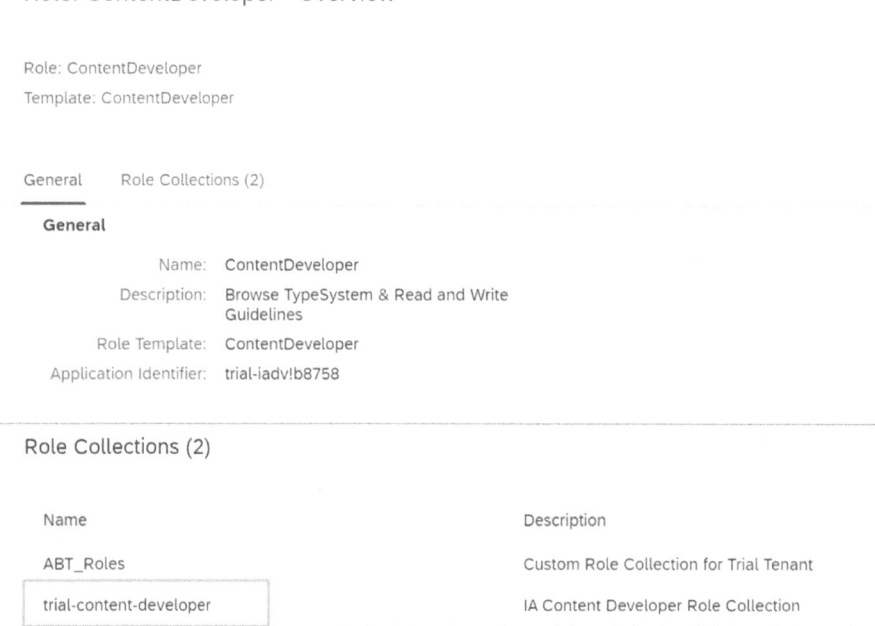

Figure 2-37. *Role collection for Integration Advisor*

Tenant administrator and content developer are the two personas linked to Integration Advisor. The roles necessary for each persona are listed in Table 2-7, along with a summary of each role's duties.

Table 2-7. *Role Collection for Integration Advisor*

Persona	Responsibilities	Roles Required	Roles Required in BTP
Tenant Admin	Controls client accounts and subscriptions Your obligation to configure IA subscription Controls users and roles in a consumer account	None	Administrator
Content Developer	Licensing	Typesystem.Read Guidelines.ReadWrite	None

See the SAP Help Portal[3] for more information on access management for Integration Advisor. The next section covers access management for Trading Partner Management.

2.1.7.10 Access Management for Trading Partner Management

There are several predefined Trading Partner Management role collections that you may assign to account users when managing users using the SAP BTP cockpit, as shown in Figure 2-38. These roles are connected to certain personas useful for integration projects based on the primary duties.

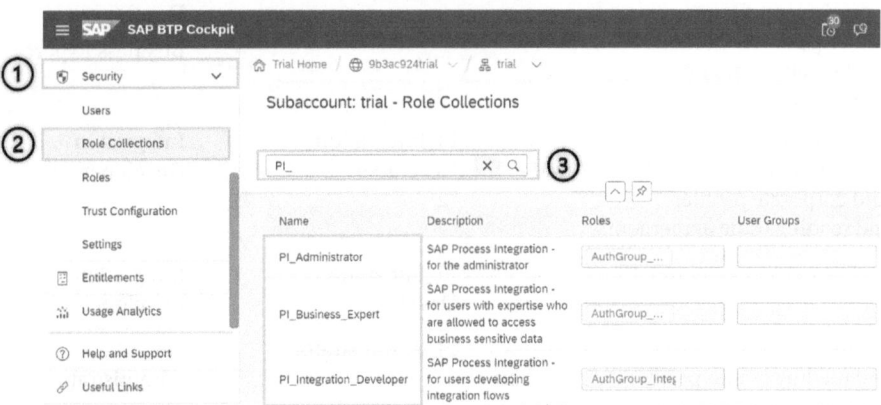

Figure 2-38. *Trading Partner Management Role Collections*

There are several predefined roles that you may assign to account users when managing users with the SAP BTP cockpit. These jobs are related to certain personas important to B2B situations based on the primary duties connected with B2B scenarios as described in Table 2-8.

---

[3] https://help.sap.com/docs/integration-suite/sap-integration-suite/ ed50e648bfea4ca08c9be8d64abc5bed.html

Table 2-8. *Role Collection for Trading Partner Management*

Persona	Role Collection	Description
Business expert	PI_Business_Expert	Enables a business expert to carry out duties related to their line of work Includes tasks such as • Tracking integration flows and integration objects' state • Examining the payload and attachments of the message
Integration developer	PI_Integration_Developer	Enables an integration developer to-carry out all configurations on trade partners, including developing and updating partner profiles, company profiles, agreement templates, and agreements and activating partner agreements Use Integration Designer to connect to a cluster to see, download, and deploy artifacts (for example, integration flows)

The responsibilities necessary to complete the different duties linked to Trading Partner Management are summarized in Table 2-9. The tasks and responsibilities' relevance to the key persona identified for Cloud Integration is also mentioned.

Table 2-9. *Trading Partner Management Tasks and Permissions*

Area	Task	Role Collection	Persona
Design	**Read-Only** View the company profile View the trading partner profile View the agreement template View partner agreement	PI_Read_Only	Integration Developer Business Expert
	TPM Configuration Customize the template for agreements, trade partner, and corporate profiles Writing and reading trade agreements	PI_Integration_Developer	Integration Developer
	Publish trading partner agreement	PI_Integration_Developer	Integration Developer
Design	**Push to Partner Directory** Writing and reading tenant partners' directories	PI_Adminstrator	Tenant administrator Technical user
	Push to Partner Directory In node management, read and deploy credentials	PI_Integration_Developer	Integration Developer
	Sensitive data management Read and write confidential corporate information Read and write private partner information	PI_Business_Expert	Business Expert
Monitor	Read monitoring data	PI_Read_Only	Integration Developer Business Expert
	Read payload data	PI_Business_Expert	Business Expert

See the SAP Help Portal[4] for more information on access management for Trading Partner Management.

The next section briefly discusses Migration Assessment.

2.1.7.11 Access Management for Migration Assessment

There are predefined Migration Assessment role collections that you may assign to account users when managing users using the SAP BTP cockpit. These roles are connected to certain personas useful for integration projects based on the primary duties.

PIMAS_Administartor and PIMAS_IntegrationAnalyst are the two personas linked to Migration Assessment. The roles necessary for each persona are listed in Table 2-10, along with a summary of each role's duties.

Table 2-10. *Role Collection for Migration Assessment*

Persona	Role Collection	Description
Administartor	PIMAS_Administartor	Administrative duties, such as configuring system settings, are carried out by PIMAS_Administrator.
Integration Analyst	PIMAS_IntegrationAnalyst	Executive duties, such as reading and executing evaluations and downloading results, are carried out by PIMAS_IntegrationAnalyst.

Visit the SAP Help Portal[5] for more information on access management for Migration Assessment.

It's time to take a closer look at how organizations can get started with this cloud-based solution. One of the best ways to do so is by setting up a trial account, which provides access to the suite's full range of features and capabilities for a limited time. By trying out SAP Integration Suite in a sandbox environment, organizations can get a feel for how the solution works and assess its potential value before committing to full deployment.

The next section explores the steps involved in setting up a trial account for SAP Integration Suite and provides tips for maximizing your trial experience.

2.1.8 Trial Account Setup: SAP Integration Suite

If you have assigned the right role collection for enterprise BTP accounts and have the subaccount viewer and administrator roles, you should be able to proceed with setting up the Integration Suite subaccount. For the trial account, you can proceed setting up BTP trial account first.

The SAP Integration Suite trial account is created inside the SAP BTP cockpit, a web-based tool allowing users to manage and monitor applications running on SAP BTP (formerly SAP Cloud Platform).

SAP offered a trial system to let users learn more about their SaaS service, and as the first move in that direction, SAP has increased the trial account expiration from 30 days to 60 days. To access SAP CI, SAP Integration Advisor, SAP API Management, SAP Enterprising Messaging, and SAP Open Connectors, Migration Assessment let's first look at how to set up a trial account in the SAP BTP for Integration Suite.

---

[4] https://help.sap.com/docs/integration-suite/sap-integration-suite/identity-and-access-management-for-trading-partner-management

[5] https://help.sap.com/docs/integration-suite/sap-integration-suite/identity-and-access-management-for-migration-assessment

2.1.8.1 Set up a BTP Trial Account

1. Navigate to SAP BTP through `https://account.hanatrial.ondemand.com/`. Sign in if you already have an SAP account.

2. If not, register for the new account with your email, provide the information, and create the account.

3. When you log in, you are asked to select your nearest region to create the account.

4. Select the nearest region, and click Create Account, as shown in Figure 2-39.

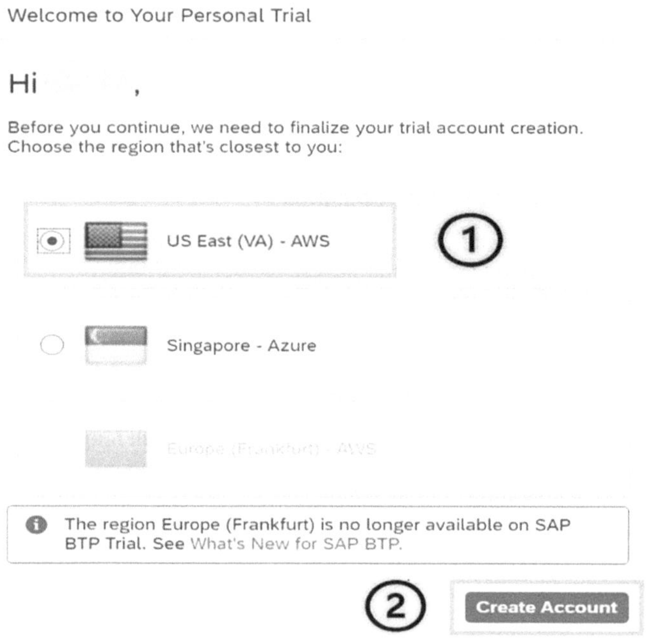

Figure 2-39. *Select data center based on location*

5. When you Create an Account, this pop-up is displayed to continue ongoing processing, which hardly takes 1-2 minutes. Press Continue to proceed further, as shown in Figure 2-40.

Welcome to Your Personal Trial

We will now set up your account. This process can take a minute.

Region:	US East (VA) - AWS
Global Account:	9b3ac924trial
Subaccount:	trial
Org:	9b3ac924trial
Space:	dev

Figure 2-40. Trial account primary information

6. On the next screen, click **Go to Your Trial Account**. The trial account has been created along with the subaccount, as shown in Figure 2-41.

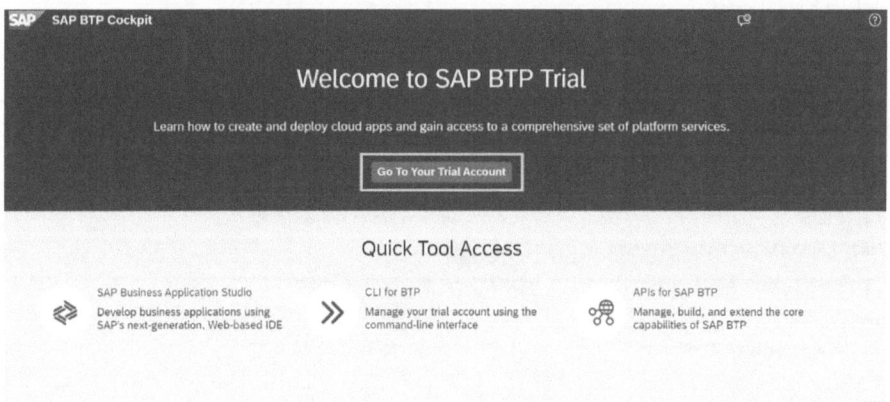

Figure 2-41. SAP BTP cockpit

7. You are navigated to the Trial Home page, where the account and subaccount have already been created. You can use this subaccount for our Integration Suite (or) you can make another subaccount dedicatedly for our purpose, as shown in Figure 2-42.

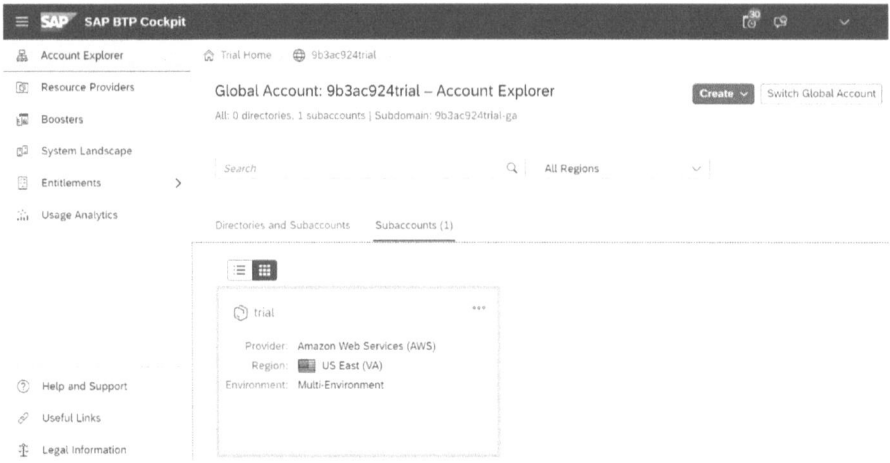

Figure 2-42. *Trial subaccount created*

8. After clicking a subaccount, you are directed to the SAP BTP cockpit, where you can see the 80 entitlements and two instances and subscriptions already created, as shown in Figure 2-43.

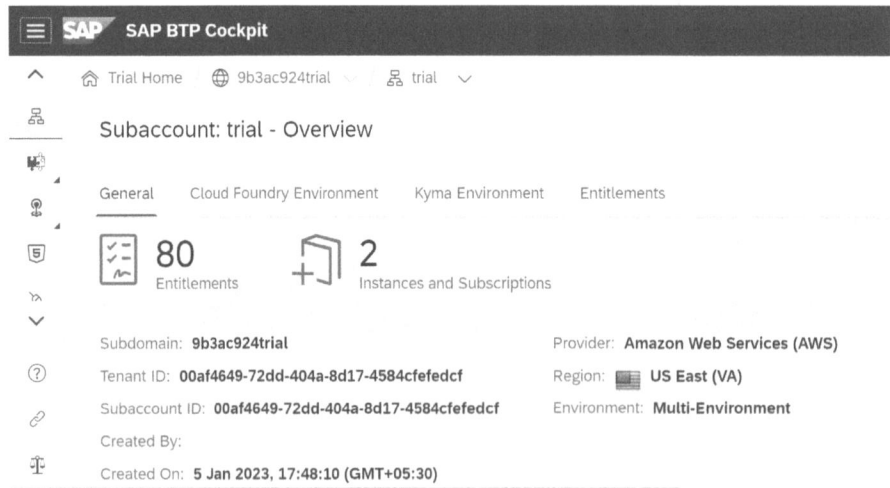

Figure 2-43. *SAP BTP cockpit*

You have created a trial account in SAP BTP. In the next section, you create a space in the Cloud Foundry environment.

2.1.8.2 Create Space: Cloud Foundry

In an enterprise account, you can create your own space in the Cloud Foundry environment or continue with by default created dev space in a trial account. The following steps explain how to create a space.

1. Navigate to **Cloud Foundry ➤ Spaces ➤ Create Space** in the left panel, as shown in Figure 2-44.

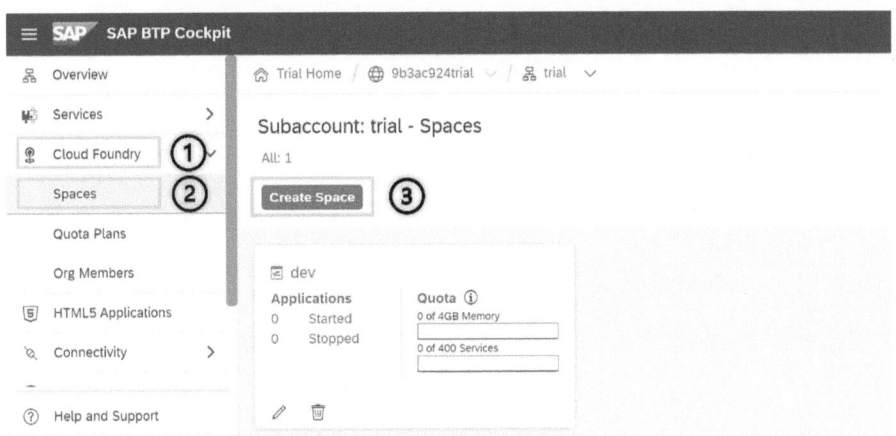

Figure 2-44. *Navigate to Cloud Foundry Create Space*

2. Give the desired space name and check all the boxes. Click **Create**, as shown in Figure 2-45.

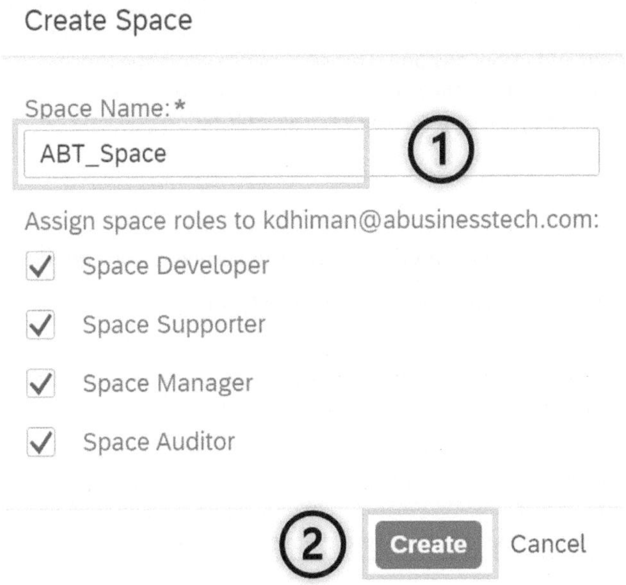

Create Space

Space Name:*

ABT_Space ①

Assign space roles to kdhiman@abusinesstech.com:

☑ Space Developer

☑ Space Supporter

☑ Space Manager

☑ Space Auditor

② Create Cancel

Figure 2-45. *Create Space in Cloud Foundry*

3. You can see the space has been created, as shown in Figure 2-46.

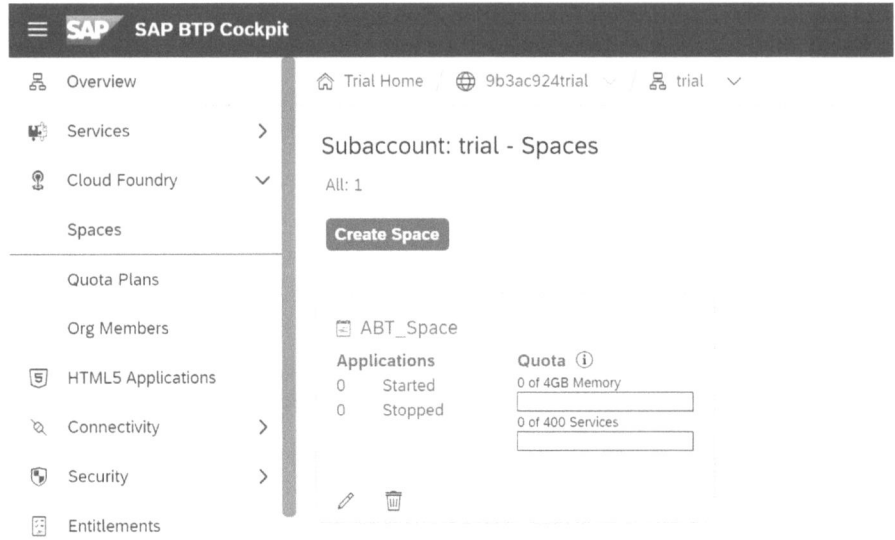

Figure 2-46. *Custom space*

This section discussed creating a space in Cloud Foundry in SAP Integration Suite. But before you can truly begin to leverage the suite's capabilities, it's important to have a clear understanding of your organization's needs and goals when it comes to enterprise integration. This understanding helps you make informed decisions about which features and entitlements to enable within your trial account, ensuring that you're getting the most out of SAP Integration Suite from day one. This section dives into the process of managing entitlements within your trial account, exploring the various options available and providing guidance on how to tailor your setup to meet your specific requirements. Let's get started.

2.1.8.3 Manage Entitlements

To avoid confusion, you should know the differences between entitlements and quotas before you begin.

Entitlements – Your right to provision and consumption of a resource is your entitlement.

Quota – The quota is the quantifiable amount that specifies the resource's maximum allowable consumption. Or how much of a service package you are allowed to utilize.

At the global account level, entitlements and quotas are handled, distributed to directories and subaccounts, and used by the latter.

Quotas and entitlements can be assigned to other subaccounts if you remove them from a subaccount and make them available again at the global account level.

You automatically receive a subaccount called *trial* when you sign up for a trial account, and all trial entitlements are, by default, assigned to it. You must manually transfer the entitlements from the default subaccount to the new subaccount if you decide to create a second subaccount and experiment there. You want to do one of the following.

- Move all trial rights

- Move some of them

- Divide a service plan's quota between two subaccounts

There are one or more service plans offered for each service. A service plan is a breakdown of the advantages and costs of a specific service type. An example would be configuring a database with numerous T-shirt sizes, each representing a distinct service package.

Some service plans include numeric quotas, allowing you to alter the quantity of units accessible in a subaccount. These units represent various things depending on the service and may limit the number of service instances, apps, or routes you can have in a subaccount.

Entitlements and quotas for subaccounts can be viewed and modified in the global account and the subaccounts.

2.1.8.3.1 Add Missing Entitlements

In Enterprise, when you buy the license for SAP Integration Suite, it becomes necessary to configure the entitlements and add the service/quota to the tenant.

Sometimes, you do not get all the service plans added to the enterprise/trial account automatically for the default trial subaccount created. Let's add some of the missing entitlements.

1. Navigate to **Entitlements ➤ Configure Entitlements** to enter the Edit mode of the subaccount, as shown in Figure 2-47.

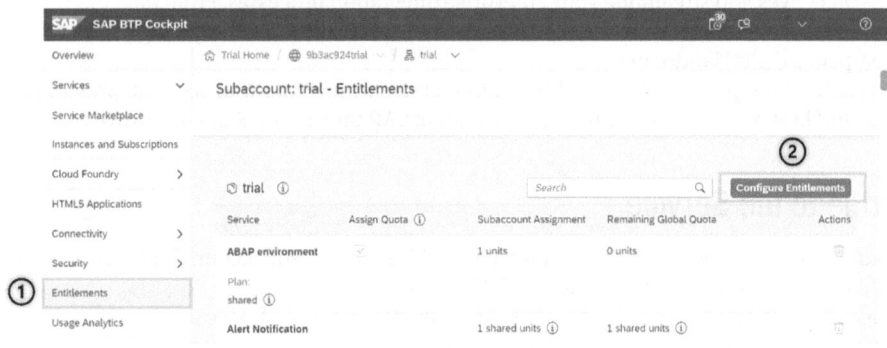

Figure 2-47. *Entitlements in SAP Integration Suite*

2. Click **Add Service Plans**.

3. Search the desired service plan, select it, and click Add Service Plan. Figure 2-48 shows all the instances already assigned in a trial account.

Figure 2-48. *Select entitlement to add to service plan*

Now that you've gained a better understanding of how to manage entitlements within your SAP Integration Suite trial account, it's time to take the next step and subscribe to the service. By subscribing to SAP Integration Suite, you'll gain access to a wide range of dynamic features and tools that can help you achieve your business objectives and streamline your enterprise integration processes. From data mapping and transformation to real-time monitoring and analytics, SAP Integration Suite has everything you need to succeed in today's fast-paced digital landscape.

The next section explores the process of subscribing to the service within your trial account, providing step-by-step guidance and best practices for getting the most out of SAP Integration Suite. Let's dive in.

2.1.8.4 Subscribe to the Service

Multiple global accounts and subaccounts can be created. Let's utilize the trial subaccount that was created by default. Clicking this trial account takes you to the SAP BTP cockpit.

The Cloud Foundry environment is default activated when you navigate this subaccount. Let's subscribe to SAP Integration Suite, but there are currently entitlements, instances, and subscriptions.

1. Select **Services ➤ Instances and Subscriptions** and click Create, as shown in Figure 2-49.

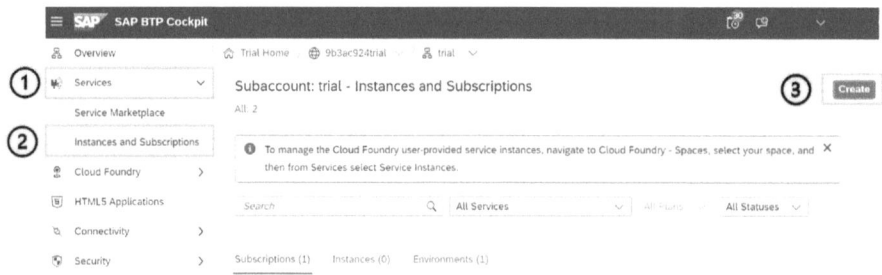

Figure 2-49. *Create new service*

2. One subscription (SAP Business Application Studio) is already active by default. The SAP Integration Suite service is created. Choose the Integration Suite service and the trial package. Select Create, as shown in Figure 2-50.

Figure 2-50. *Basic information for creating new service*

3. The Service Integration Suite is created in the SAP BTP cockpit, as shown in Figure 2-51.

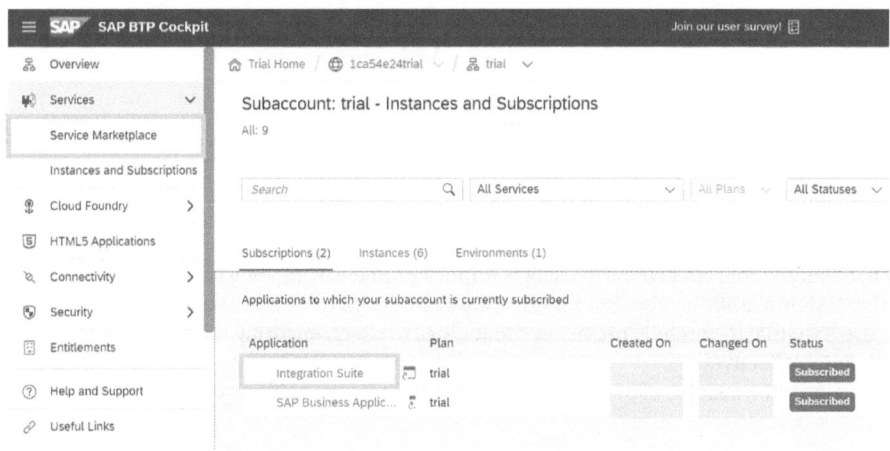

Figure 2-51. *Integration Suite*

The next section assigns role collection to the users in SAP Integration Suite.

2.1.8.5 Assign Role Collection

1. Select **Security ➤ Users**. Choose the entry against your name. In the Role Collections section, choose **Assign Role Collection**, as shown in Figure 2-52.

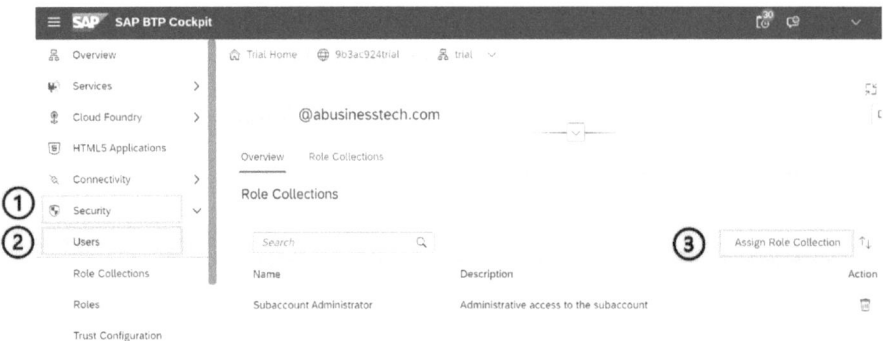

Figure 2-52. *Assign role collection to user*

2. Select **Integration Provisioner** as the role collection and assign it to the user, as shown in Figure 2-53.

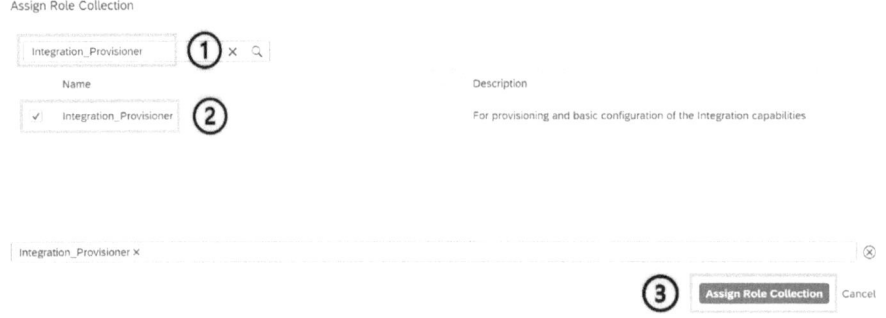

Figure 2-53. *Integration Provisioner Assign Role Collection*

3. You must assign the role collection per the capabilities discussed in sections 2.1.7.7, 2.1.7,8, 2.1.7.9, 2.1.7.10, and 2.1.7.11.

As you have seen, assigning role collections to users is important in ensuring they have access to the right capabilities within SAP Integration Suite. But it's not enough to assign roles and assume everything works smoothly. It's also essential to carefully provision capabilities to users, ensuring they have the tools and resources to do their job effectively.

The next section explores the process of provisioning capabilities to users within your SAP Integration Suite trial account.

2.1.8.6 Provisioning Capabilities

SAP Integration Suite allows the Cloud Foundry environment. Although the Neo environment is still supported in Cloud Integration, any new capabilities will only be found in the Cloud Foundry environment.

1. In the SAP BTP cockpit dashboard, navigate to the SAP Integration Suite UI, as shown in Figure 2-54. To use the services of Integration Suite, you must add the capabilities of SAP Integration Suite, which you learn while moving through the chapter.

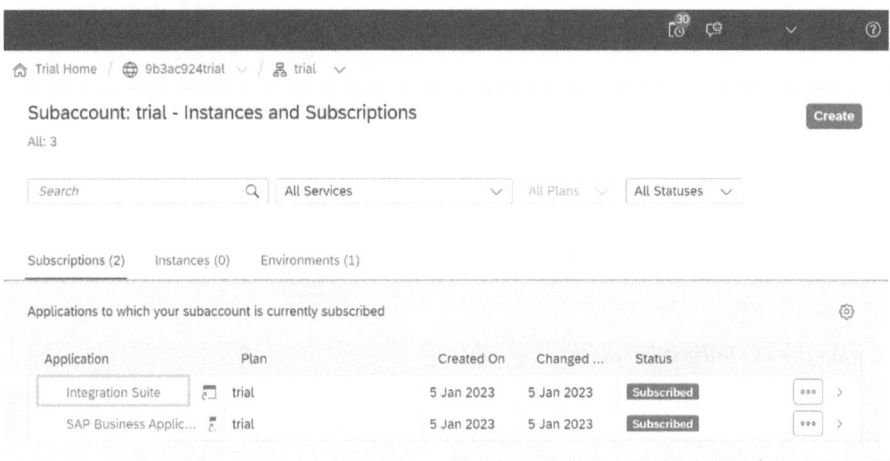

Figure 2-54. *Open SAP Integration Suite UI*

2. In the Integration Suite Launchpad, choose **Add Capabilities** to activate the capabilities offered by Integration Suite, as shown in Figure 2-55.

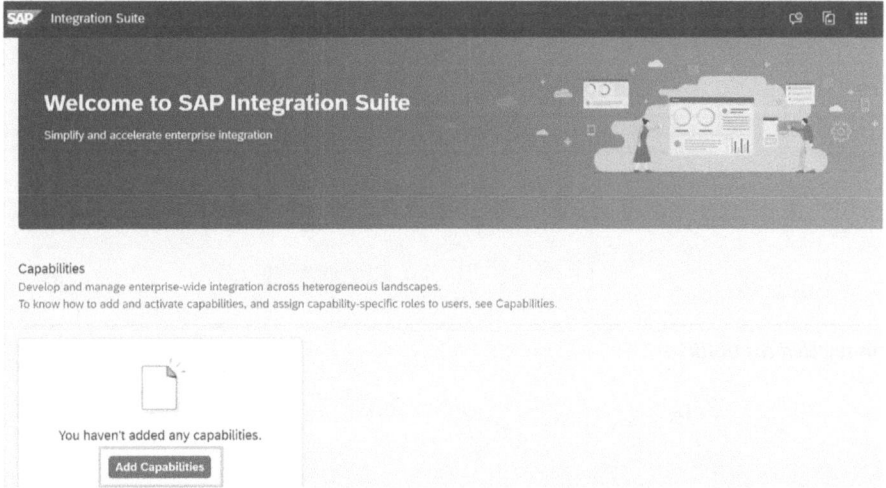

Figure 2-55. *Add capabilities to Integration Suite*

3. Select the capabilities according to the need and requirements of the organization. Select all the capabilities of Integration Suite to proceed further. After selecting the capabilities, click **Next**, as shown in Figure 2-56.

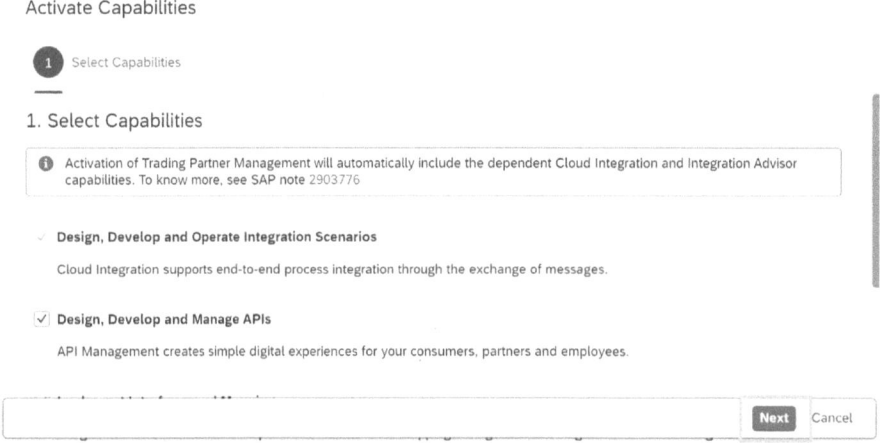

Figure 2-56. *Select Capabilities in Integration Suite*

4. When you click Next, a pop-up asks you to confirm the capabilities. Confirm the capabilities, and click **Next**, as shown in Figure 2-57.

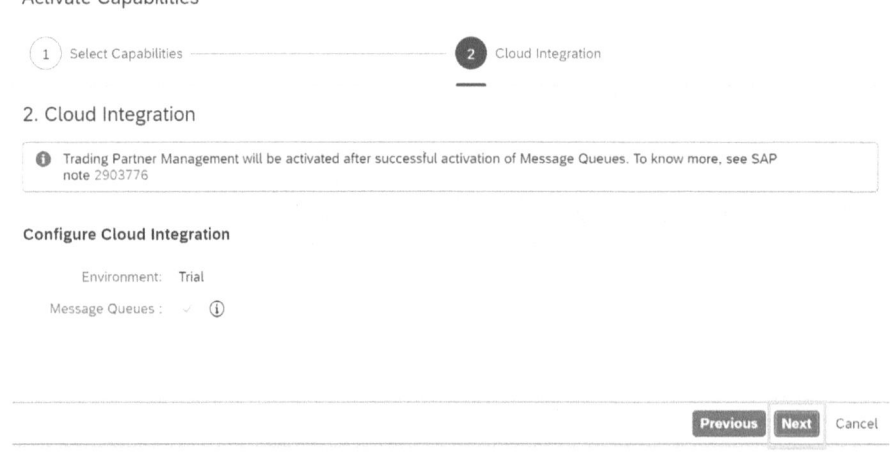

Figure 2-57. *Confirm selected capabilities*

5. After confirming all the capabilities, select **Activate** to provision the selected capabilities, as shown in Figure 2-58.

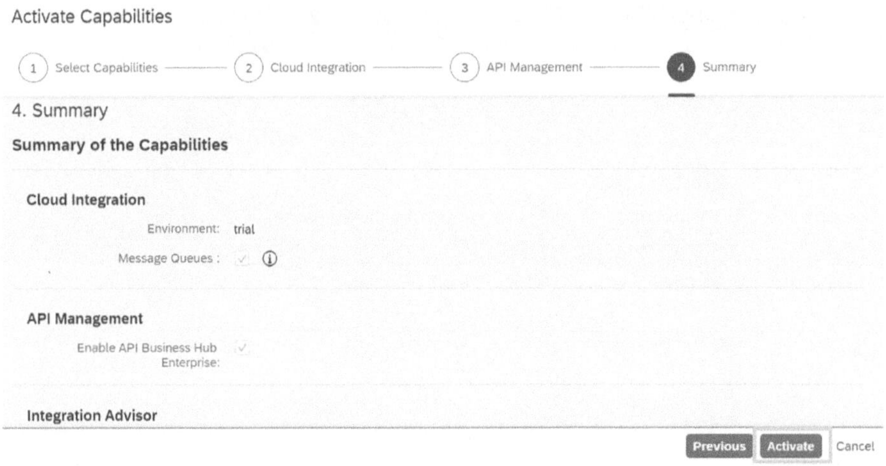

Activate Capabilities

(1) Select Capabilities ——— (2) Cloud Integration ——— (3) API Management ——— (4) Summary

4. Summary

Summary of the Capabilities

Cloud Integration

Environment: trial

Message Queues :   ✓ ⓘ

API Management

Enable API Business Hub   ✓
Enterprise:

Integration Advisor

Previous   Activate   Cancel

Figure 2-58. *Activate capabilities*

6. The processing for adding capabilities starts and takes about 60 minutes to complete and add capabilities, as shown in Figure 2-59.

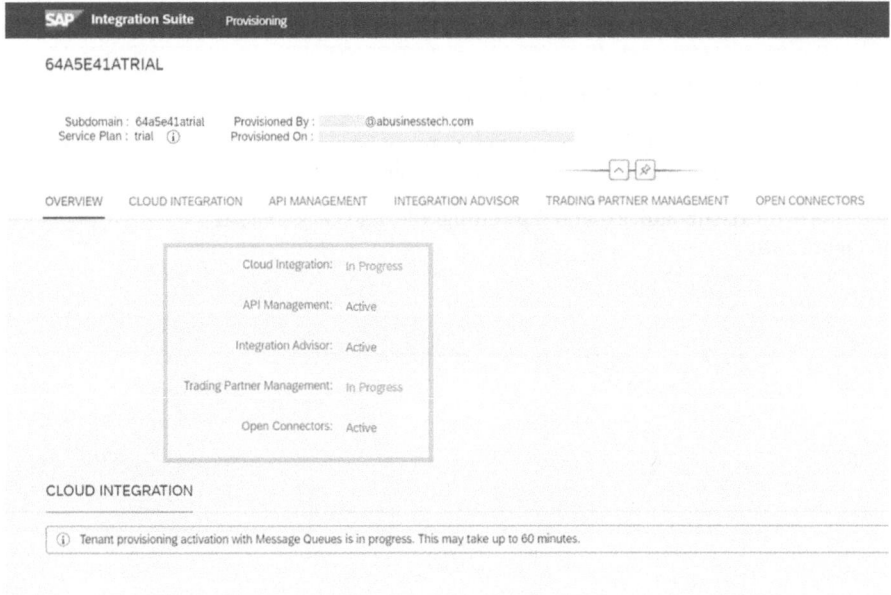

Figure 2-59. *Provisioning capabilities*

7. After provisioning all the selected capabilities, you see the tiles of capabilities in the SAP Integration Suite dashboard, as shown in Figure 2-60.

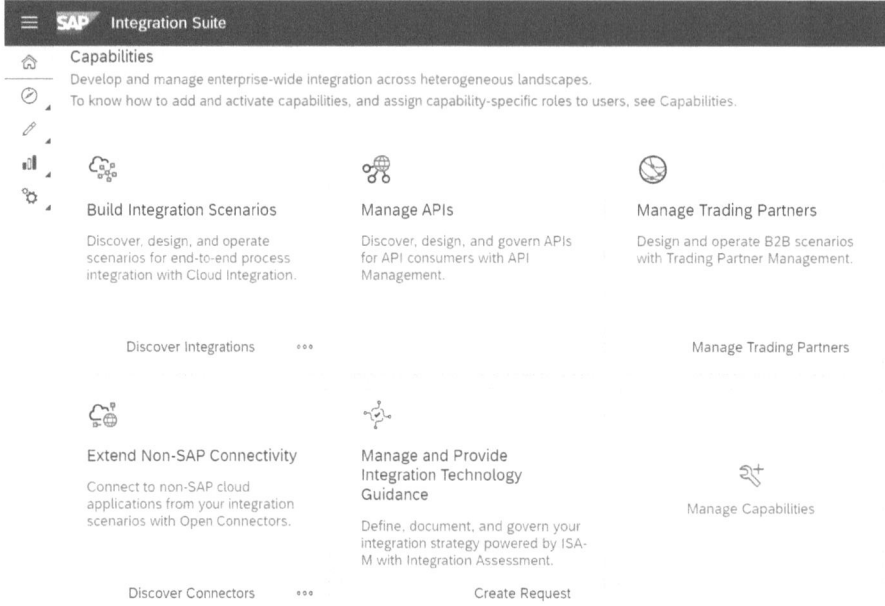

Figure 2-60. *SAP Integration Suite dashboard with all capabilities*

2.1.8.6.1 Provisioning SAP API Management

After successfully creating the trial account for our tenant, you may not be able to see the Discover or Design tabs in the Manage APIs tile. In such a situation, take the following steps.

1. On the left side of the dashboard, expand the Settings tile, and click APIs, as shown in Figure 2-61.

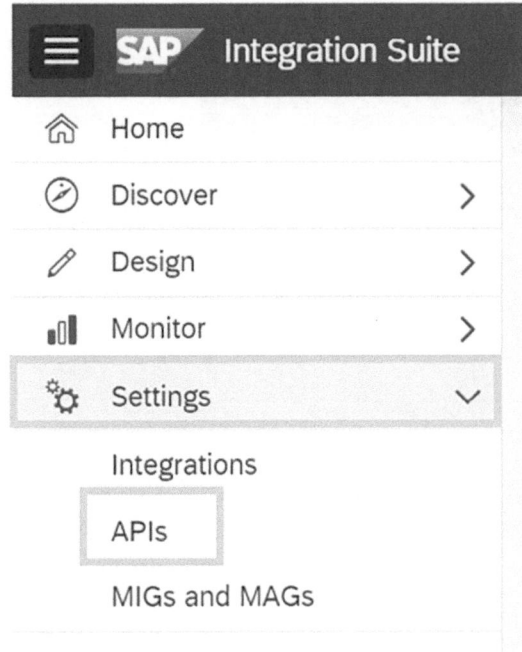

Figure 2-61. *APIs to set up API Management*

2. After clicking APIs, you see your account is provisioned. This might take some time to set up, as shown in Figure 2-62.

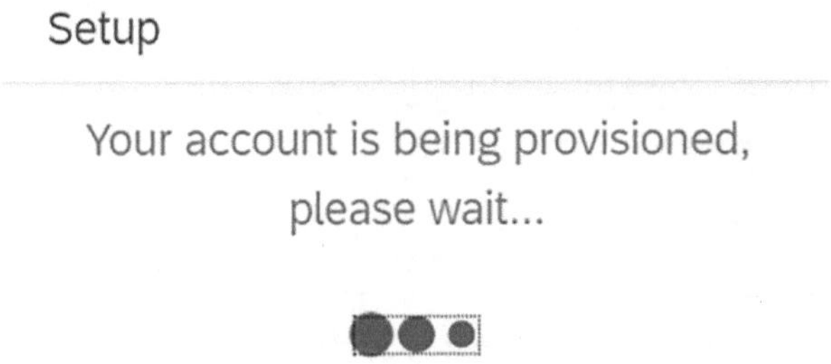

Setup

Your account is being provisioned,
please wait...

Figure 2-62. *Set up SAP API Management*

3. After successfully setting up, you are logged out. Click the link to log in to SAP Integration Suite, as shown in Figure 2-63.

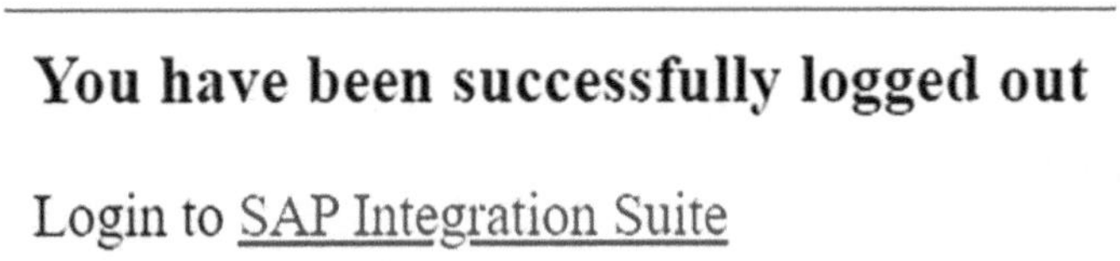

You have been successfully logged out

Login to <u>SAP Integration Suite</u>

Figure 2-63. *Successful logout from SAP Integration Suite*

4. Once you log in to SAP Integration Suite, you can access all the capabilities and find more features in the left-hand side tile, as shown in Figure 2-64.

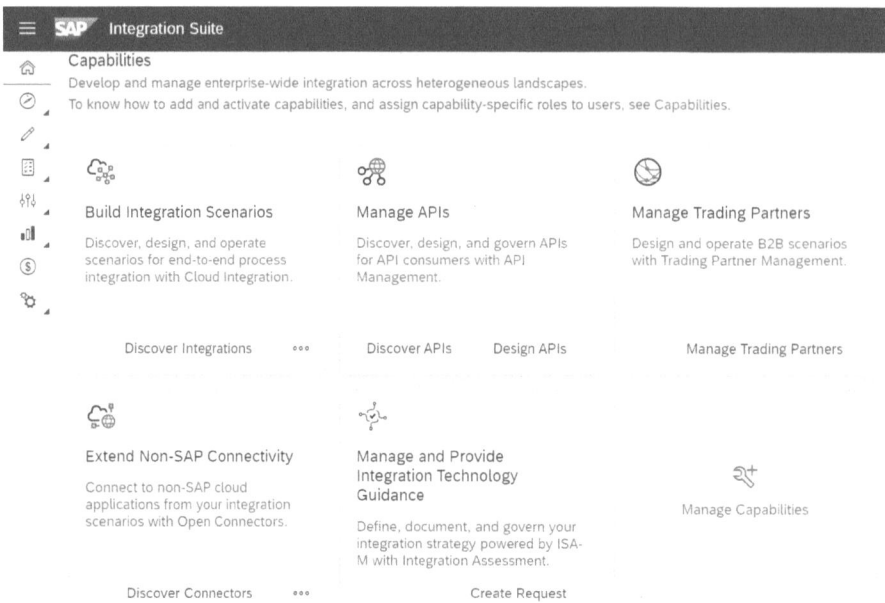

Figure 2-64. *Provision capabilities in SAP Integration Suite*

5. After provisioning all the capabilities, the SAP Integration Suite dashboard looks like Figure 2-65.

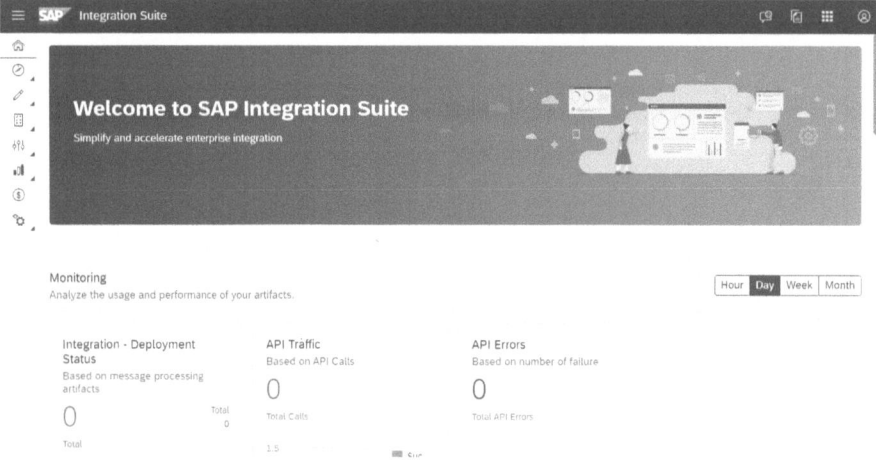

Figure 2-65. *SAP Integration Suite dashboard*

6. All the capabilities are provisioned to the trial tenant. You can manage capabilities to build service instances and assign roles automatically provisioned in the tenant, as shown in Figure 2-66.

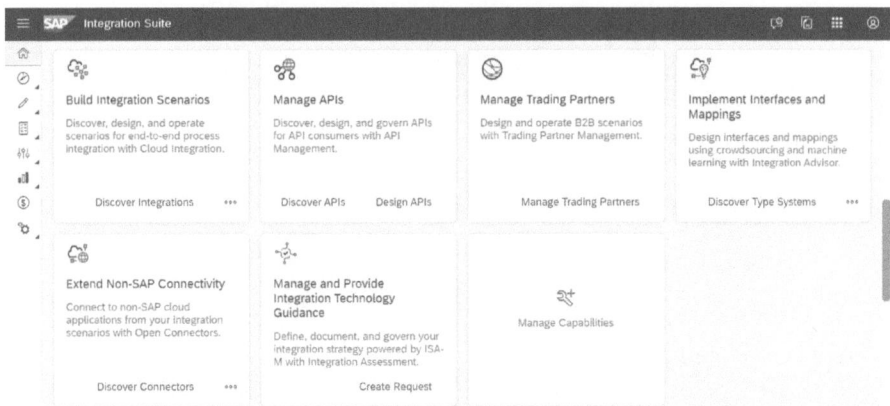

Figure 2-66. *Manage Capabilities in SAP Integration Suite*

This section discussed provisioning capabilities, a key component of starting with SAP Integration Suite. But as your needs grow and you begin to scale your integration efforts, manually provisioning services and assigning roles can quickly become a time-consuming and error-prone process. That's where boosters come in. This efficient tool allows you to automate the process of building service instances and assigning roles within your SAP Integration Suite trial account, streamlining the provisioning process and freeing up valuable time and resources for other critical tasks.

This section explores how to use boosters to automatically build service instances and assign roles within your trial account, providing step-by-step guidance and best practices for maximizing your efficiency and accuracy.

2.1.8.7 Boosters: Automatically Build Service Instances and Assign Roles

Boosters are a series of guided and interactive actions that let you choose, set up, and use services on SAP BTP to accomplish a particular technical objective. In this situation, SAP Integration Suite boosters assist you in creating service instances and assigning roles.

In our example, a service instance specifies how a remote component can call the Process Integration Runtime service. The SAP BTP client is defined as a service instance in the context of cloud integration. The credentials and other data needed later to invoke the integration flow are contained in the service key generated from the service instance.

1. Navigate to the overview page of your SAP BTP global account.

2. Choose **Boosters** on the left navigation pane.

3. Look for the Enable Integration Suite tile in the list. Start the execution by choosing **Start** on the tile, as shown in Figure 2-67.

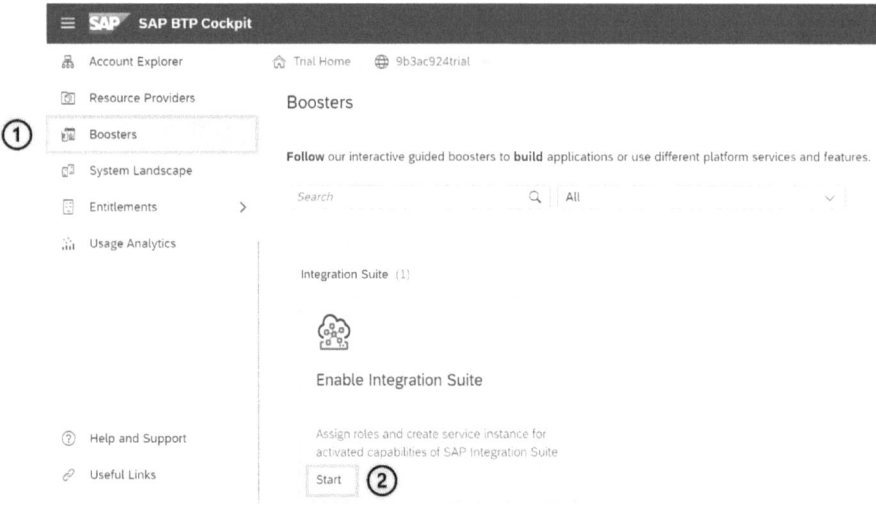

Figure 2-67. *Start boosters to assign roles and create service instances*

4. The first step is to configure the subaccount to enable Integration Suite. Provide the following information to proceed further.

    • Subaccount: trial

    • Org: the default

    • Space: dev (For different environments, you can name your space accordingly, for example, dev, test, prod)

Click **Next** and move to the next screen, as shown in Figure 2-68.

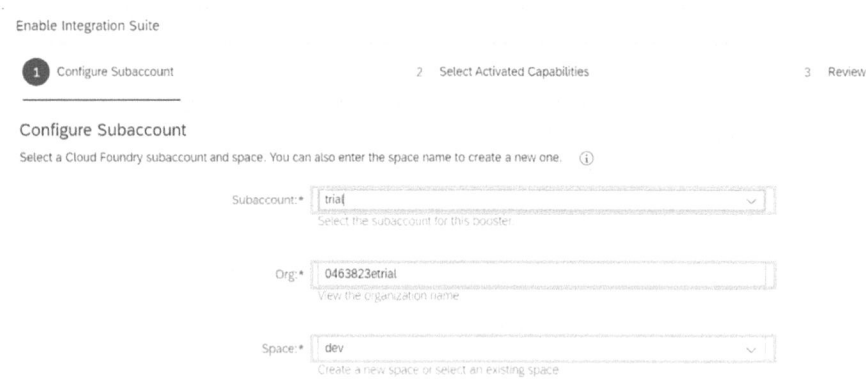

Figure 2-68. *Configure Subaccount*

5. For provisioning the capabilities, select the capabilities, and click **Next**, as shown in Figure 2-69.

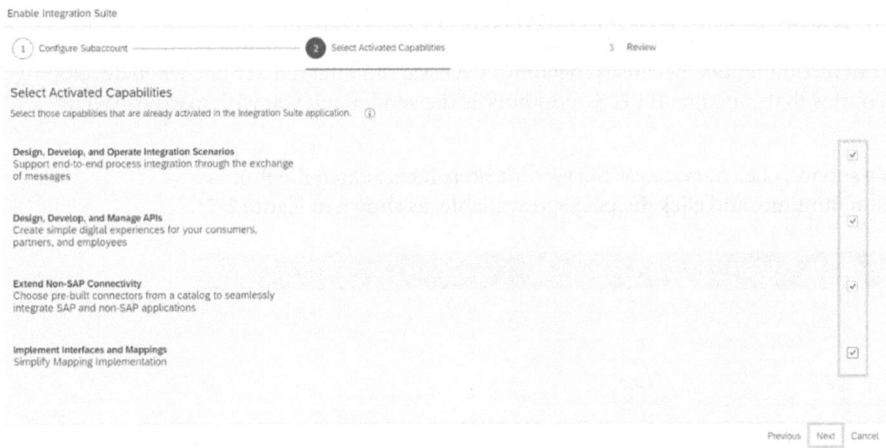

Figure 2-69. *Select Activated Capabilities*

6. You see the progress status and ongoing processing. Within a few minutes, you see the Success status for Booster, in which the roles and services instances are available for activated capabilities in the subaccount.

7. You can navigate to the subaccount after successfully assigning the roles and instances, as shown in Figure 2-70.

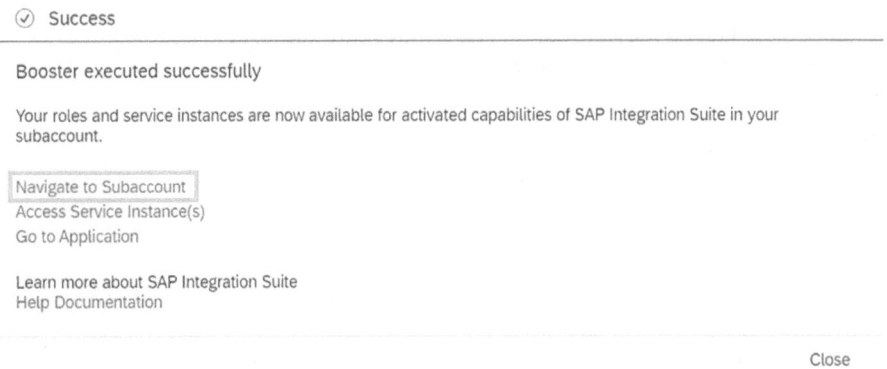

Figure 2-70. *Navigate to Subaccount*

This section briefly discussed boosters for automating the setup process and streamlining your trial account configuration. But to truly harness the power of SAP Integration Suite, you must have a robust Process Integration Runtime environment. This runtime environment provides the backbone for enterprise integration, enabling seamless connectivity and data exchange across your systems and applications.

The next section explores setting up Process Integration Runtime within your trial account, providing guidance and best practices for configuring this critical component of SAP Integration Suite. Although this is an optional topic, you can start development after completing all other steps mentioned in earlier sections.

2.1.9 Setup Process Integration Runtime (Optional)

This step is optional but recommended because it becomes necessary in Integration Suite when developing cloud integration scenarios that consume HTTPS endpoints as the sender and you wish to expose web services.

1. To enable service, click **Services ➤ Service Marketplace**. Search the Process Integration Runtime, and click the package available, as shown in Figure 2-71.

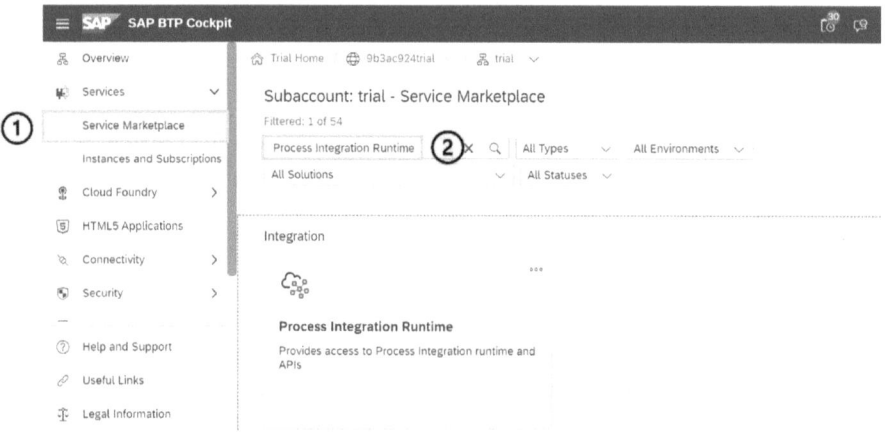

Figure 2-71. *Process Integration Runtime*

2. Go to the Process Integration Runtime tile, and click the **Create** button to subscribe to the package, as shown in Figure 2-72.

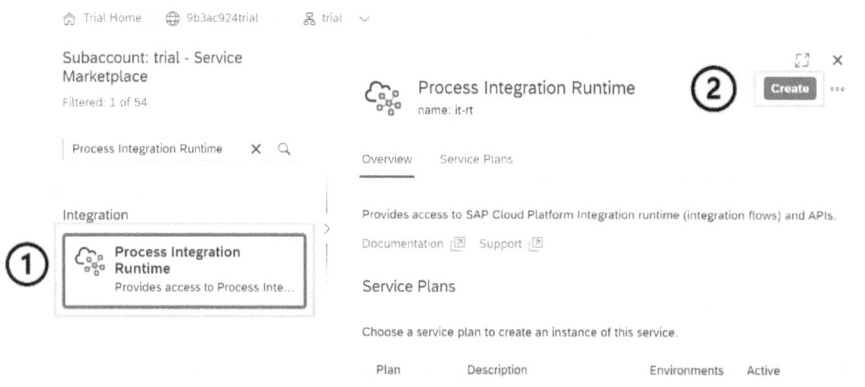

Figure 2-72. *Subscribe to Process Integration Runtime*

3. Provide the following information for a new instance in or subscription to the Process Integration Runtime package.

- Service: Process Integration Runtime

- Plan: integration flow

- Runtime Environment: Cloud Foundry

- Space: dev or the custom-created Space

- Instance Name: ProcessIntegrationRuntime

Click **Next** after giving all the details, as shown in Figure 2-73.

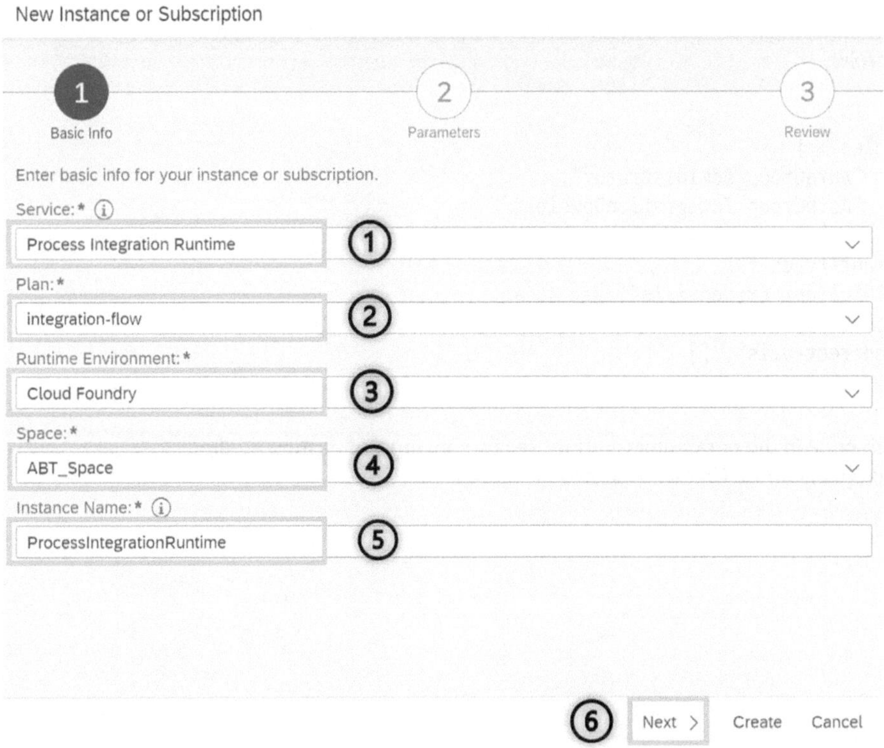

New Instance or Subscription

Figure 2-73. *Create new instance for Process Integration Runtime*

4. The next step is very important, in which you have to add the JSON format code. Specify the parameter value as follows.

```
{
 "roles":[
   "ESBMessaging.send"
 ]
}
```

For SAP Integration Advisor. While creating the new instance, select **api** as the plan; the rest of the information remains the same.

The following role is used for SAP Integration Advisor.

```
{
"roles":[
"WorkspacePackagesEdit"
]
}
```

For **SAP API Management**, while creating the new instance, select **api** as the plan and provide the same information as for Cloud Integration.

Paste the following JSON codes in the Provide Parameters section to assign a specific role.

```
{
    "roles": [
        "AuthGroup_Administrator",
        "AuthGroup_IntegrationDeveloper"
    ],
    "grant-types": [
        "client_credentials"
    ],
    "redirect-uris": []
}
```

Enter this code in the parameters editing area by copying and pasting it. After selecting **Next**, choose Create, as shown in Figure 2-74.

New Instance or Subscription

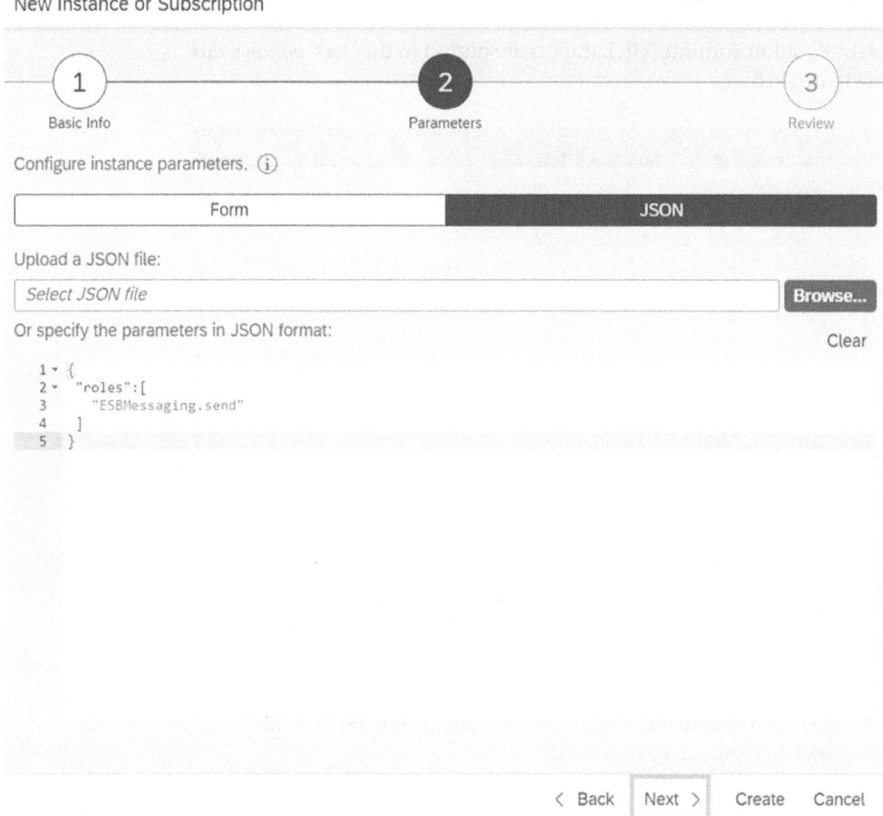

Figure 2-74. *Specify parameters in JSON format*

5. The dashboard displays the instance created, as shown in Figure 2-75.

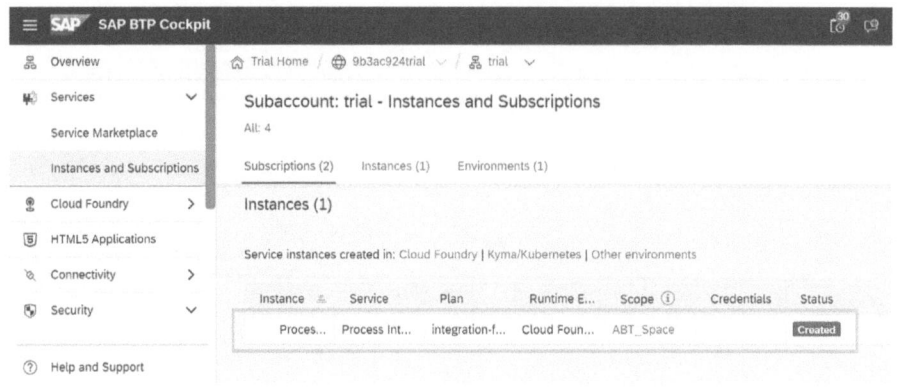

Figure 2-75. *New instance created*

2.1.9.1 Create a Service Key

1. In ProcessIntegrationRuntime, click the Create button in the Service Keys tab, as shown in Figure 2-76.

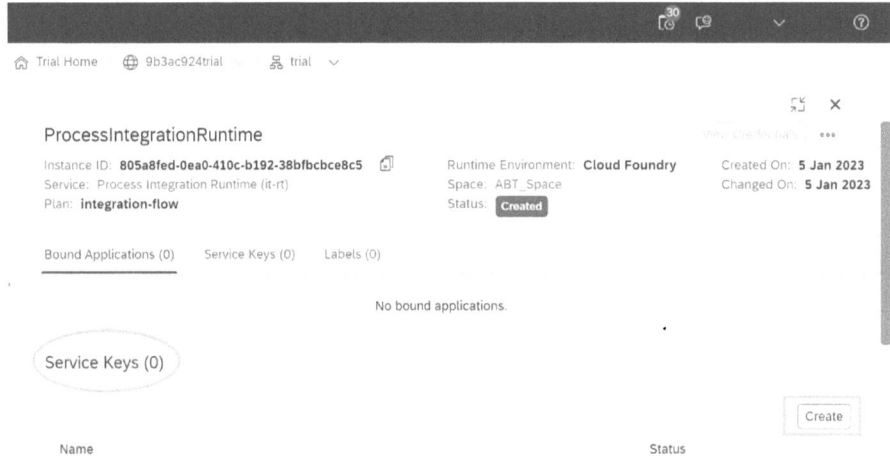

Figure 2-76. *Create service key*

2. You see the pop-up to name the service key without adding JSON code. Click Create, as shown in Figure 2-77.

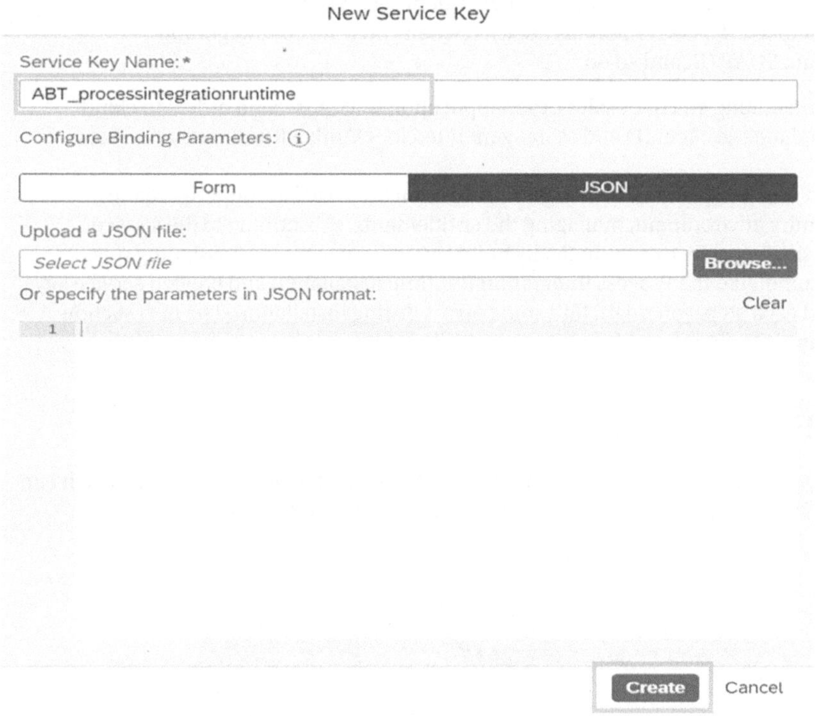

New Service Key

Service Key Name: *

ABT_processintegrationruntime

Configure Binding Parameters: ⓘ

| Form | JSON |

Upload a JSON file:

Select JSON file   Browse...

Or specify the parameters in JSON format: Clear

1 |

Create   Cancel

Figure 2-77. *Name service key*

> 3. To see the value of the key, click the service key. The JSON code is shown in Figure 2-78.

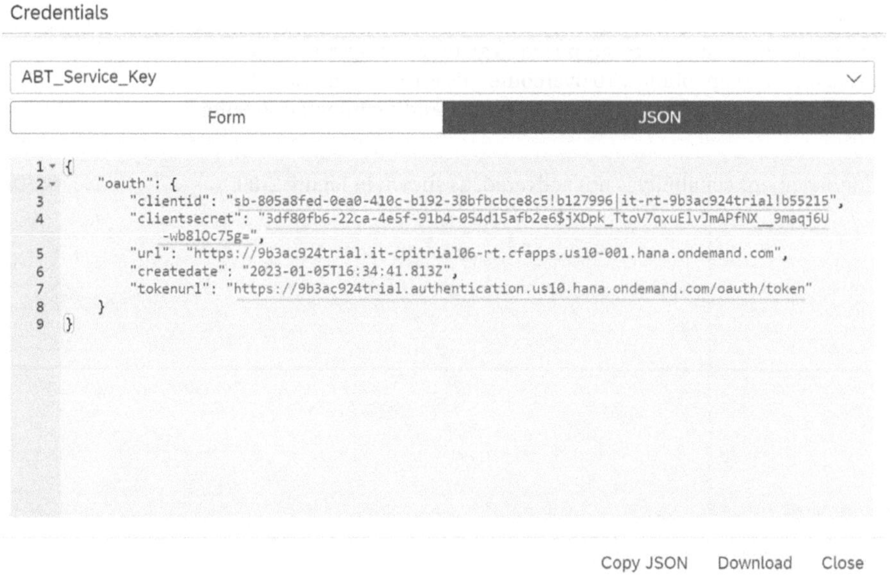

Credentials

ABT_Service_Key   ⌄

| Form | JSON |

```
1 ⏷ {
2 ⏷     "oauth": {
3           "clientid": "sb-805a8fed-0ea0-410c-b192-38bfbcbce8c5!b127996|it-rt-9b3ac924trial!b55215",
4           "clientsecret": "3df80fb6-22ca-4e5f-91b4-054d15afb2e6$jXDpk_TtoV7qxuElvJmAPfNX__9maqj6U
                -wb81Oc75g=",
5           "url": "https://9b3ac924trial.it-cpitrial06-rt.cfapps.us10-001.hana.ondemand.com",
6           "createdate": "2023-01-05T16:34:41.813Z",
7           "tokenurl": "https://9b3ac924trial.authentication.us10.hana.ondemand.com/oauth/token"
8       }
9  }
```

Copy JSON   Download   Close

Figure 2-78. *Trial tenant ClientID and client secret*

4. This ClientID is the OAuth 2.0 username, and the Client secret is the password to fetch the bearer token. These credentials are used when you test your integration flows in Postman, SOAP UI, and so on.

For example, you are integrating an HR or sales order application that must send orders to cloud integration via HTTPS. You share the client ID and secret generated for OAuth 2.0 authentications of your web services/API.

You have gone through each step for creating an SAP Integration Suite trial account by creating the space in the Cloud Foundry environment, managing the entitlements, subscribing to the service, provisioning capabilities, using boosters to automatically build service instances, and assigning role collections. You have also completed the Process Integration Runtime installation and created service keys. Although you have followed each step, you might find some errors during installation. The next section covers common errors during installation.

2.1.10 Common Errors (Installation)

Users may experience frequent issues throughout SAP Integration Suite's installation procedure, which can be complicated. The following explains the most typical errors and methods to fix them.

1. You do not have access to the requested page, as shown in Figure 2-79.

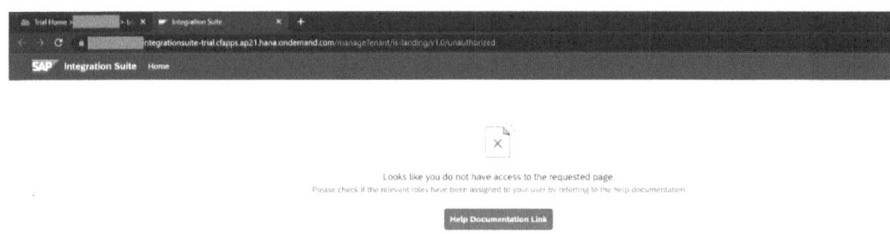

Figure 2-79. *Error while adding capabilities*

This error may arise when you are supposed to go to the Integration Suite dashboard to add the capabilities. To overcome this error, assign the role collection integration provisioner to the user before adding Integration Suite capabilities.

2. SAP API Management capability is not activated, as shown in Figure 2-80.

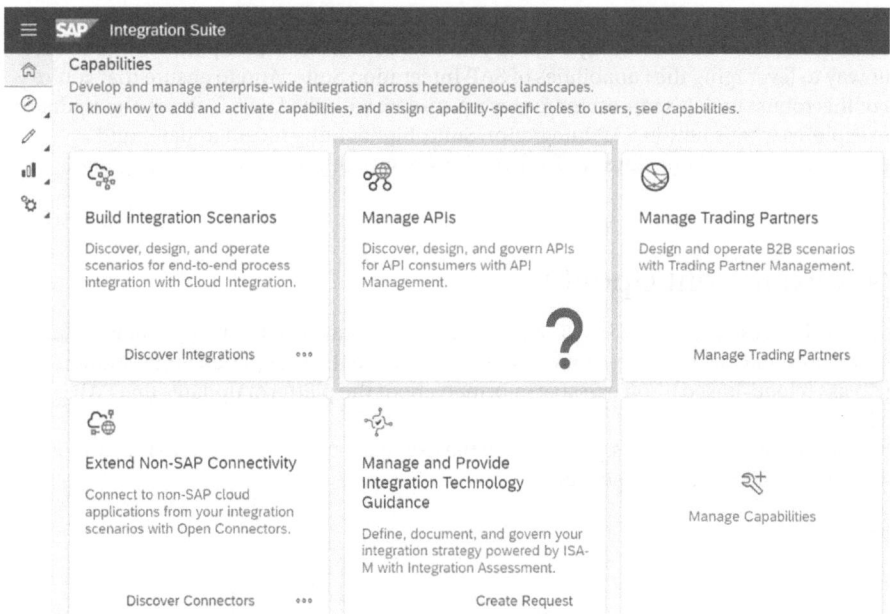

Figure 2-80. *API Management not provisioned*

On the left side of the Integration Suite dashboard, Expand the Settings tile, click the APIs, and follow the screen prompt to resolve, as shown in Figure 2-81.

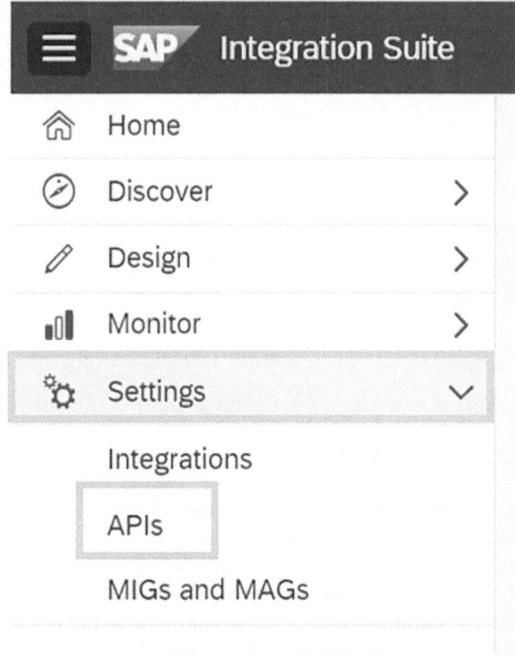

Figure 2-81. *APIs to set up API Management*

This section briefly discussed the common errors you get during the installation of the trial account in SAP Integration Suite. But once you've successfully created your trial account and set up the components, you'll be well on your way to leveraging the capabilities of SAP Integration Suite. And to ensure that you're getting the most out of this robust tool, it's important to stay up-to-date with the latest features and updates.

The next section explores what's new in SAP Integration Suite, highlighting recent updates and enhancements to help you achieve your business objectives and streamline your enterprise integration processes.

2.1.11 What's New: Recent Updates

You can often learn about changes/updates to SAP Integration Suite. You can also learn about new features and capabilities delivered periodically every six months or a year because the Integration Suite solution is SAP-managed true iPass (cloud-based). You can stay informed about the platform updates that SAP is delivering periodically.

The pop-up screen shown in Figure 2-82 is presented to the user each time they log in to Integration Suite, and It can be disabled by clicking the checkbox.

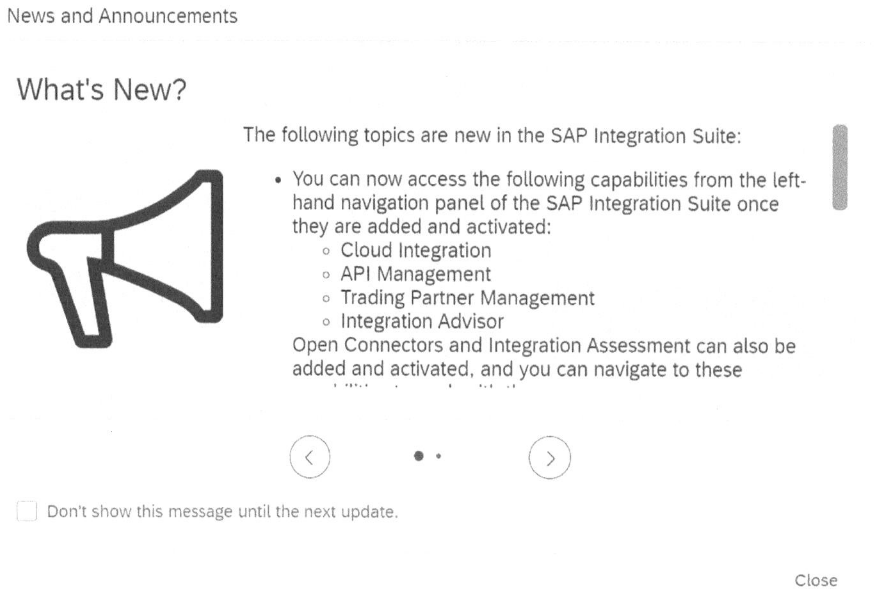

Figure 2-82. *What's New? in SAP Integration Suite*

2.1.12 Access Integration Suite: Bookmark URLs

If you have followed all the steps for the trial account setup correctly and assigned the right role collection, your SAP BTP Integration Suite is fully provisioned and ready to use for further learning.

After successfully installing SAP BTP, you can access all the capabilities over the browser.

Depending upon your roles, personas, and job description, you may have access to one or many of the capabilities of SAP Integration Suite. If you have access to Integration Suite, you can navigate to all the capabilities from the dashboard. But if your role is defined to one of the capabilities like Cloud Integration or API Management, you can navigate to the capabilities separately by sharing the link to the application with your team individually. Every organization has different URLs depending on the org, space, and service instance created.

For example, the trial account has generated the following Integration Suite capability URLs.

- Cloud Integration

    - Discover Integrations
      `https://4d5b5bf7trial.integrationsuite-trial01.cfapps.us10-001.hana.ondemand.com/shell/integration.`

    - Create Integrations
      `https://4d5b5bf7trial.integrationsuite-trial01.cfapps.us10-001.hana.ondemand.com/shell/design`

- API Management

    - Design APIs
      `https://4d5b5bf7trial.integrationsuite-trial01.cfapps.us10-001.hana.ondemand.com/shell/develop`

    - Discover APIs
      `https://4d5b5bf7trial.integrationsuite-trial01.cfapps.us10-001.hana.ondemand.com/shell/api`

- Trading Partner Management
  `https://4d5b5bf7trial.integrationsuite-trial01.cfapps.us10-001.hana.ondemand.com/shell/tpm`

- Integration Advisor

    - Discover Type Systems
      `https://4d5b5bf7trial.integrationsuite-trial01.cfapps.us10-001.hana.ondemand.com/shell/typesystems`

- Create MIGs
  `https://4d5b5bf7trial.integrationsuite-trial01.cfapps.us10-001.hana.ondemand.com/shell/migs`

    - Create MAGs
      `https://4d5b5bf7trial.integrationsuite-trial01.cfapps.us10-001.hana.ondemand.com/shell/mags`

- Cloud Connector
  `https://4d5b5bf7trial.integrationsuite-trial01.cfapps.us10-001.hana.ondemand.com/shell/typesystems`

- Integration Assessment
  `https://4d5b5bf7trial.intas-cpitrial06.cfapps.us10.hana.ondemand.com/app/index.html`

2.1.13 Summary

This chapter covered the main characteristics and advantages of SAP Integration Suite, including its support for hybrid integration situations, management and security of APIs, and end-to-end visibility and monitoring of integrations. The many capabilities of SAP Integration Suite, such as the Cloud Platform Integration, API Management, Integration Advisor, Trading Partner Management, Migration Assessment, Open Connectors, and Integration Assessment.

The chapter also covered the functionalities and the role collections which are necessary to assign to the user for functioning the capabilities. It went deep into the roles and role collections in SAP BTP cockpit, and you learned how to set up a trial account.

The upcoming chapters explore each of the Integration Suite capabilities in detail, starting with SAP API Management.

CHAPTER 3

SAP Integration Advisor

This chapter examines the SAP Integration Advisor, a cloud-based service that facilitates the integration process. Organizations may link their numerous systems and applications more easily thanks to SAP Integration Advisor's extensive range of tools for creating, managing, and testing integration scenarios.

Each section covers a distinct feature of the SAP Integration Advisor and is organized into a separate section. You start by learning a general overview of the tool, along with the important terms and vocabulary that go along with it. The first setup procedure, which entails setting up the library of type systems and developing unique message types, is then covered.

You learn all about message implementation guidelines and mapping guidelines, the export of documentation and runtime artifacts and mapping artifacts in SAP Cloud Integration, how to configure OAuth client credentials in a Cloud Foundry context, and the significance of data security and privacy.

3.1 Overview of SAP Integration Advisor

A cloud-based tool called SAP Integration Advisor may assist you in simplifying and speeding up the implementation flow of your B2B/A2A and B2G integration process. It makes it simple to produce integration content using a crowd-based machine-learning technique. According to tests, Integration Advisor may accelerate the production to the deployment of content by roughly 60%. Additionally, you may manage and distribute your artifact and use the artifact supplied by other application users with comparable commercial requirements.

The main issue with B2B/A2A/B2G integration is that several business partners use various industry standards, like UN/EDIFACT, SAP IDoc, and ASC X12, to mention a few. Integration Advisor resolves this issue. It takes much effort to manually design a new interface for every new standard created to enable this integration. As a place to start, Integration Advisor provides you with a library of type systems. Use the messages in this type of system library to produce new message implementation guidelines, mapping guidelines, and runtime artifacts with various integration solutions, such as SAP Cloud Integration and SAP Process Orchestration.

The application's dashboard is shown in Figure 3-1.

© Jaspreet Bagga 2023
J. Bagga, *Introduction to Integration Suite Capabilities*, https://doi.org/10.1007/978-1-4842-9630-1_3

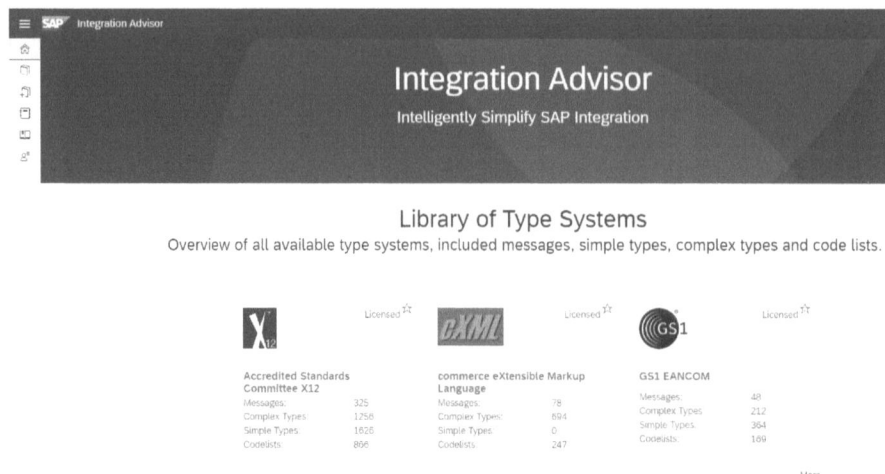

Figure 3-1. *Integration Advisor dashboard*

Next is an overview of Integration Advisor.

3.1.1 Overview of SAP Integration Advisor

SAP Integration Advisor has various sections: type systems, custom type systems, message implementation guidelines, and message application guidelines. We will go through each sections one by one.

3.1.1.1 Type System Overview

The B2B/A2A/B2G standard maintenance organizations have made several message templates available in the library of type systems. The organization that owns each type of system creates and maintains it. For example, SAP SE develops and maintains the type of system SAP IDoc. Similarly, ANSI ASC X12 maintains the ASC X12 type system. If you and your business partners use these industry standards for your business messages, this is a fantastic place to start when establishing your B2B interface. Figure 3-2 shows the Library of Type Systems dashboard in SAP Integration Suite.

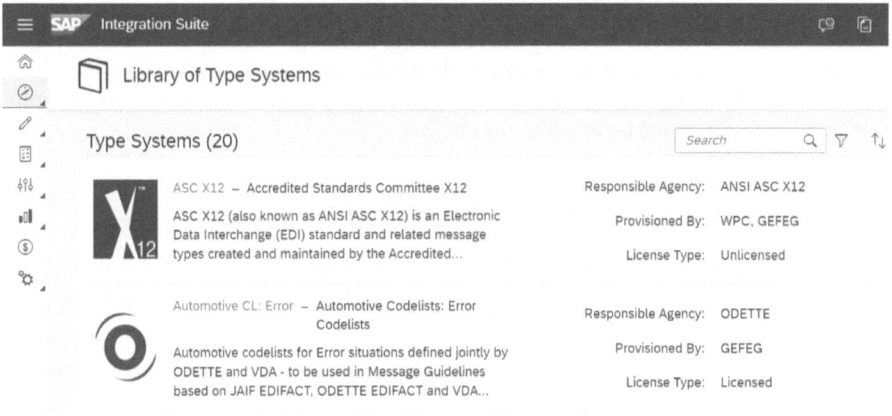

Figure 3-2. *SAP Integration Advisor Type Systems*

3.1.1.2 Custom Type Systems

You can submit your messages to the library of custom type systems, allowing you to build MIGs and MAGs according to these custom message structures. B2B standards, proprietary communications not provided by Integration Advisor, or standard messages you have extended are very common scenarios for using a library of custom type systems. This feature allows you to do the following.

- Upload XSD files for custom interfaces

- Build MIG and MAG

- Make your MIG unique

- View a list of your personalized messages

3.1.1.3 Message Implementation Guidelines

A MIG (Message Implementation Guidelines) is used as a source or target in a mapping guideline. This starts with a message from the type system that applies to your case.

The MIG has all the details needed to create a unique message interface. To make sure you don't need to consult any extra documentation to implement the interface, Integration Advisor uses the message in a type system as a starting point. By offering a MIG, you can guarantee that all users involved in implementing the interface know the rules. The upcoming sections of the chapter look at creating and working with MIGs.

3.1.1.4 Mapping Guidelines

A *mapping guideline* (MAG) is a feature of SAP Integration Advisor that provides guidelines for mapping data between source and target messages. The mapping stage of an integration application like SAP Cloud Integration uses a MAG runtime artifact. This is a guide or reference for implementing mapping in the integration application. Users who utilize messages that follow the A2A/B2B type system standards find it easier to map messages when you provide such an artifact.

Based on source and target message implementation guidelines, a MAG is created. It shows the mapping of the specified nodes at either side, describes each mapping component in depth with definitions or notes, and offers further guidance for the transformation, such as functions or code value mappings.

After a brief introduction to SAP Integration Advisor, let's look at terms associated with SAP Integration Advisor so that they are familiar to you going further into the chapter.

3.1.2 SAP Integration Advisor Terminology and Glossary

With the growing complexity of business applications and systems, it is essential to have a common language and understanding of the terms used in integration processes. This section describes common acronyms used in SAP Integration Advisor.

- **Type system**: An independent framework for the design of reusable messages and data types. Its foundations include the following.

  - Responsible organization

  - Using syntax rules to represent syntax

  - Definition and use of schema

  - Modeling, naming, and organizing whole communications through a method

  - Template repository

- **Message Implementation Guideline (MIG)**: A thorough description of an A2A/B2B message type that considers only certain factors and limitations to satisfy the user's business purpose requirements in a specific business context.

- **Mapping Guideline (MAG)**: An artifact that you may utilize directly in the mapping process of an integration application is a MAG. Based on source and destination message implementation guidelines, a MAG is created. It shows how the defined nodes on either side are mapped, describes every mapping component in depth with definitions or notes, and offers additional guidance for the transformation, like functions or code value mappings.

- **Business context**: The formal description of a particular business situation as indicated by the values of several context categories, allowing for the distinctive differentiation of various business situations. The business context represents a very relevant part. It marks the MIG for the system to present you with an appropriate recommendation based on the specified business context. It is utilized for statistical data supplied in subsequent IA releases. As a result, it is necessary to set at least one business context value. Integration Advisor offers five separate business context categories with values for setting the owner's and the business partner's business context.

  - **Business process**: When defining a business scenario, the business activity that is being carried out is typically the most significant part of that circumstance. The business process context provides a method for clearly identifying the business activity.

  - **Product classification**: Refers to the elements of a business scenario that have to do with the products or services being traded, used, or otherwise involved in the business process.

  - **Industry classification**: This information describes the industry or sub-industry in which the business process is conducted.

- **Geo political**: Permits the depiction of business context elements relating to the area, nationality, or cultural elements with a geographic basis. The country is the starting point for the geopolitical context category.

- **Business process role**: This term refers to the characteristics of a business scenario unique to one or more players involved in the business process.

- **Message**: Describes the electronic exchange of business documents between two business partners is described by a message.

- **Complex types**: Complex reusable building blocks that are put together to form a message. For expressing the attributes of a complex item (e.g., Partner, Location, or Address), each complex type has a complex structure of child elements.

- **Simple types**: The behaviors and properties of leaf components used to convey data in an instance are defined by reusable simple types.

- **Qualifiers**: Integration Advisor's qualifier reference technique effectively supports the semantic qualification. The semantically generic components and the leaf elements, which add additional meaning through code values, are combined by the qualifier references. This innovative method is very effective since it raises the business meaning the level of understanding of customized interfaces in MIGs and, in addition, considerably decreases the effort required for mappings.

- **Notes**: A node's definition and significance are further explained in Notes. The numerous types of notes that include usage scenarios, additional restrictions, a comment, or usage examples can be used to categorize these details.

Next, let's look at SAP Integration Advisor's initial setup.

3.1.3 Initial Setup of SAP Integration Advisor

Let's look deeper into Integration Advisor, starting with the library of type systems.

3.1.4 Library of Type Systems

The B2B/A2A/B2G standard maintenance organizations have diligently curated numerous message templates, accessible in the library of type systems. Each type system is owned and maintained by its respective organization; for example, SAP SE oversees SAP IDoc, while ANSI ASC X12 manages the ASC X12 type system. Utilizing these industry standards for your business messages in collaboration with your partners provides an intelligent foundation for establishing a robust B2B interface.

Messages, associated complex types, basic types, and codelists comprise a typical type system library. The following are B2B standards.

- ASC X12

- UN/EDIFACT

- Odette

- cXML

- EANCOM

- Automotive EDIFACT

- JAIF EDIFACT

- VDA EDIFACT

- ODETTE EDIFACT

- ISO codelists

- UN/CEFACT recommendations

- Automotive codelists (JAIF, ODETTE, VDA, UNC, Error)

- GS1 codelists

Now that you have explored the basic functionalities of the library of type systems, let's move on to custom type systems. The SAP Integration Advisor allows users to define custom types that can be used to define message structures. This is particularly useful when the standard types provided by the library of type systems are insufficient for the user's needs.

3.1.5 Custom Type Systems

A custom type system is a set of message types and associated data structures that have been created especially to meet the particular integration needs of an organisation. A Custom Type System, as opposed to conventional type systems offered by B2B/A2A/B2G maintenance organisations, enables companies to develop and maintain own message formats and structures, offering flexibility and adaptation to their unique integration scenarios. Now that you know what custom type systems may offer you, let's have a practical experience creating one example.

You require a custom type system to upload and use custom messages in SAP Integration Advisor. To achieve the same, adhere to the steps in the following section.

3.1.5.1 Add a Custom Message

In SAP Integration Advisor, Add Custom Message is a feature that allows users to create custom messages using custom type systems. Custom messages can be created from scratch, or they can be based on existing messages. The following steps add a custom message in SAP Integration Advisor.

1. Go to the Custom Type Systems link in your application's left pane, as shown in Figure 3-3.

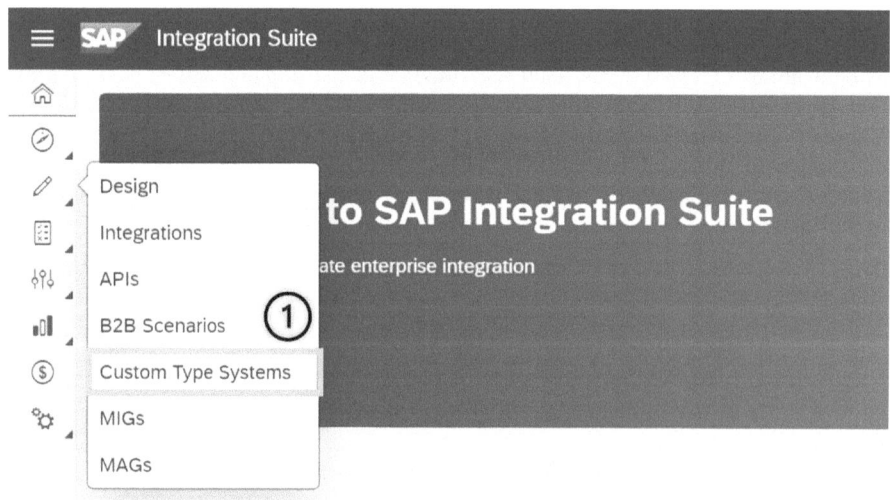

Figure 3-3. *Custom Type Systems*

2. Go to the Messages tab after selecting and opening the custom type of system, as shown in Figure 3-4.

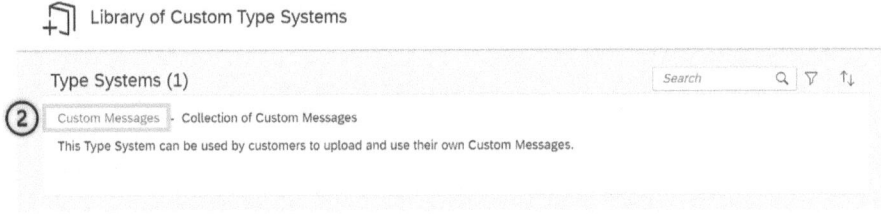

Figure 3-4. *Custom Messages*

3. The list of uploaded custom messages is shown on this tab. Select **Add** to add a custom message, as shown in Figure 3-5.

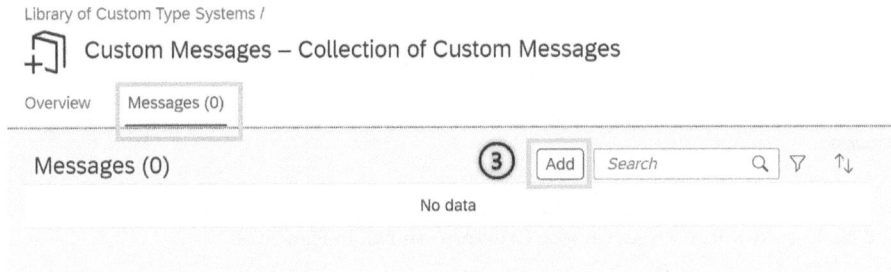

Figure 3-5. *Add custom message*

4. In the Upload XSD File stage, select the file and click **Browse**, as shown in Figure 3-6. Your back-end team provides this file.

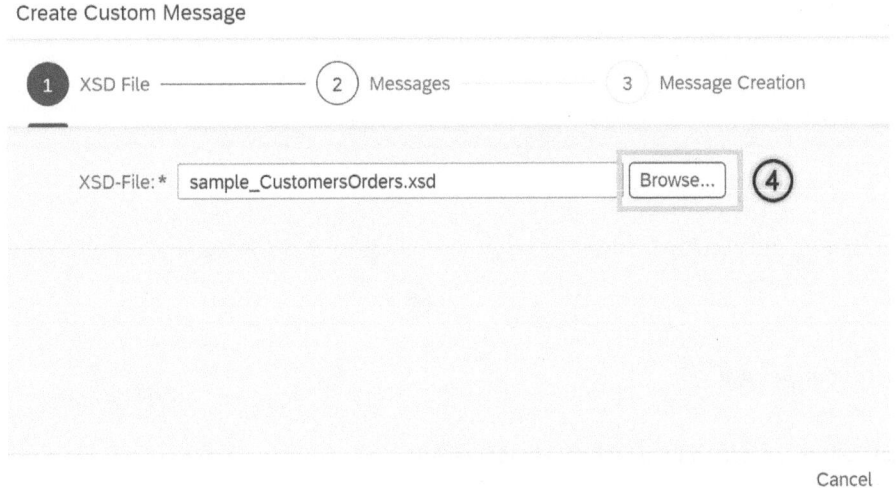

Figure 3-6. *Browse XSD file*

5. You see every message found in your XSD. Select the necessary message in Messages, as shown in Figure 3-7.

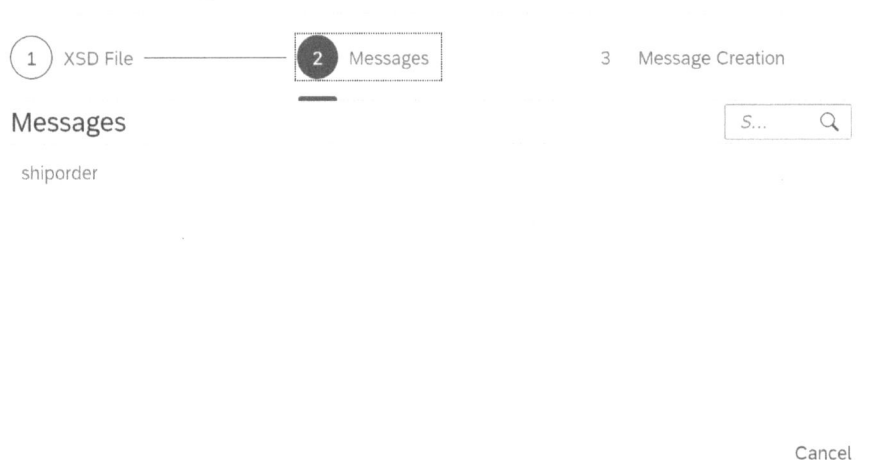

Figure 3-7. Select message

6. The Message Creation phase automatically fills in the message's details. You can check and change the information. Choose **Create**, as shown in Figure 3-8.

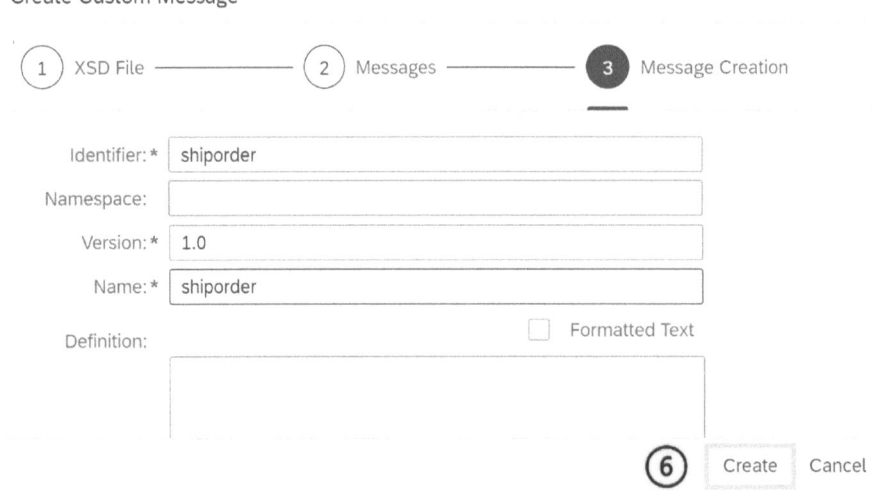

Figure 3-8. Create a custom message

7. The final custom message added is shown in Figure 3-9.

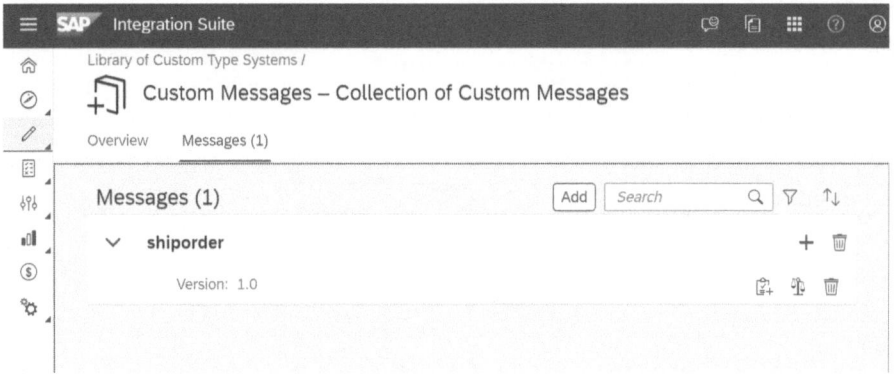

Figure 3-9. *Custom message*

Now that you have seen adding the custom messages in SAP Integration Advisor, it's important to understand how to manage them effectively. As with any other element in an integration flow, custom messages may need to be modified or removed as business requirements change or system updates are implemented. The next section explores how to delete custom messages in SAP Integration Advisor. Doing so ensures that only the necessary messages are in your flow, leading to improved performance and easier maintenance over time.

3.1.5.2 Delete a Custom Message

Users can delete a custom message that is no longer required or is disrupting the integration flow by using the Delete option. This can be helpful when the custom message is no longer needed or when the integration flow has changed, and the custom message needs to be removed.

The following steps delete a custom message.

1.   From the left pane, select **Custom Type Systems**.

2.   Go to the Messages tab after selecting and opening the custom type system.

3.   The number of uploaded custom messages for the type of system is shown on this tab. For the individual custom message you want to remove, select Delete, as shown in Figure 3-10.

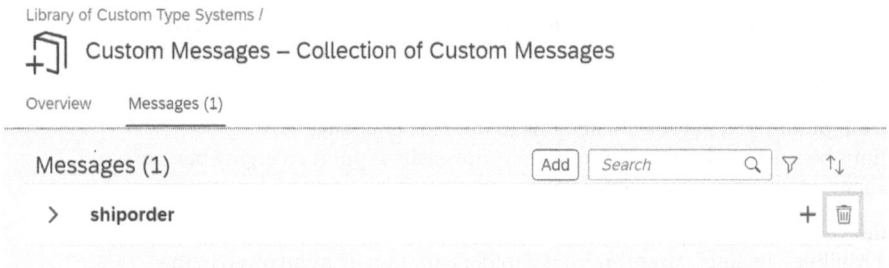

Figure 3-10. *Delete a custom message*

4. Click **Delete** after checking the appropriate boxes for the message, version, and revision, as shown in Figure 3-11.

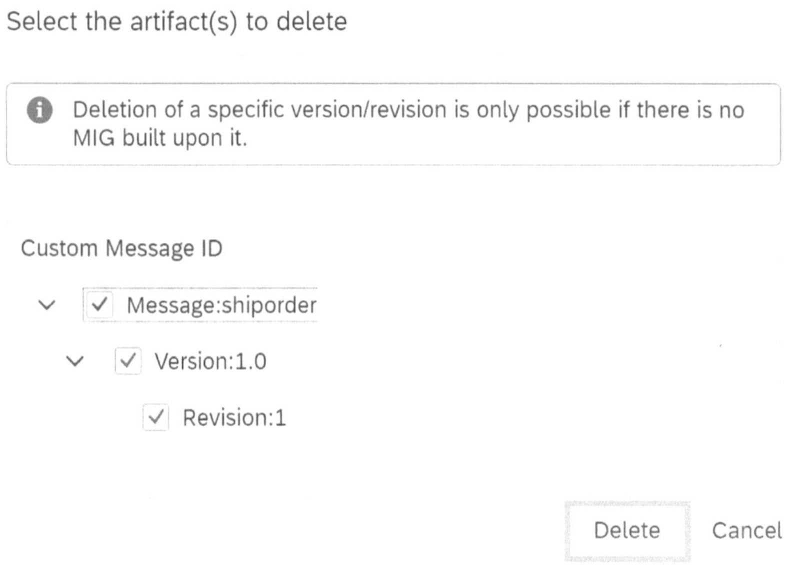

Select the artifact(s) to delete

> ℹ️ Deletion of a specific version/revision is only possible if there is no MIG built upon it.

Custom Message ID

⌄ ☑ Message:shiporder

  ⌄ ☑ Version:1.0

    ☑ Revision:1

Delete   Cancel

Figure 3-11. *Delete versions of custom message*

You have added custom messages to SAP Integration Advisor and looked at how to remove those messages from SAP Integration Advisor. Yet, there are some disadvantages to custom messages with which you need to be familiar. The disadvantages of custom messages in SAP Integration Advisor are covered in the next section.

3.1.5.3 Disadvantages of Custom Messages

Users should be aware of several disadvantages of using custom messages in SAP Integration Advisor. The following are some of these disadvantages.

- **Limited scalability**: Custom type systems may encounter performance and stability problems when dealing with big data sets and complicated data structures.

- **Lack of connection with other SAP tools**: Customized solutions might not work together easily with other SAP products, such as SAP Process Orchestration or SAP HANA, necessitating extra development work.

- **Limited adaptability**: It may be challenging to upgrade or change existing integrations because custom type systems may not easily support changing business requirements or adapt to new technologies.

- **Dependence on IT expertise**: Because custom type systems need specialist technical abilities, business users and stakeholders may not be as involved in the integration process.

- **Risk of data loss or corruption**: Because custom type systems cannot have strong error handling and data validation methods, there is a higher chance that data is lost or corrupted during integration.

Now that you've covered custom type messages in SAP Integration Advisor, it's time to dive into message implementation guidelines. MIG is a set of best practices and guidelines for designing and implementing messages in your integration flow. By following MIG, you can ensure that your messages are consistent, reliable, and optimized for performance.

The next section discusses MIG's key principles and walks through a step-by-step process for implementing messages in SAP Integration Advisor. You'll learn to design messages that meet your business requirements, map data between different systems, and troubleshoot common issues.

3.1.6 Message Implementation Guidelines (MIGs)

A MIG (Message Implementation Guidelines) is the source or target of SAP Integration Advisor mapping guidelines. This starts with a message from the type system that applies to your case.

The MIG has all the details needed to construct a unique message interface. To ensure you don't need to consult any other documentation about developing the interface, SAP Integration Advisor uses the message in a type system as a starting point. By offering a MIG, you can guarantee that all users involved in implementing the interface know the rules.

3.1.6.1 Searching MIGs

The list of all the MIGs you've developed is shown on the Message Implementation Guidelines screen. MIGs are found and shown using filters that consider a variety of factors. To specify your search criteria, use the free text filter search or the extended filter at the top of the screen. You can filter and display MIGs that match your search text using the free text filter search option. It can be utilized as a speedy filter to compare your search string to several MIG properties.

You can also select Extended Filter to filter MIGs according to requirements. You can list MIGs that match your query by using the necessary filter fields. You can further modify the filter bar by selecting Filters, as shown in Table 3-1.

Table 3-1. *Filter Criteria*

Filter	Description
MIG Name	The MIG's name is listed in the Overview tab's General Information section.
Summary	The MIG summary is available in the Overview tab's Documentation section.
MIG Version	The MIG version is listed in the Overview tab's General Information section.
Status	This is in the Overview tab's General Information section and includes the MIG status.
Type System	The MIG type system is described in the Overview tab's General Information section.
Type System Version	The type system version of the MIG is available in the Overview tab's General Information section.
Message Type	The Overview tab's General Information section lists the MIG's message type.
Created By	The distinctive ID of the user who made the MIG. It is accessible in the Overview tab's Administrative Data section.
Created Between	The time frame that the MIG was produced.

(continued)

Table 3-1. (*continued*)

Filter	Description
Modified By	The distinctive ID of the person who made the MIG modifications. It is accessible in the Overview tab's Administrative Data section.
Modified Between	The timestamp that the MIG was changed.
Last Imported By	The distinctive ID of the person who most recently imported the MIG. It is accessible in the Overview tab's Administrative Data section.
Last Imported Between	The timestamp in which the MIG was brought in.

This section discussed MIGs and the filter criteria. Next, let's look at a practical example of a MIG.

3.1.6.2 Create MIGs: Practical Example

The source and target used in a mapping guideline are known as MIGs. You create a MIG based on your requirements using a template from one of the type systems offered in the collection as the foundation. The application also makes recommendations on which nodes would be best for you, using the current MIGs as a point of reference.

1. Select the MIGs symbol in the left pane. Create a MIG by selecting **Add** on the detailed screen, as shown in Figure 3-12.

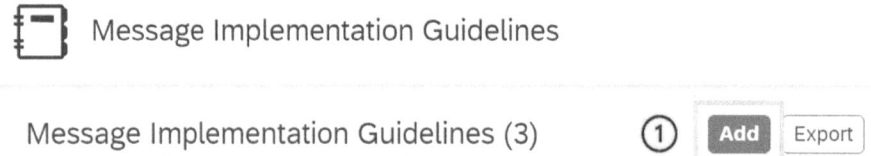

Figure 3-12. *Create new MIG*

2. In the wizard, you see the choices presented in the Type System step. You can select Custom and Standard MIGs. In Custom, you must configure all the details. In Standard, you can use the predelivered, as shown in Figure 3-13.

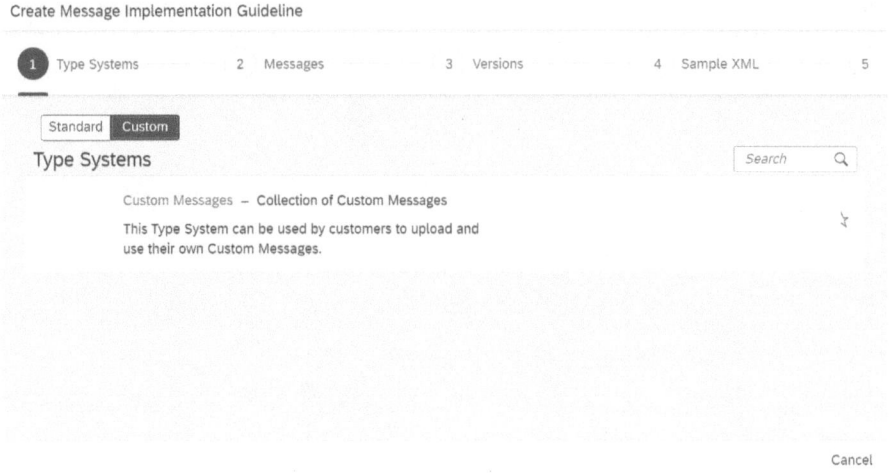

Figure 3-13. *Configure MIG*

3. In the Messages step, a list of all messages for the selected type system is visible. A message can be chosen from the list.

4. The message's available versions are shown in the Versions step. From the list, select the desired version.

5. Update the following fields in the MIG creation step.

   - Name: The MIG's name.

   - Direction: To improve the proposal service's accuracy, direction is employed in conjunction with your own and your partners' business contexts. From a B2B engagement perspective, the values provided describe the direction.

     - In: A business partner sends you a message, which is described in the MIG.

     - Out: In the MIG, a message you deliver to a business partner is called *out*.

     - Both: The MIG refers to a message that may be sent to or received from a partner.

   - Summary: You can enter a text description in Message Implementation Guideline (MIG). In the MIG overview list, this explanation is a brief text document.

   - Applied to Business: Choose the business context you want to include. You are given more drop-down list options based on the business context that you add.

6. Click Create.

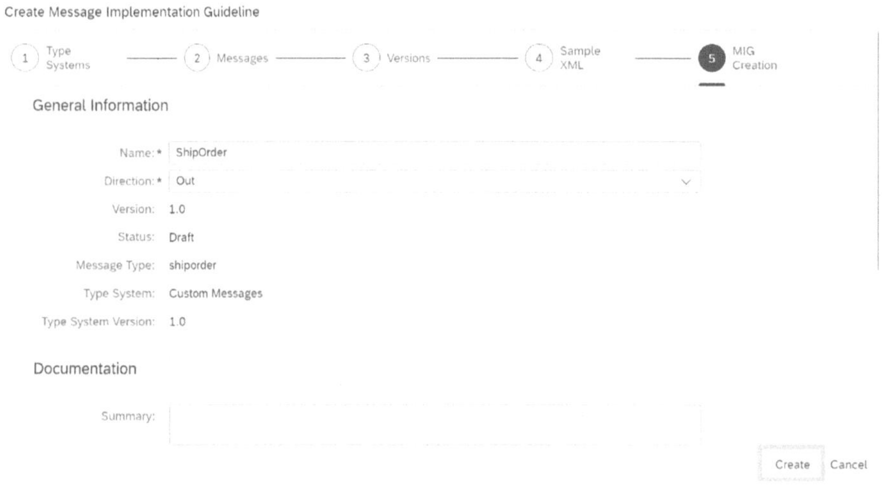

Create Message Implementation Guideline

①   Type
     Systems   ——   ② Messages ——   ③ Versions ——   ④ Sample
                                                         XML   ⑤ MIG
                                                                        Creation

General Information

Name: *	ShipOrder
Direction: *	Out
Version:	1.0
Status:	Draft
Message Type:	shiporder
Type System:	Custom Messages
Type System Version:	1.0

Documentation

Summary:

Create   Cancel

Figure 3-14. *Create MIGs*

You work with MIGs in the next section and examine nodes, codelists, and qualifying nodes.

3.1.6.3 Working with MIGs

By generating a new MIG or selecting one from the MIG area of the SAP Integration Advisor dashboard, you gain access to the MIG that you want to work with.

1. Open your MIG in SAP Integration Advisor and Click Edit.

2. You can use the arrows to expand and collapse the nodes.

3. When you select Get Proposals, a proposal indicator shows which fields might be most appropriate. This is computed using the various MIGs that are readily available.

4. Select the checkbox next to the node and then pick the element to edit the node's properties. By doing this, you may inspect and modify the parameters in the Details pane.

5. Set a review status for the Message Implementation Guideline or compare the MIG when reviewing it.

   • To indicate a node's property editing state, set the Review Status property.

   • Several MIG specifics can be compared, including documentation, properties, code values, status, and example values.

3.1.6.3.1 Working with Nodes

The fundamental units of integration scenarios are nodes. They stand in for distinct processing operations carried out on messages as they move through an integration scenario. Nodes can perform a wide range of tasks, including data transformation, message routing to other systems, and message content validation.

- A node's properties are editable.

- You can edit and change the nodes' characteristics after creating a MIG for the structure.

- The Edit option allow you to change the MIG.

- Select the Nodes tab, then select the node you want to change. The node details are shown to the right of the structure (see Figure 3-15).

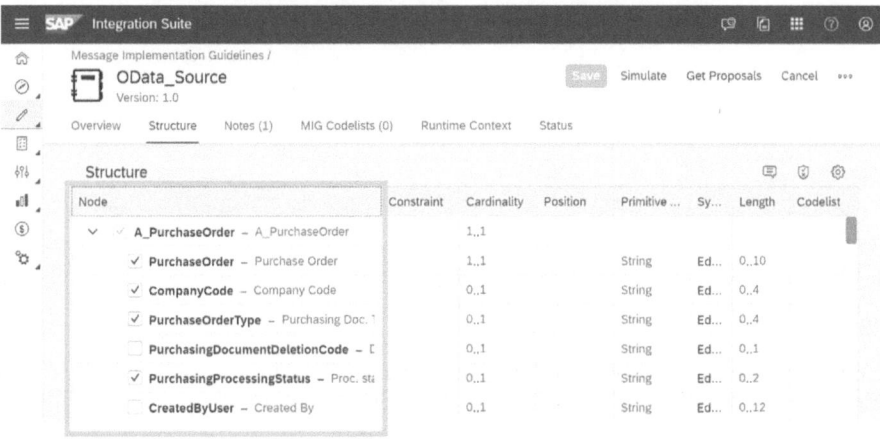

Figure 3-15. *Nodes in SAP Integration Suite*

3.1.6.3.1.1 Notes Tab

You can add notes specific to that node using this tab, as shown in Figure 3-16.

- You can add notes specific to that node using this tab.

- Assign the note a Category. Various kinds of notes can be added to a node, including the following.

  - Example

  - Usage

  - Comment

  - Constraint

  - Technical Information

- In the Number section, enter a number for the note type you are creating.

- Text should be entered in the Text area. By checking the box next to "Formatted Text," you can format the text.

- Select Apply.

- Using the button, you can also sort the notes inside each category according to quantity.

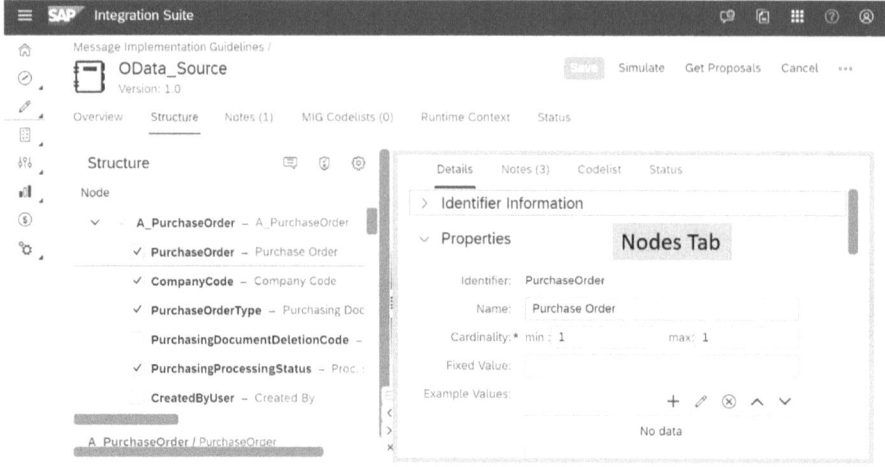

Figure 3-16. *Nodes tab*

3.1.6.3.1.2 Status Tab

1. Setting a node's review entry can be done using the Status tab. This aids in locating the nodes you wish to focus on later, as shown in Figure 3-17.

2. Multiple users can review documents using this tab as well.

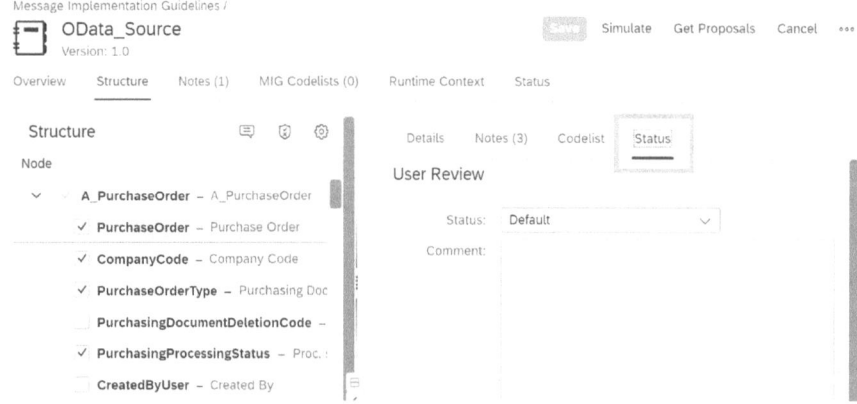

Figure 3-17. *Status tab in SAP Integration Advisor*

The next section examines codelists.

3.1.6.3.2 Codelists

A codelist is a list of abbreviations with meaningful definitions, which helps business partners comprehend the code values and lowers the risks of working together.

Codelists are created and updated by multinational Standards Developing Organizations (SDO). For instance, the names of countries and other geographical entities can be identified using values from the International Organization for Standardization (ISO) codelist ISO 3166-1.

Elements that are semantically general in a message implementation guideline are given a more specific meaning by code values (MIG). For instance, using a code value to specify whether a generic element like "Party" refers to a "seller party" or a "buyer party" is an example of how code values might be utilized.

You can view every codelist made available by a type system.

The library of type systems contains all the codelists a type system has made public. Next to any chosen codelist, there are brackets with the total code values. When you choose a specific codelist, the whole list of code values and their descriptions are displayed in the Code Values tab.

Four main types of codelists are now supported by SAP Integration Advisor.

- Codelists based on the type system come from the type system itself and only apply to MIGs that share the same type system.

- Reusable codelists are maintained by organizations like ISO and the UN/ECE (United Nations Economic Commission for Europe). They can be utilized in MIGs built on any kind of system. Consider the UN/ECE-based codelist for measure unit codes or the ISO-based codelists for nation, language, and currency codes. They are portrayed as distinct type systems that are limited to codelists.

- MIG codelists are defined locally by users within a MIG.

- Local codelists are defined specifically for using that element and cannot be used elsewhere. They are frequently employed in connection with custom messages.

The associated checkmarks in the Codelist column let you know which MIG elements are references to codelists. The Details panel for the selected element opens when a codelist-referencing element is selected. The code values and other information about the codelist are displayed on the Codelist tab of the Details panel. Any code value is selectable or deselectable.

The next section focuses on how to create a MIG codelist in SAP Integration Advisor.

3.1.6.3.2.1 Create MIG Codelist

A MIG codelist is a message implementation guide used in SAP Integration Advisor to specify a list of acceptable values for a specific message field or element. Codelists can be used to guarantee that messages follow certain standards and norms and to lower the possibility of mistakes or inconsistencies in message data.

In SAP Integration Advisor, you can do the following.

- Add code values manually to MIG codelists

- Create a MIG codelist by submitting a CSV file

Next, let's briefly look at creating a MIG codelist in SAP Integration Advisor.

3.1.6.3.3 Add Code Values Manually to MIG Codelists

If the reusable or type system–based codelists do not satisfy your needs, create new MIG codelists.

1. To add a new codelist, select Add on the MIG Codelists tab.

2. In the relevant fields in MIG Codelists, enter the codelists ID, name, and definition, as shown in Figure 3-18.

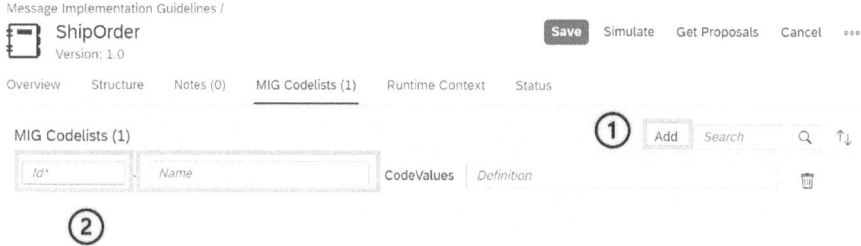

Figure 3-18. *Insert codelist manually*

3. Click CodeValues, as shown in Figure 3-19.

4. Select Add in the resulting panel on the Code Values tab to input the necessary values, as shown in Figure 3-19.

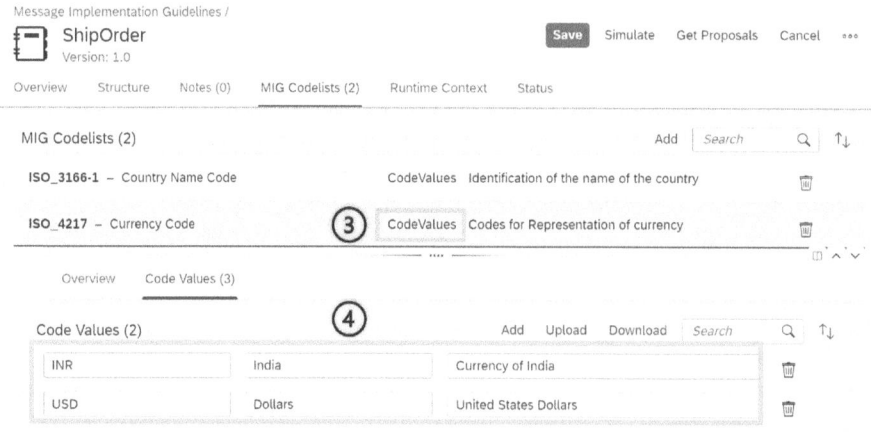

Figure 3-19. *Define code values*

You have created the MIG codelist by creating the code values manually to the MIG codelist. Next, you learn to create the MIG codelist using a CSV file.

3.1.6.3.4 Create MIG Codelists by Submitting a CSV File

In SAP Integration Advisor, you can create MIG codelists by submitting a CSV file. This can be a useful way to quickly create codelists with many values or to import values from external sources such as spreadsheets or databases.

The following steps create a MIG codelist by submitting a CSV file in SAP Integration Advisor.

1. To add a new codelist, select **Add** on the MIG Codelists tab.

2. In the relevant fields in **MIG Codelists,** enter the codelist's identification, name, and definition.

3. Click **CodeValues**.

4. To import the CSV file containing the code values, select **Upload** in the resultant panel's Code Values tab, as shown in Figure 3-20.

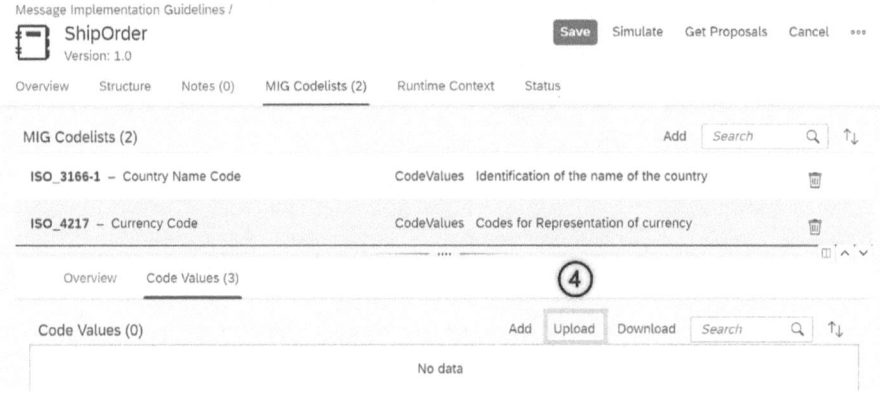

Figure 3-20. *Code Values Upload*

5. Select the CSV file holding the code values in the Code Value Upload dialog box, then browse for and upload it.

6. If the CSV file has any code values that include escape characters, a list of those code values is shown when the file is uploaded. Click Remove after selecting the rows in which you wish to remove escape characters.

7. Select **Skip All** if you wish to skip this stage.

8. If you wish to add more code values to the uploaded values, select Add and fill in the necessary fields in the new entry.

9. Select **How to Download the template** from the Code Value Upload dialog box.

10. Save your changes after editing the downloaded template file CodeValueDownload.csv.

11. Choose **Select** to upload the modified file with code values.

Once the codelist has been created and published, it can be used in integration scenarios to validate message data against the predefined list of values. If a message contains a value not in the codelist, an error is generated, and the message is rejected.

Creating MIG codelists by submitting a CSV file can quickly and efficiently define large lists of valid values. It also allows values to be imported from external sources, making it easier to maintain consistency across different systems and partners.

3.1.6.3.5 Provide Leaf Nodes with Codelists

Codelists can be assigned to any leaf node of your choosing.

1. Choose the leaf node to which you wish to apply a codelist on the Structure tab.

2. In the next panel, select the Codelist tab. To choose a codelist from one of the following options, select Add, as shown in Figure 3-21.

   - MIG Codelists allows you to give the leaf node a codelist.

   - If you want to give the leaf node a codelist based on a type system, select from Type System and then the desired codelist from the drop-down box.

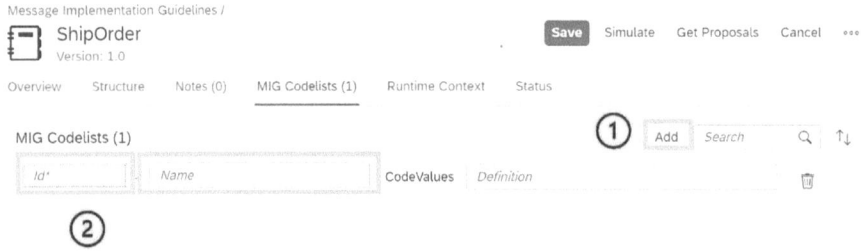

Figure 3-21. *Assign codelist to the leaf node*

3. The code values can be selected or deselected as necessary.

3.1.6.3.6 Qualifying Nodes

When creating your Message Implementation Guideline, you can qualify nodes to give the context of the semantically generic element.

The message structure consists of generic data segments that include logically connected information. The context of generic elements in your MIG message structure can be changed and precisely specified to provide the remaining data. For instance, it is unclear at the semantic level which date or address the message interface refers to when using a generic element like date/time reference or address. As a result, more details are needed to clear up any confusion about the semantic function of a segment. A given instance's qualifiers can be used to describe the type of address stored there.

The fields street, city, state, postal code, and country are typical components of an address segment. A delivery address, a contact address, or a billing address are examples of addresses that may appear more than once. In these circumstances, you may use a standard address segment structure and then use a qualifier to designate the kind of address kept inside a specific instance of a data segment.

The next section examines creating the qualifier marker in SAP Integration Advisor.

3.1.6.3.6.1 Creating a Qualifier Marker

Using the qualifier feature, you can set a qualifier marker that indicates the element qualified by another data element. Qualifier markers can be generated independently or automatically provided by message templates from type systems to suit your business context. To construct a qualifier marker, follow the steps outlined next.

You must go through the Codelist section before jumping on this chapter.

1. The leaf node you intend to utilize as the qualifying node should be selected on the **Structure** tab, as shown in Figure 3-22.

2. Select (**Add**) in the Qualifiers field of the resultant pane's Details tab, as shown in Figure 3-22.

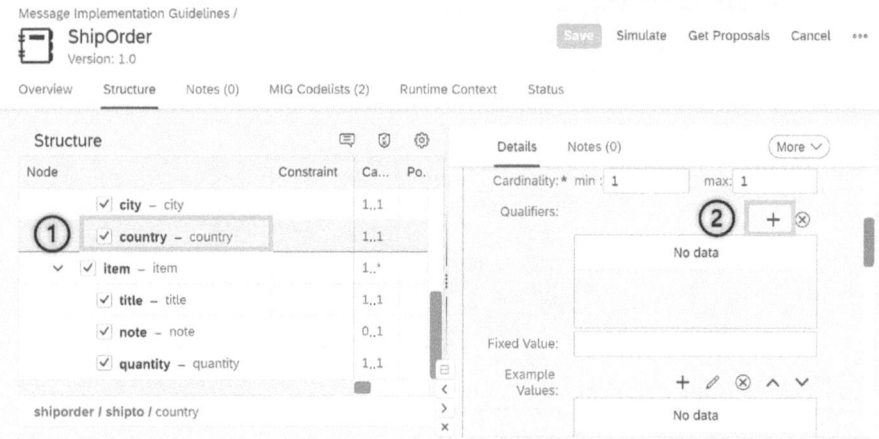

Figure 3-22. Add qualifier in MIG

3. Choose the node you want to qualify from the list in the Qualifiable Nodes dialog box, as shown in Figure 3-23.

4. Select **Add**, as shown in Figure 3-23.

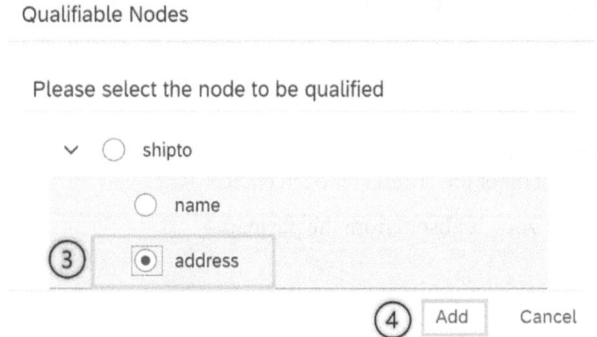

Figure 3-23. Insert qualifiable node

The next section examines creating a qualified instance.

3.1.6.3.6.2 a Creating Qualified Instance

In SAP Integration Advisor, a *qualified instance* is a specific instance of a message that meets predefined criteria or conditions. In the context of qualifying nodes, creating a qualified instance involves defining the criteria or conditions a message must meet to be processed by a particular node.

Qualifying nodes selectively process messages based on predefined criteria like content or metadata. For example, a qualifying node may only process messages that contain a specific value in a certain field or messages that meet certain security or compliance requirements.

The following steps create a qualified instance in a qualifying node.

1. Choose the node you want to qualify by right-clicking it on the **Structure** tab, as shown in Figure 3-24.

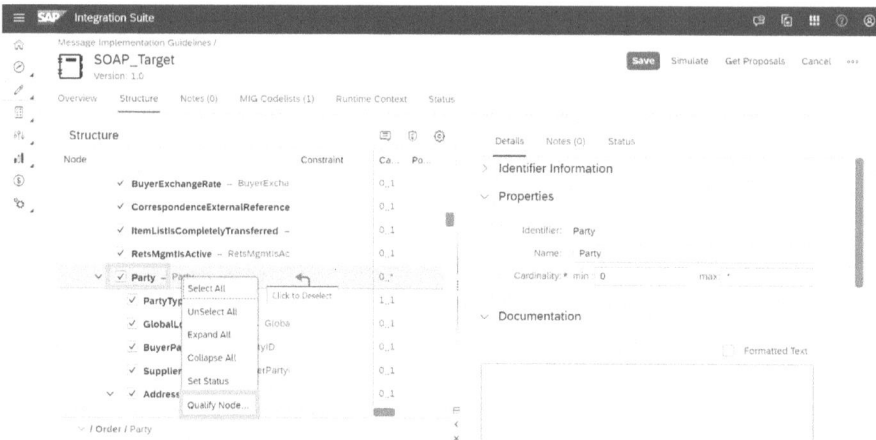

Figure 3-24. *Qualify Node…*

2. The context menu allows you to select **Qualify Node**.

3. This brings up the dialog box for **Creating Qualified Nodes**.

    •   The qualifiers list is displayed in the dialog box's Qualifiers section.

4. In the Selected Qualifier Marker drop-down menu, select the qualifier numbered 1.

5. This shows a list of possible values for the chosen qualifier marker. Select a value from the list by checking the box, as shown in Figure 3-25.

    •   In the list, you can select more than one qualifying value. Selecting more than one value indicates that the node must meet the criteria for each choice.

    Select **No Value** in the checkbox if the node is absent from the payload.

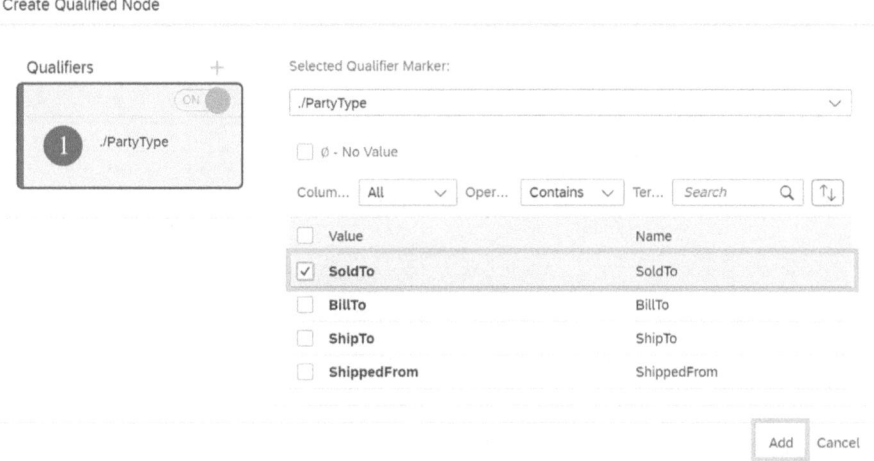

Figure 3-25. *Create Qualified Node*

6. To qualify the node, you have the option to add additional qualifiers. For a node to be qualified, it must have multiple qualifier markers. Once you decide to add a qualifier, the list of qualifiers will increase by one. However, if a qualifier already has only one qualifier marker, the button to add more qualifiers will be disabled.

7. Steps 4 and 5 must be completed to add the qualifying values for the newly added qualifier, which is designed as 2 in the list.

8. The specified qualifying value causes the qualifier marker arrow to turn blue. Details about the qualifying are shown in the right pane in the Qualifier tab.

9. The qualified node's original name is followed by the readable name of the qualifier value.

3.1.6.3.6.3 Additional Qualification Options

You have created a qualifier marker and a qualified instance in SAP Integration Advisor MIG from the context menu of the qualified node. You can now carry out the following operations. These operations can be applied to the items that you have created.

- You can alter the qualifier values of the node using **Change Qualified Node**, as shown in Figure 3-26.

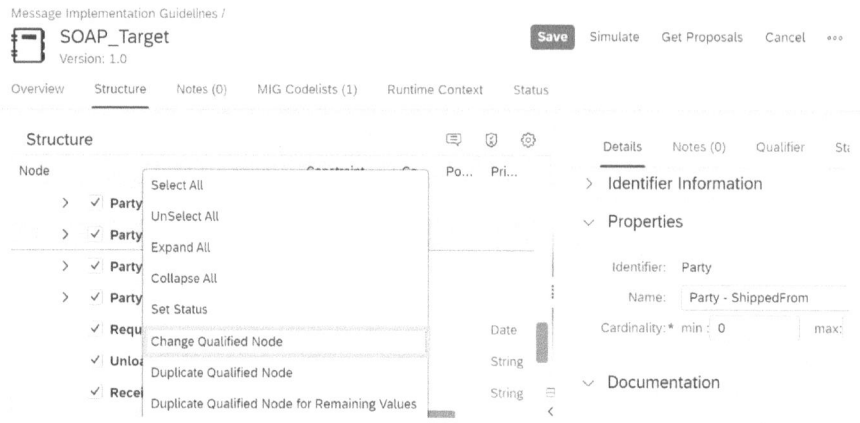

Figure 3-26. *Change Qualified Node*

- You can alter the qualified nodes' order by selecting **Change Order of Qualified Node**.

- Duplicate Qualified Node can make a second instance of the qualified node with a different qualifier value, as shown in Figure 3-27.

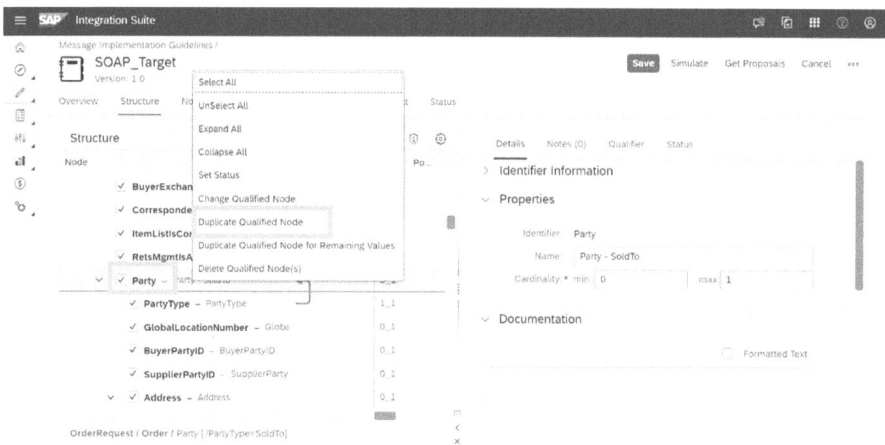

Figure 3-27. *Duplicate Qualified Node*

- The **Duplicate Qualified Node for Remaining Values** function generates a duplicate qualified node that contains values not present in other instances of the same element group.

- Another qualifier can be added to the node using Add Qualifier. This choice also allows you to convert a simple qualification into a compound qualification.

- You can eliminate one or more qualified nodes using the delete qualified node(s) command. You have two choices if you choose this option.

- **Delete this Qualified Instance**: If the chosen node has other qualified instances, selecting this option only removes this one and keeps the node qualified.

- **Delete All Qualified Instances**: Selecting this option causes a list of all nodes that are qualified for the specified node to be shown. Select the qualified node from the list if you want to keep the information for that node. This transfers the selected qualified node's information to the unqualified node.

3.1.6.4 Simulating a Message Implementation Guideline

This feature refers to the recommended best practices for simulating messages in the SAP Integration Advisor tool.

Simulating messages involves testing how a specific message is processed in the integration flow without sending the message through the entire flow. This allows for testing and troubleshooting before actual implementation.

1. Go to the Structure tab after opening your MIG.

2. To run an embedded simulation, select Simulate. The MIG can be simulated in two different ways.

   - **Simulate with MIG Example Data** duplicates the message implementation guideline using the sample MIG data already available.

   - **Simulate with Payload Data** requests that you supply an example XML payload file for the MIG simulation.

   The button that displays above the MIG structure after the simulation can also be used to re-run the simulation according to the option you selected. Figure 3-28 shows the options available in the Simulate in MIG.

   If there is a payload, the simulation is restarted using the payload file that was previously uploaded. Select **Simulate** and upload the payload file in Simulate with Payload data if you wish to simulate with a new or changed payload file.

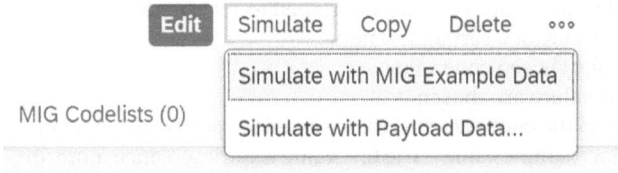

Figure 3-28. Simulate MIG

3. A new column has been added. The message structure contains simulation data filled with the following information.

   - The input values from the payload data are represented by the rows containing values in the structure.

   - The navigation buttons ‹› indicate multiple values in the payload if there are numerous instances of each. You can navigate to the additional instances of the payload values using these buttons.

- A group node additionally has the navigation button shown if it appears at least once in the simulation data.

- The leaf node is indicated in the structure by the icon ⌐✓, even if it is empty.

- The absence of the leaf node in the structure is indicated by the icon ⌐ˣ.

3.1.6.4.1 Simulate with MIG Example Data

When selecting this option, the application creates a complete payload file that matches the MIG definition and is utilized as input during simulation.

Cardinality

When simulating with the example data, the maximum cardinality determines the cardinality of the group node and leaf node.

- If maxOccurs=1, a node is only created once.

- If maxOccurs>1. The node is produced twice, regardless of the maxOccurs value. In doing this, it is possible to examine the node's repetition without producing a lot of payload files.

Each leaf node's data is retrieved in the order shown in Table 3-2.

Table 3-2. *Leaf Node Value*

Value	Description
Example Value	These values are those specified as typical values for a specific node. Each value from the list is assigned to a specific instance of a leaf node during simulation in the order supplied. The first value is retrieved once more for the following occurrence once the last value in the list has been reached. The same value is obtained for every instance of the leaf node if one example value is specified.
Fixed Value	The simulation then retrieves the fixed value for all instances of the leaf node if it does not have an example value but instead has a fixed value applied to it.
Code Value	The simulation then considers the data from the code values assigned to a leaf node if there is neither an example value nor a fixed value for a certain leaf node. The only values that are fetched are any chosen code values. All code values are utilized if no code values are chosen. The sequence in which the values from the code value list are assigned to specific instances of leaf nodes is the same as that used for example values. The first value is retrieved once more for the following occurrence once the last value in the list has been reached.
Dummy Value	A dummy value is generated for the node based on the node identification if nothing is defined.

3.1.6.4.2 Simulate with Payload Data

By selecting this option, you can view and upload local system data as a payload.

Uploading the payload data in XML format is recommended. There may be extra nodes or nodes in the payload you're attempting to import that aren't declared in the MIG. These nodes' information is shown in a dialog box after the payload data has been imported, but they won't be considered during the simulation. Select OK.

3.1.6.5 Version Management: Message Implementation Guidelines

Version Management is an important aspect of the MIGs in SAP Integration Advisor. It refers to the recommended best practices for managing different versions of the integration flow and messages within the Integration Advisor tool.

In SAP Integration Advisor, users can create different versions of the integration flow and messages. This is important because it allows for versioning and tracking changes made to the integration flow over time. This means that users can revert to previous versions of the integration flow or message in case of any issues or errors.

This section describes the stages a message implementation guideline goes through, the function of each phase, and how to activate a MIG.

An MIG can have one of three status types.

- Draft

- Active

- Deprecated

The Status column is set to Draft, and the version is set to 1.0 when MIG is created. This updated version of MIG can now be changed to suit your business needs. An existing MIG can also be copied. The Version field is set to 1.0, and the word *copy* is appended to the name of the copied MIG. The MIG's name can be changed later.

A MIG's status is changed to Active when it is activated. This MIG version is finalized once enabled and cannot be changed afterward. However, you can still export runtime artifacts and documentation from the MIG. By enabling a MIG, the message implementation guideline also be anonymously added to the knowledge graph, helping to advance and enhance the proposal service.

Open the active MIG version, then select Edit to create the subsequent draft version. Additionally, you can select Edit from the MIG overview's action button. You then be asked if you want to create a draft version. Select OK. The MIG's subsequent version is now created with a Draft status. For instance, if the MIG's active version is 2.0, the new draft MIG's version is set to 3.0.

A MIG becomes deprecated and has its Status field set to Deprecated after it has been successfully activated.

The following explains how to activate a custom message.

1. Choose ⭕ ⭕ ⭕ and then select Activate Message from your MIG, as shown in Figure 3-29.

Integration Suite
Message Implementation Guidelines /
SOAP_Target
Version: 1.0
Save
Simulate
Get Proposals
Cancel
Activate
Export
Overview
Structure
Notes (0)
MIG Codelists (1)
Runtime Context
Status
Structure
Node
Constraint
Cardinality
Position
Primitive ...
Sy...
Length
Codelist
OrderRequest – OrderReque
1..1
MessageHeader – Messa
1..1
Order – Order
1..1
ActionCode – ActionC
0..1
Token
xs...
2..2
PurchaseOrderID – Pu
0..1
Token
xs...
1..35
PurchasingDocumentT
0..1
Token
xs...
1..5
PurchasingOrganizatio
0..1
String
xs...
1..4

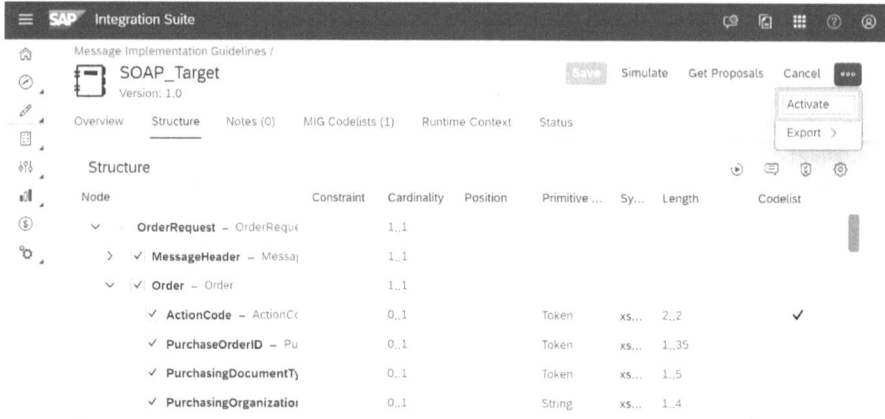

Figure 3-29. *Activate MIG*

2. To indicate the state of the activation, a message is presented. Additionally, you can see the status upon activation in the MIG's Overview tab's Message Status field.

3. After activating the customized message on which the MIG is based, you may activate your MIG, as shown in Figure 3-30.

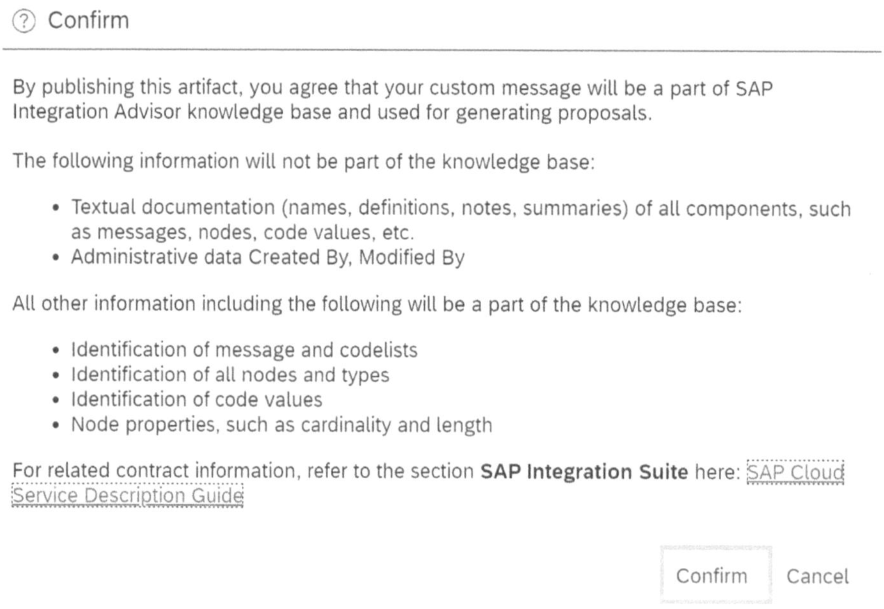

Confirm
By publishing this artifact, you agree that your custom message will be a part of SAP Integration Advisor knowledge base and used for generating proposals.
The following information will not be part of the knowledge base:
• Textual documentation (names, definitions, notes, summaries) of all components, such as messages, nodes, code values, etc.
• Administrative data Created By, Modified By
All other information including the following will be a part of the knowledge base:
• Identification of message and codelists
• Identification of all nodes and types
• Identification of code values
• Node properties, such as cardinality and length
For related contract information, refer to the section SAP Integration Suite here: SAP Cloud Service Description Guide
Confirm
Cancel

Figure 3-30. *Activate custom message*

Once the MIG version management has been created and justified, the next step is to Migrate Message Implementation Guidelines.

3.1.6.6 Migrate Message Implementation Guidelines

Your MIG can be transferred to a different (newer or older) version of the same type of system.

1. Select the MIG you wish to migrate on the Message Implementation Guidelines screen by selecting (More Options), and then select **Migrate**, as shown in Figure 3-31.

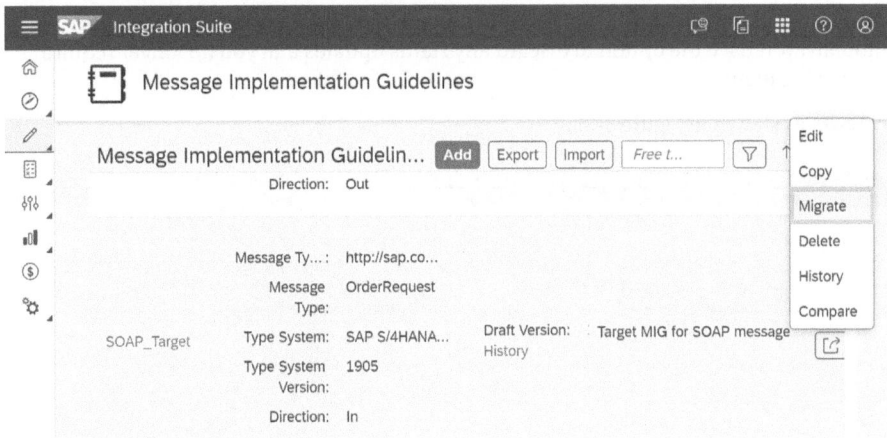

Figure 3-31. *Migrate MIGs*

2. Choose the version you want to migrate your MIG from the Versions tab of the Migrate Message Implementation Guideline dialog box.

    The Migrate Message Implementation Guideline dialog box shows the MIG's specifics, including its new version number.

    Choose **Create**, as shown in Figure 3-32.

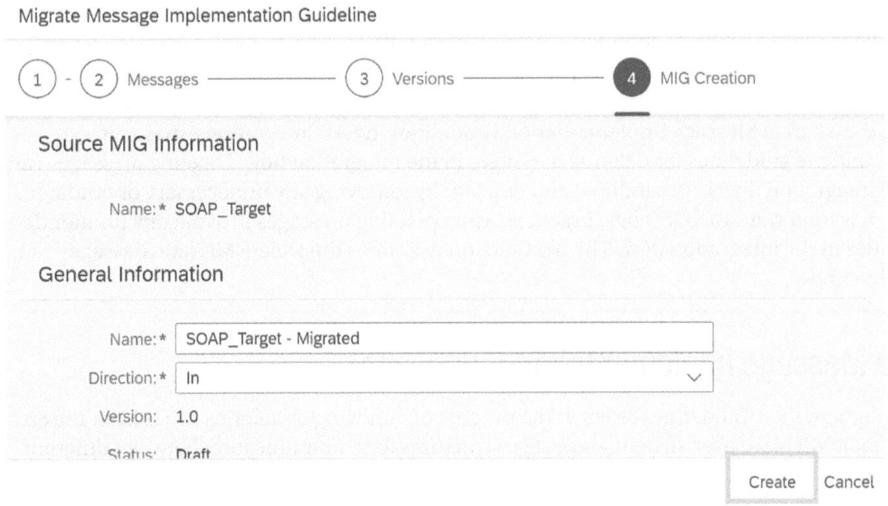

Migrate Message Implementation Guideline

(1) - (2) Messages ——————— (3) Versions ——————— (**4**) MIG Creation

Source MIG Information

    Name:* SOAP_Target

General Information

    Name:* | SOAP_Target - Migrated |

    Direction:* | In ⌄ |

    Version: 1.0

    Status: Draft

| Create | Cancel |

Figure 3-32. *Create new version of MIG*

3. There is a new MIG created.

4. Review the message structure on the Structure tab, then filter the Status column for any notifications to see the number of errors, informational messages, warnings, and comments.

   a. To display further information in the Notifications section, select the necessary cell in the Status column.

   b. To add comments and change the node's status in the User Review section, select Edit. Additionally, you have the option to discard any status updates that you no longer require after reviewing them.

   c. Select **Save**, as shown in Figure 3-33.

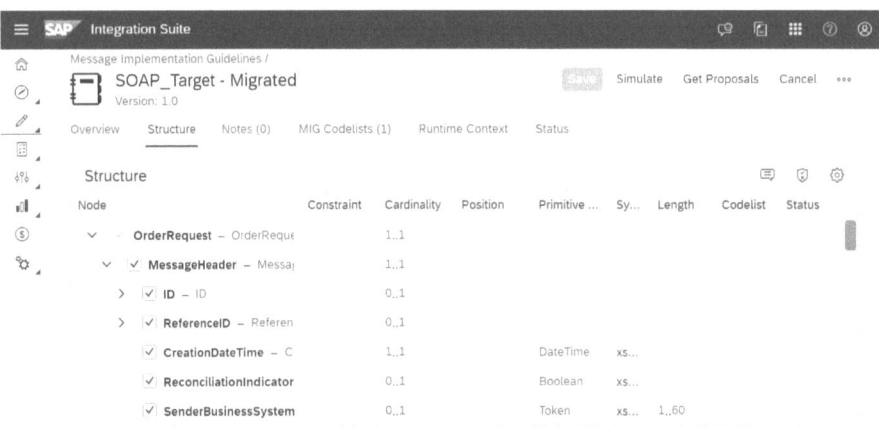

Figure 3-33. *Migrated MIGs*

5. For a comparison of the MIG and the underlying message template, select Show Changes Mode.

6. To alter your MIG to suit your company's needs, select Edit.

7. Select **Save**.

After reviewing the Migrate Message Implementation Guidelines in SAP Integration Advisor, it is important to understand the guidelines for deleting messages in the integration flow. Deleting messages can help ensure that the integration flow is streamlined and efficient by removing any unnecessary or outdated messages. However, it is important to follow best practices when deleting messages to avoid any unintended consequences or issues in the integration flow. The next section examines the Delete Migrate Message Implementation guidelines.

3.1.6.7 Delete Message Implementation Guidelines

Delete Message Implementation Guidelines refers to the process of removing or deleting MIGs from the SAP Integration Advisor tool. MIGs are used to define message structures for communication between different systems and applications.

Deleting MIGs can be necessary if a MIG is no longer relevant or if it needs to be updated or replaced. There are different options for deleting MIGs depending on their status.

- **Delete MIGs having Draft, Active and Deprecated Versions**: This option allows you to delete all versions of a MIG that are in draft, active, or deprecated status. This permanently remove the MIG from the system.

- **Delete Specific Version of a MIG**: This option allows you to delete a specific version of a MIG while keeping other versions intact. This can be useful if there is a mistake in a specific version or if it is no longer needed.

3.1.6.7.1 Delete MIGs with Draft, Active, and Deprecated Versions

Deleting MIGs with draft, active, and deprecated versions is a process in SAP Integration Advisor that allows users to remove MIGs that are no longer needed or relevant.

MIGs can have different versions, including Draft, Active, and Deprecated. When a new version of a MIG is created, the previous version is marked as Deprecated, indicating that it should no longer be used.

Users can choose to delete MIGs with any of these three versions, depending on their needs. Deleting a MIG removes it from the SAP Integration Advisor system, including all versions and associated data.

It is important to note that deleting a MIG cannot be undone, so it is recommended to carefully review and consider the implications before proceeding with deletion. Additionally, if a MIG is referenced by a Mapping Implementation Guideline (MAG), the MAG should be updated before deleting the MIG to avoid any potential errors or issues.

The following steps delete the MIG.

1. Select the MIG you want to delete on the Message Implementation Guidelines screen by selecting (More Options), and then select Delete, as shown in Figure 3-34.

2. Select the Delete entire Message Implementation Guideline (including history) option in the Delete Message Implementation Guideline dialog box.

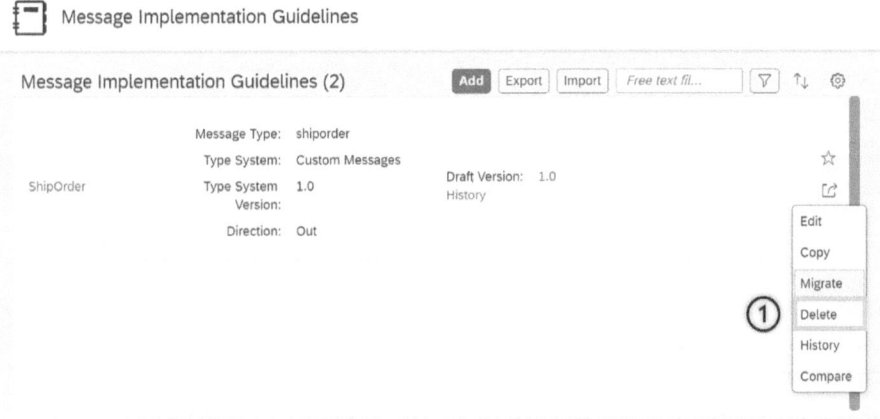

Figure 3-34. *Delete Message Implementation Guidelines*

3.1.6.7.2 Delete Specific Version of a MIG

Delete specific version of a MIG is a feature in SAP Integration Advisor that allows users to remove a specific version of a MIG that is no longer needed or relevant.

MIGs can have multiple versions, including Draft, Active, and Deprecated. Each version of the MIG contains unique data and may be associated with different Mapping Implementation Guidelines (MAGs) or other artifacts.

By using the "Delete specific version of a MIG" feature, users can select a specific version of the MIG to delete while keeping the other versions intact. This feature allows users to clean up their SAP Integration Advisor system and remove unnecessary versions of MIGs without losing all of the data associated with the MIG.

Using the screen for the MIGs

1. Select the MIG you want to delete on the Message Implementation Guidelines screen by selecting (More Options), and then select Delete.

2. Select the Delete this (latest) version option in the Delete Message Implementation Guideline dialog box.

Using Overview Tab

1. Pick the precise MIG version you want to remove by using the Overview tab on the Message Implementation Guidelines page.

2. Choose **Delete** or (More) and then choose Delete on the Overview page, depending on the size of your screen, as shown in Figure 3-35.

3. To confirm, select OK in the Confirm dialog box.

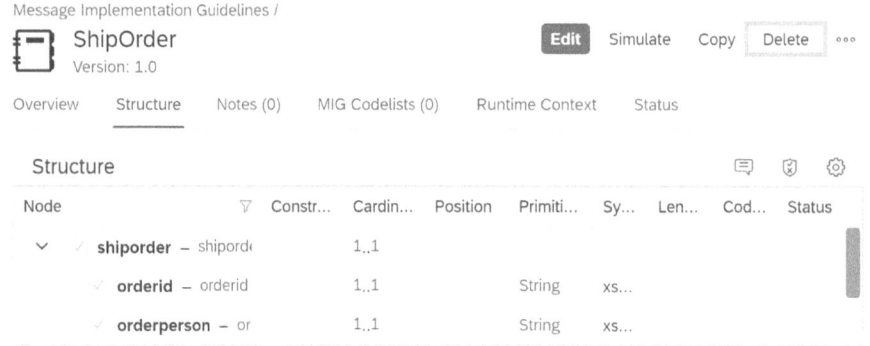

Figure 3-35. *Delete MIG*

The following explains the process using the History button.

1. Select **History** for the MIG you want to delete on the Message Implementation Guidelines screen, as shown in Figure 3-36.

2. Choose (action menu) on the version you want to delete from the Message Implementation Guideline History Versions table. Choose OK to confirm.

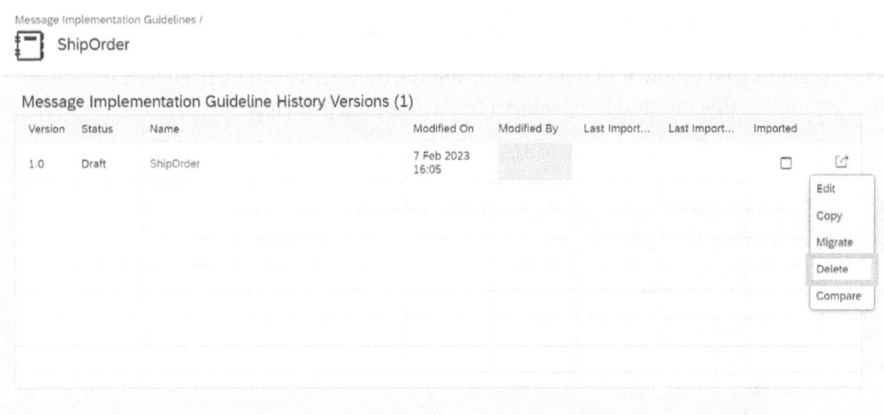

Figure 3-36. *Delete MIG using History link*

After creating a MIG, the most important step is to create the MAGs, as discussed in the next section.

3.1.7 Mapping Guidelines (MAGs)

An integration application's mapping step uses a MAG as a runtime artifact. This serves as a guide or reference for putting mapping into the integration application. For users who utilize messages that follow the A2A/B2B type system standards, you ease the mapping work by supplying such an artifact.

Based on source and destination MIGs, a MAG is created. It shows how the defined nodes on either side are mapped, describes every mapping component in depth with definitions or notes, and offers additional guidance for the transformation, like functions or code value mappings.

3.1.7.1 Code Value Mapping

The global code value mappings that you can use in your MAG are listed on the Code Value Mapping tab. To view the source and target code values, select and open a code value mapping. The free text filter (extended filter) offered at the column level allows you to further refine your search for a particular code value. When there are many code values, this is an effective approach to filter through them.

If you wish to change the target code values, select Edit.

Only the code values chosen in the source MIG are included in these global code value mappings. From the underlying codelist, select Add to add more source code values.

- If you want to add one source value from the list of source codes, do so.

- If you want to add all the values from the source codelist that were previously unselected, choose the remaining source values.

To learn more about MAGs, let's look at a practical example.

3.1.7.2 Create a Mapping Guideline (MAG): Practical Example

When mapping between systems that follow B2B type system standards, a MAG can be used as a source, a reference, or guidance. You utilize this method to develop a fresh mapping principle.

1. From the application's left pane, choose MAG, as shown in Figure 3-37.

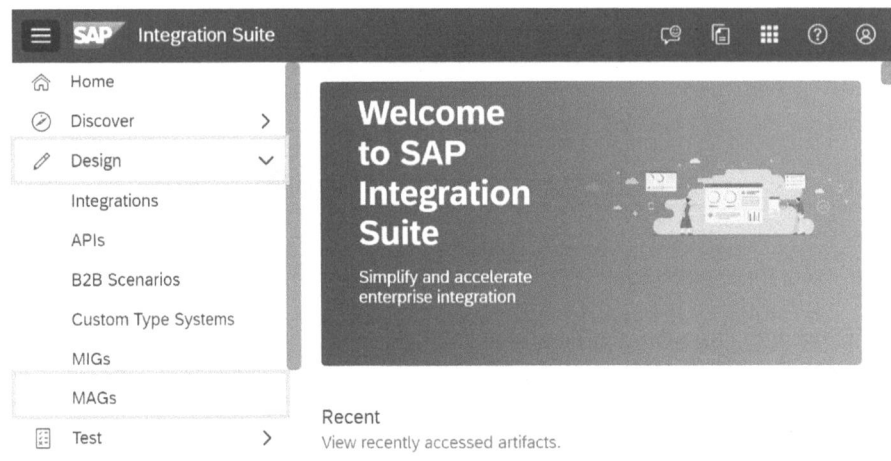

Figure 3-37. *Create MAGs*

2. Click **Add** in the Mapping Guidelines tab, as shown in Figure 3-38.

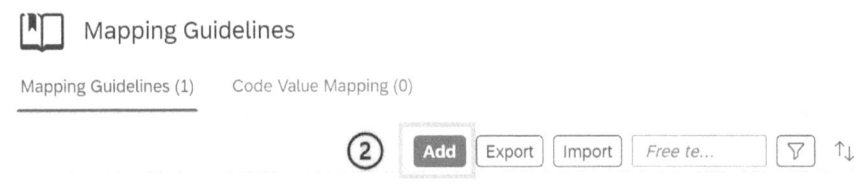

Figure 3-38. *Add MAG*

3. Choose **Next** after selecting the MIG you want to utilize as the source in your MAG, as shown in Figure 3-39.

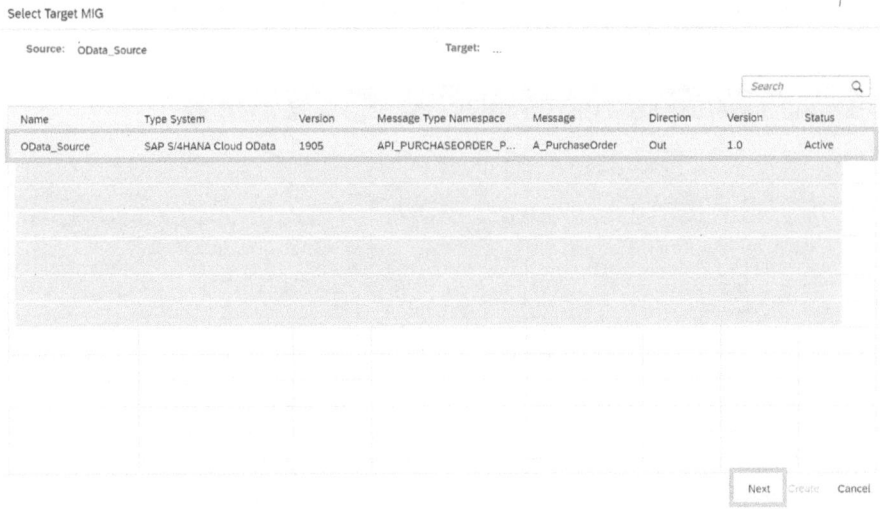

Figure 3-39. Source MIG

4. Choose **Create** after selecting the MIG you wish to use as your MAG's target, as shown in Figure 3-40.

Figure 3-40. Select Target MIG

5. A duplicate of an existing MAG can also be used to generate a new one.

6. Select **Copy** by clicking the Action button on the Overview page. There are two choices for the copy feature.

- Copy MAG Only: This choice only generates a duplicate of the MAG.

- Copy MAG and Referenced MIGs: By selecting this option, a copy of the MAG and all referenced MIGs is made.

7. The MAGs are created by selecting the source and the target MIGs, as shown in Figure 3-41.

Figure 3-41. *Created MAG dashboard*

Once the MAG has been created in SAP Integration Advisor, the main step is to enhance the MAG by working with the mapping source and the target nodes. Let's use the different functions in MAG.

3.1.7.3 Working with Mapping Guidelines (MAGs)

A MAG that you've already generated can be edited to have the mapping specifics changed to fit your scenario. You can also include documentation that gives the MAG's users further details about its applicability.

There may be situations where the MIGs differ significantly, such as various node hierarchies, different message structures, etc. In these situations, mapping becomes challenging. To get around this, SAP Integration Advisor gives you a method that helps modify a source MIG's structure so that its content may be readily mapped with the tragic MIG structure.

1. Choose **Edit**.

2. To modify the MAG's mapping information, select the Mapping tab.

3. Use the Proposal Service, drag the field from the Source to the field on the Target you want to map to, or both.

4. By choosing the mapping entity line, more information or features can be added to or managed for each piece.

3.1.7.3.1 Pretransformation of Message Implementation Guidelines

Pretransformation is a way to transform the structure of a MIG. It allows a business user to define, simulate and pretransform a source MIG including the source message instance so that it fits better to the structure of the target MIG. Pretransformation applies a structural transformation to the hierarchy of the incoming payload and its corresponding MIG defined by a pretransformation script (PTS). PTS also needs to provide a corresponding runtime artifact.

Two key operations make up pretransformation.

- The MIG structure on the left side of a MAG is replaced with a MIG structure that has undergone MIG transformation, which is carried out at design time to produce the associated changed message structure.

- Payload transformation that takes place during an integration flow execution. In order to accomplish this, the application would export a second XSLT script, which is still part of the same XSL file as the primary XSLT script.

3.1.7.3.1.1 Create a Pretransformation

Let's imagine that you have a MAG that has two MIGs and that you want to change one of the MIGs to continue the mapping.

1. Choose **Edit** when your MAG is open.

2. Open the Overview tab. The Pretransformation field is in Source and Target MIGs. The value is None if there is no transformation. Select the **(+)** Add button, as shown in Figure 3-42.

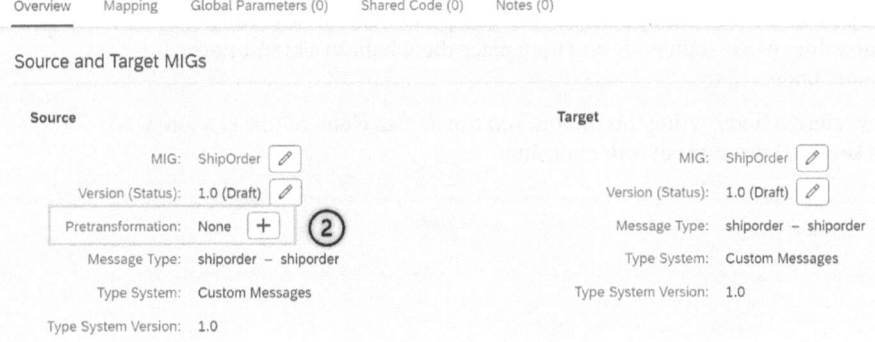

Figure 3-42. Pretransformation

3. This launches the Mapping tab with the Pretransformation feature. The mapping structure's source MIG is selected by default.

4. There are three sections on the page.

- MIG Structure: The MIG structure that was selected for pretransformation. This would always display the original, unaltered source MIG structure.

- Pretransformation/Main Transformation: The button that allows you to switch between the pretransformation and the main transformation MAG. You can define your transformation stages and the relevant parameters in the Pretransformation tables.

- Pretransformed Structure: This structure is subjected to the pretransformation you specify, and the resulting modified MIG is then known as the pretransformed MIG. You can review the outcomes and then proceed to the main transformation for mapping, as shown in Figure 3-43.

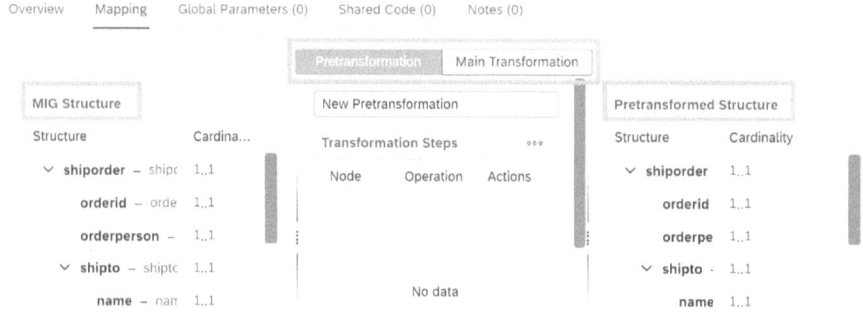

Figure 3-43. *Pretransformation sections*

5. The name of the transformation is displayed in the New Pretransformation text field. You can change it and give it a meaningful name based on your business's objectives.

6. Select **Add** next to Transformation Steps. These are the choices it offers, as shown in Figure 3-44.

    • Group by Key: Using this option, you can group the elements according to the values of a certain node and then place them behind a brand-new wrapper node.

    • Copy referred node: Using this option, you can add an element that is referenced by a key or ID to a node of your choosing.

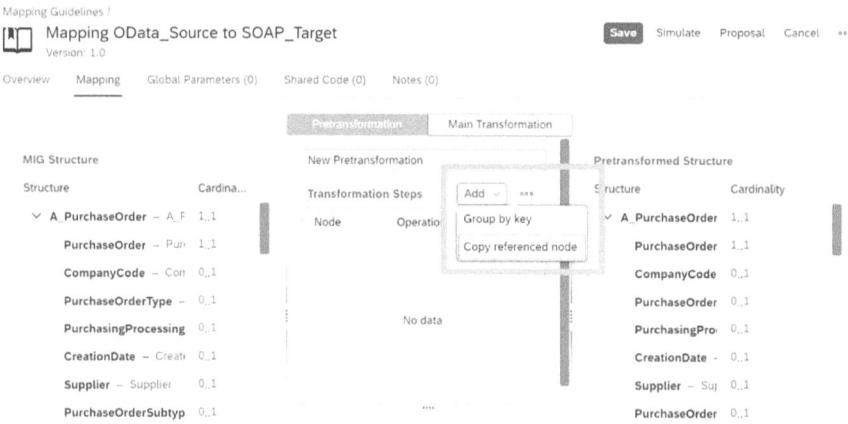

Figure 3-44. *Pretransformation Add*

7. The parameters presented in Table 3-3 should be kept in the Operation Parameters table if you selected Group by Key.

Table 3-3. *Group by Key: Transformation Steps*

Parameter	Value
Grouped node	This is the group node that groups the instances.
Key node	This is the leaf node containing the grouping key. This node needs to be the grouped node's child node.
Wrapper name	This is the name of the node in the group that hold the grouped nodes. This node takes on the role of the grouped node's parent node.
Group key node	The group key for each instance is contained in a child node of the wrapper node.

8. If you selected Copy referenced node, keep the Operation Parameters values, as shown in Table 3-4.

Table 3-4. *Operation Parameter: Transformation Steps*

Parameter	Value
Receiving node	This is the group node that receives a copy of the referred node.
Node with reference ID	It contains the reference ID for the node cloned in a leaf node. This node needs to be the receiving node's child node.
Referenced Node	Group node that is replicated and referenced.
Referenced ID Node	Contains the referenced ID in a leaf node. The referenced node's child node must be this node.

9. Select **Execute** from the menu that appears next to the Transformation Steps table after maintaining all the items. Pretransformed Structure is shown on the right after the operation has been applied to the MIG structure.

10. The outcomes can be examined, and your procedure can be adjusted as necessary. The Delete button is in the Transformation Steps table and can also be used to remove your transformation step.

The next section explores mapping a source and the target nodes.

3.1.7.3.2 Mapping Source and Target Nodes

The MAGs editor can be used to define a mapping between a source and target node. The Hierarchical Structure of the Source and Target Message Implementation Guidelines is displayed in the Mapping tab (MIGs). You can define a data transformation by applying functions based on your needs and dragging a source node to a target node to specify a mapping.

The structure of the final target document varies depending on.

- The kind of mapping established between source and target nodes.

- The target node's cardinality.

The key words used in node mapping are defined in Table 3-5.

Table 3-5. *Node Mapping Terminology*

Term	Description
Group Node	This is a node without any data that has subnodes. It may also function as a group node's subnode.
Leaf Node	This is a data-containing subnode.
Cardinality	It describes how many times an element may appear in a document. There are minimum and maximum values.
Mapping Line	The creation of the target node is determined by mapping, which links certain components of the source structure with one or more of the target structure's elements. Nodes and subnodes are connected from source to target by a mapping line.
Mapping Entity	Using a mapping entity, one or more source elements are mapped to a single target element.

3.1.7.3.2.1 Mapping Cardinality

Mapping cardinality represents the number of nodes associated with the source and target document structure, and the root node sets the context for interpreting the cardinalities.

Let's focus on leaf mapping by group. A target node is produced when a source group node is mapped to a target leaf node; the target node's cardinality is defined by the actual cardinality of the source group node. To give information for the target instances, you must map the pertinent source leaf nodes with the newly generated target node, which is empty initially.

- Within this mapping, the maximum target occurrence is 1.

    - The target leaf node is formed if at least one genuine occurrence of the source group node exists.

    - The target leaf node is not formed if the actual occurrence of the source group node and the minimum occurrence of the target leaf node are both zero.

    - The target leaf node is formed if there is no real occurrence of the source group node and at least one instance of the target leaf node.

- The target leaf node's cardinality matches that of the mapped source group node if the target maximum occurrence is greater than 1. Independent of the target cardinality, each instance of the source group node creates exactly one instance of the group node in the target.

- Within this mapping, the destination node is only mapped to one leaf node.

    - Each instance of the target node's content is filled with information from the associated instance of the source node if only one instance of this node is suitable for the particular target node.

    - If multiple instances of this particular leaf node are pertinent, the values are concatenated for each instance of the source group node separately and applied to the associated instance of the target leaf node.

The following Mapping options come when you map the source and the target.

- 1:1 Mapping
    - It represents a mapping between a single element from the source and a single element from the target.
    - A target node is produced for each instance of the source node if the target node's maximum occurrence value is greater than one.
- N:1 Mapping
    - It symbolizes a relationship where many source items are mapped to a single target element.
    - The contents of the source nodes are concatenated and put onto the target node if the maximum occurrence value of the target node is equal to one.
- 0:1 Mapping
    - A target leaf node can be given a constant value directly without a specified mapping.

The supported mapping nodes are listed in Table 3-6.

Table 3-6. *Supported Mapping Cardinality and Mapping Types*

Mapping Cardinality	Mapping Type
1:1	Leaf to Leaf
1:1	Group to Group
1:1	Leaf to Group
N:1	Leaf to Leaf
0:1	None to Leaf (Constants)
1:1	Group to Leaf
N:1	Group to Leaf

The mapping options not supported in SAP Integration Advisor are shown in Table 3-7.

Table 3-7. *Mapping Cardinality and Types Not Supported*

Mapping Cardinality	Mapping Type
N:1	Group to Group
N:1	Leaf to Group

The next section examines conditional mapping.

3.1.7.3.2.2 Conditional Mapping

In a group-to-group node mapping, you can put criteria on leaf nodes to limit the number of target group node instances created and their cardinality.

A leaf node is inserted as a condition node in group-to-group node mapping.

- By default, the mapping list's leaf node has the condition checkbox selected.

- The mapping pane's Condition tab becomes accessible.

- You can enter the condition as an XPath expression in the XPath Code editor that appears.

Only when the sequence isn't empty does the default XPath expression for the mapping, boolean ($nodes in/*), return true. All conditional node instances pertinent for one instance of the target node are contained in the sequence $nodes id/*. The default expression can be modified to suit your company's needs.

The next section explains deleting mapping lines in MAGs.

3.1.7.3.2.3 Delete Mapping Lines

You can remove unnecessary mapping entities and lines. To delete a selected object or mapping line, hit the Delete key, as shown in Figure 3-45.

- Alternatively, you can delete the mapping line or object from the context menu by performing right-clicking it.

- Delete any mapping line or object in the Source column by selecting the Delete icon next to it in the mapping pane's Mapping List tab.

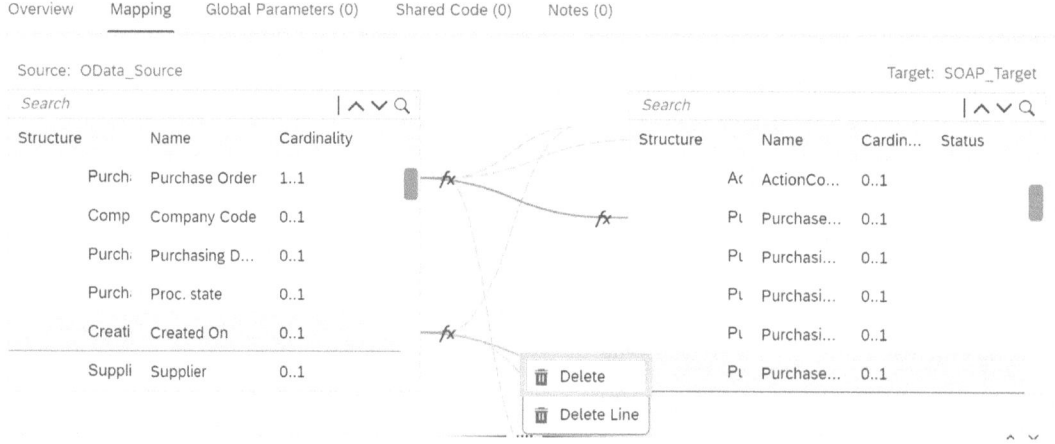

Figure 3-45. *Delete mapping line*

A mapping entity also allows you to remove a single mapping line. You can choose Delete Line from the context menu when right-clicking a specific mapping line while selecting the mapping entity.

The next section discusses the data and time conversion in MAGs.

3.1.7.3.3 Date and Time Conversion

In the message structure of a MAG, the source and target nodes of the date, time, and datetime types can be mapped.

According to the kind of format, the DateTime Conversion feature can be configured to include any of the following features.

- When mapping, the source node's format can be changed to the target node's format if the nodes have different formats. You must set the nodes' format to accomplish this. Go to the Functions tab, select Switch to DateTime Mapping, and then select the element, as shown in Figure 3-46.

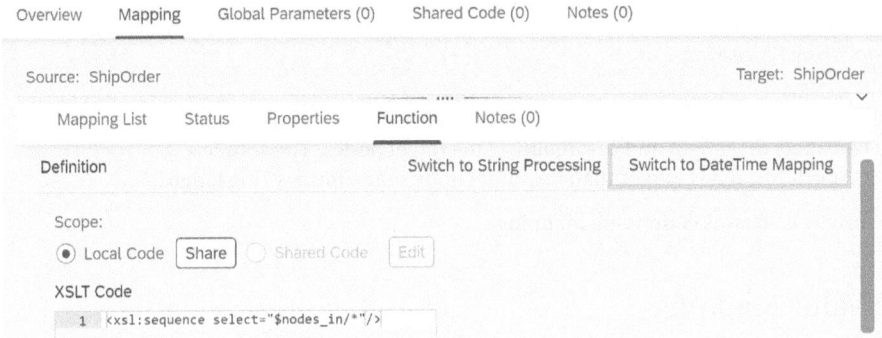

Figure 3-46. *Switch to DateTime Mapping*

- Set the values for the Source DateTime Format and the Target DateTime Format on the DateTime Conversion tab, then click Save.

- The DateTime Conversion tab automatically shows if the source and target nodes are Date, Time, or DateTime fields. When setting the appropriate message implementation rules, the default DateTime format of the source and target nodes is set.

The next section covers the string processing type.

3.1.7.3.4 String Processing Type

Using the string processing type, you can map multiple source nodes of the type string and token to one or more target nodes.

- The source node must be selected before switching to string processing may be done on the Functions tab, as shown in Figure 3-47.

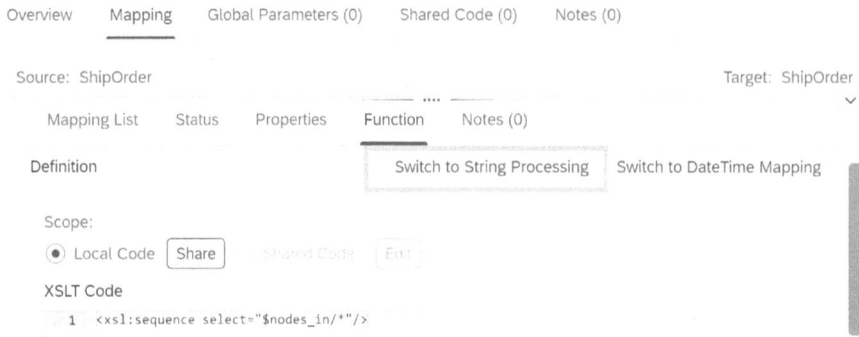

Figure 3-47. *Switch to String Processing*

- To distribute a long source message across the target nodes, the string processing function considers the length and cardinality of the target nodes. The source message is shortened if it is longer than the target node's maximum string length.

The next section briefly discusses code value mapping.

3.1.7.3.5 Code Value Mapping

Code value mappings show how a code value set at the source element is changed into a code value needed at the target element. The definitions and semantics of the two items should be the same, even though the strings for each element may change.

The Code Value Mapping tab appears only when the source and target items in a mapping contain code values. MAG's proposal service proposes the target structure's mapped code values. Depending on your needs, you can adjust the proposal or add code value mappings. The following explains how to do this.

1. Select **Code Value Mapping** after selecting the source node.

2. Click **Copy** and choose a value mapping from the list if you want to copy a value mapping to another mapping element.

3. Typically, runtime artifacts are generated first, followed by creating a code value mapping at the semantic level, which is then converted into an XSLT function. However, select Switch to Function if you need to transform your code value mapping into a function. As a result, the Functions tab's Code Value Mapping XSLT function is automatically generated.

4. By selecting the value help and picking the appropriate value from the list, you can modify the code value in the target node.

If you have previously used a standard codelist, use the following options.

- If you want to link a global mapping to a specific code value, select Assign Global Mapping.

- By selecting Create Global Mapping, you can also turn the local code value mapping into a global mapping. Other mapping standards can take advantage of this global mapping.

3.1.7.3.5.1 Deprecated Code Values

A code value is marked as deprecated by the icon ◇ next to it.

- When a value is still in the codelist but not selected in the MIG, it is a deprecated code value.

- The value is changed or dropped from that specific codelist version.

- The MIG changes or replaces the code value. In this instance, MAG displays the domain GUID of the missing value.

The tooltip located beneath the icon ◇ can be used to differentiate between these values.

You can delete the deprecated values if they are listed in Source Code Value by using the supplied symbol 🗑 next to each value.

You can swap out any deprecated values in Target Code Value with an existing value by using the symbol ⬚.

In the appropriate MIG, you can pick the deselected values once more.

The next section discusses functions.

3.1.7.3.6 Functions

Functions specify a data transformation between source and target nodes within the MAG editor. The Function tab on the mapping pane is available only when leaf nodes are mapped, or a function is assigned to a target leaf node with no corresponding source node. The XSLT Code editor in the Function tab is used to formulate expressions based on XSLT V2.0 or XPath 2.0. The shared code feature enables users to construct frequently used code snippets and reuse them as needed. Parameters can be used to provide values to functions.

Look at the following example.

- **Source Data as an Input**: A function receives and transforms an input value as an argument. The substring function extracts the first three characters from the given data.

```
<xsl:sequence select="substring($nodes_in,1,3)"/>
```

- **Function as an Input**: A function that extracts the year component from the current-dateTime function is supplied the function as an input.

```
<xsl:sequence select="year-from-dateTime(current-dateTime())"/>
```

- **Parameter as an Input**: A function receives a parameter as an argument. Concatenating the source field with the global parameter PARAM, defining and giving the value 01, and assigning it to the destination field.

```
<xsl:sequence select="concat($PARAM,$nodes_in)"/>
```

3.1.7.3.6.1 Use Functions

The MAG editor uses functions to express a data transformation between source and target nodes.

1. Choose the target leaf node to which the function t should be applied.

2. The code snippet should be edited in the XSLT Code editor's Function tab. The function's default scope is local. Additionally, you can write a function to reuse it throughout the mapping.

3. In the XSLT Code editor, reuse a function from shared code as follows.

    a. Choose **Shared Code** from the menu.

    b. Select **Replace**.

    c. Choose the desired shared code from the drop-down menu.

4. Do the following to pass an argument to a function.

    d. Select **Add new parameter** from the Function tab.

    e. Enter the parameter's name in the Function Parameter Name field of the Create New Function Parameter dialog box. Select Create.

    f. If you want to link it to a function parameter, pick a global parameter.

5. From the list of global parameters, choose the one you need. As an alternative, Search can be used to filter for a parameter.

The next section discusses how to create a shared flow.

3.1.7.3.6.2 Creating the Shared Flow

You can create frequently used code snippets and reuse them as required within your MAG.

1. Select **Add** in the Shared Code tab to generate a suitable code snippet.

2. The code snippet's name should be entered in the Name field of the Create New Snippet dialog box.

3. You can fill up the Summary section with a summary of the code snippet.

4. To generate a fresh code snippet, select Create.

5. Enter any supporting data in the Documentation area of the Overview tab.

6. Fill out the Implementation field on the Definition tab with the function definition.

7. Click **Save**.

8. The final Shared Code is generated, as shown in Figure 3-48.

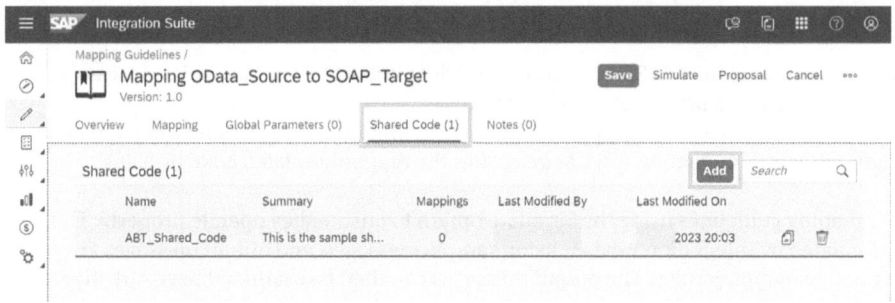

Figure 3-48. *Shared Code*

The next section defines the global parameters.

3.1.7.3.6.3 Define Global Parameters

To define reusable parameters for your Mapping Guideline, use the Global Parameters tab (MAG).

1. To add a new parameter, select Add on the Global Parameters tab.

2. Enter a name for the parameter in the Parameter Name field of the Create New Parameter dialog box and a default value in the Default Value field.

3. Click **Create**.

4. The global parameter is created, as shown in Figure 3-49.

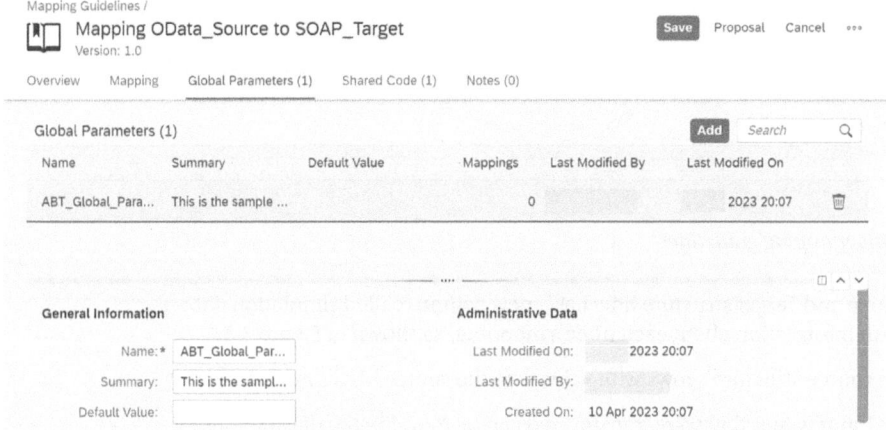

Figure 3-49. *Global Parameters*

The next section discusses simulating a MAG.

3.1.7.4　Simulate Mapping Guideline: Practical Example

The simulate mapping guidelines feature in SAP Integration Advisor is used to validate and test message mappings before deploying them in a production environment.

You can specify a set of mapping rules that spell out how the mapping should be done when creating a mapping in Integration Advisor. Integration Advisor generates the mapping-related code by using these rules.

You can test the mapping guidelines using the Simulate option to ensure they operate properly. The mapping process is simulated by Integration Advisor using sample messages, and output messages are generated depending on the mapping rules. The output messages can then be examined to ensure they correspond to the anticipated outcomes.

The following steps simulate the mapping guidelines.

1. Go to the Mapping tab after opening your MAG.

2. Expand every node to check if the mapping is complete, and select Simulate to run an embedded simulation inside the MAG. The MAG can be simulated in two ways, as shown in Figure 3-50.

   - Simulate with MIG Example Data

   - Simulate with Payload Data

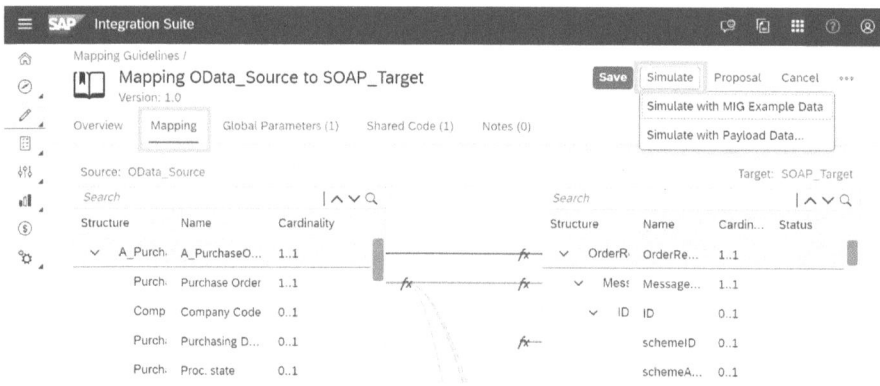

Figure 3-50.　*Simulate mapping guidelines*

3. The Source and Target structure now has a new column called simulation data filled with information about each node's mapping, as shown in Figure 3-51.

   - The source structure's rows with values are the source MIG's input values.

   - The input values that were transformed can be found in the destination structure.

   - The navigation buttons indicate multiple leaf node instances, if there are any, in the simulation data. You can navigate to the additional instances of the payload values using these buttons.

   - A group node additionally has the navigation button shown if it appears at least once in the simulation data.

- The minus sign (-) indicates that the transformed or input value on the target message is empty.

- The icon indicates that the leaf node is not in the simulated data.

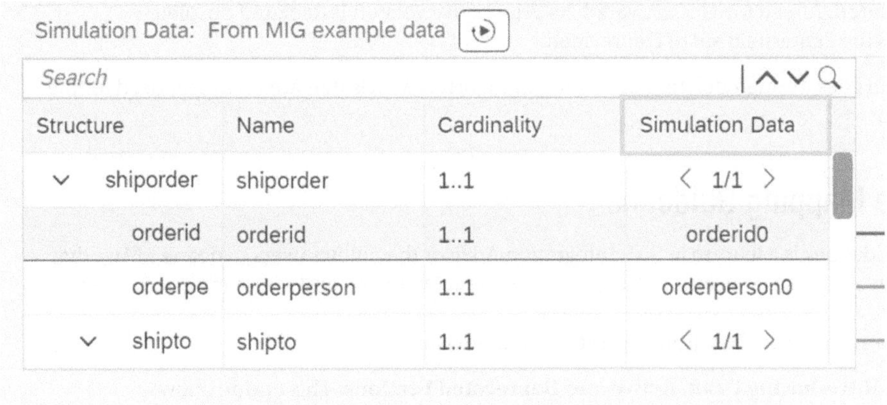

Simulation Data: From MIG example data ⊙			
Search			| ∧ ∨ Q
Structure	Name	Cardinality	Simulation Data
∨ shiporder	shiporder	1..1	⟨ 1/1 ⟩
orderid	orderid	1..1	orderid0 —
orderpe	orderperson	1..1	orderperson0 —
∨ shipto	shipto	1..1	⟨ 1/1 ⟩ —

Figure 3-51. *Simulatation Data*

4. Pick a mapping entity from the source or target structure to inspect the mapping after the simulation.

5. In the Condition tab of that mapping entity, you can view the following details of that conditional mapping if you've set any conditions to a source leaf node to regulate the instances of a target group node.

   - The conditional mapping nodes

   - The stage of creating a target node according to the input given in the mapping

   - The mapping structure icon indicates the target nodes that the conditional mapping prevents from being formed.

6. The Simulation Data column only shows instances of the target group nodes that meet the condition conditions for conditional mapping for target group nodes. Choose if you want to see every node occurrence, whether or not they satisfy the requirement.

The next section discusses MAG version management.

3.1.7.5 Version Management: Mapping Guidelines

Once you have modified the mapping information in a MAG to fit your situation, you can turn on the MAG to put it to good use. The following describes the three types of status that can be specified for a MAG.

- **Draft**: When a new MAG is produced, the newly created MAG has the Version field set to 1.0 and the Status field set to Draft. This updated version of MAG is now adaptable to your business needs.

- **Active**: The Status of the MAG is changed to Active after activating a MAG. This version of the MAG cannot be changed once activated because it is a final version. However, you can still export runtime artifacts and documentation from the MAG.

Open the active MAG version, then select Edit for the subsequent draft version. Additionally, you can select Edit from the MAG overview's action button. You then be asked if you want to create a draft version. Select OK. The next MAG version is now developed with a Draft status. For instance, if the MAG's active version is 2.0, the new draft MAG's version is set to 3.0.

- **Deprecated**: After a MAG is activated, its prior active version is declared obsolete and has the Status field set to Deprecated.

The next section goes through deleting unwanted or incorrect MAGs that are no longer used for the integration scenarios.

3.1.7.6 Delete Mapping Guideline

Delete Mapping Guidelines is a feature in SAP Integration Advisor that allows users to delete MAGs that are no longer needed. This can be useful for keeping the system organized and removing outdated or unused MAGs.

There are two ways to delete MAGs in SAP Integration Advisor.

- **Delete MAGs having Draft, Active and Deprecated Versions**: This option allows users to delete all versions of a MAG, including draft, active, and deprecated versions. When a MAG is deleted using this option, all its associated versions and data are permanently removed from the system.

- **Delete Specific Version of MAG**: This option allows users to delete a specific version of a MAG. This can be useful if a specific version of a MAG is no longer needed, but other versions of the MAG should be kept.

3.1.7.6.1 Delete MAGs Having Draft, Active, and Deprecated Versions

In SAP Integration Advisor, MAGs can have multiple versions, including draft, active, and deprecated versions. Deleting MAGs having draft, active, and deprecated versions means deleting all the versions of a specific MAG, including any draft or deprecated versions.

When you delete a MAG, all associated mappings, functions, and shared flows are also deleted. Therefore, before deleting a MAG, it's important to ensure that it's no longer needed and that there are no dependent objects. If a MAG is still needed, but only certain versions are no longer required, you can delete a specific version of a MAG instead of all versions.

The following explains how to delete a MAG.

1. Select the **MAG** you wish to eliminate by selecting (More Options) on the Mapping Guidelines tab of the Mapping Guidelines screen, and then select Delete.

2. Choose the **Delete** Mapping Guideline (with history) option in the dialog box to delete MAGs, as shown in Figure 3-52.

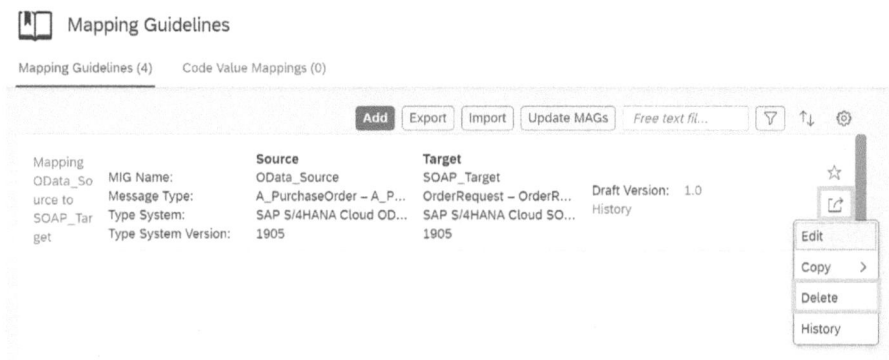

Figure 3-52. *Delete MAG*

3.1.7.6.2 Delete Specific Version of MAG

In SAP Integration Advisor, MAGs can have multiple versions, including draft, active, and deprecated versions. If you want to delete only a specific version of a MAG, you can use the "Delete Specific Version of MAG" option.

Deleting a specific version of a MAG removes only that particular version and its associated mappings, functions, and shared flows. The other versions of the MAG remain unaffected. This option can be helpful when you need to delete an older or deprecated version of a MAG while keeping the latest or active version.

The specific version of MAG can be deleted.

3.1.7.6.2.1 Using the Mapping Guidelines Tab

1. Select the MAG you wish to eliminate by selecting (More Options) on the Mapping Guidelines tab of the Mapping Guidelines screen, and then select Delete.

2. Select the Delete this (latest) version option in the dialog box for the Delete Mapping Guideline command.

3.1.7.6.2.2 Using the Overview Tab

1. Select the MAG version you want to delete from the Mapping Guidelines screen's Mapping Guidelines tab.

2. Depending on the screen size, select **Delete** or More and then select Delete from the Overview tab, as shown in Figure 3-53.

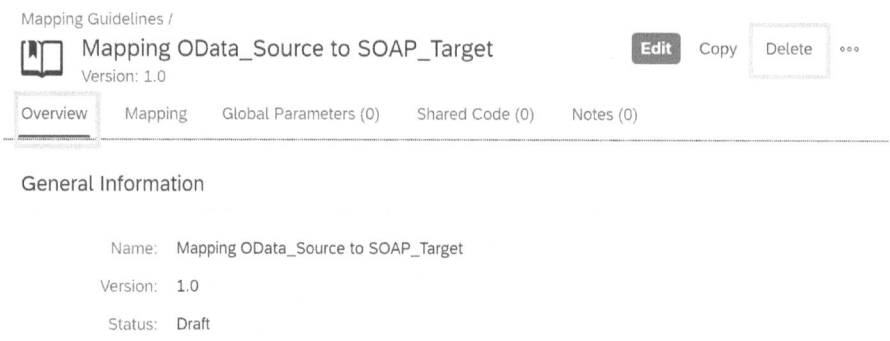

Figure 3-53. Delete MAG using the Overview tab

3. Select **OK** in the Confirm dialog box.

3.1.7.6.2.3 Using the History Link

1. Select History in the MAG that you wish to delete on the Mapping Guidelines tab of the Mapping Guidelines screen.

2. Select the version you want to delete from the Mapping Implementation Guideline History Versions table by selecting it from the action menu, as shown in Figure 3-54.

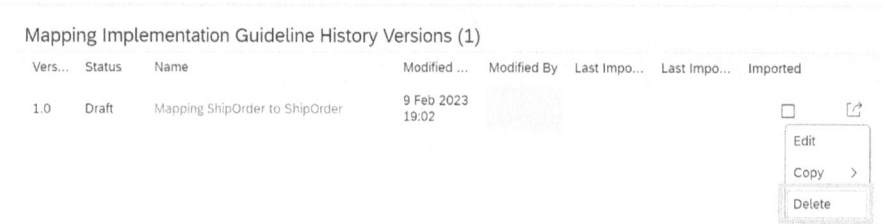

Figure 3-54. Delete using History link

3. Select OK in the Confirm dialog box.

The next section examines how to import and export MAGs and MIGs in SAP Integration Advisor.

3.1.8 Import and Export in SAP Integration Advisor

You can export runtime artifacts, documentation, MIG, and MAG using SAP Integration Advisor. MIGs and MAGs can also be imported using the program.

3.1.8.1 Export MIG/MAG

In SAP Integration Advisor, Export MAG (Mapping Artifacts and Guidelines) is a feature that allows you to export the mapping artifacts and MAGs for a particular interface. This includes the mapping data structures, the message mapping code, and associated MAGs. Export MIG (Mapping Implementation Guidelines) is a feature that allows you to export the implementation guidelines for a particular mapping. These guidelines specify how the mapping should be implemented in a particular environment, such as the system settings, parameter values, and other implementation details.

1. Select Message Implementation Guidelines or Mapping Guidelines from the left pane after logging in to your application.

2. Go to the overview page and select **Export**, as shown in Figure 3-55.

Figure 3-55. *Export*

3. Decide which MIG or MAG you want to export. Additionally, you may select multiple items from the list. To find a specific MIG/MAG, use the available filter option beneath each column header, as shown in Figure 3-56.

4. Select **Export** to download the files in ZIP format, as shown in Figure 3-56.

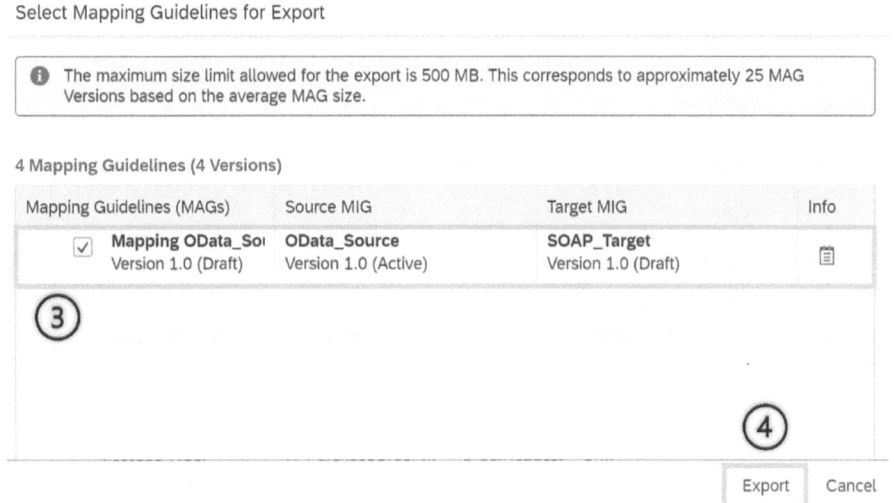

Figure 3-56. *Export MAG*

The next section discusses importing MIGs and MAGs in SAP Integration Advisor.

3.1.8.2 Import MAG/MIG

Importing MIGs and MAGs allows users to reuse or migrate existing artifacts from one system to another. It is also useful when users want to collaborate with other integration teams or share guidelines with other companies.

1. Select Message Implementation Guidelines or Mapping Guidelines from the left pane after logging in to your application.

2. Select **Import** from the overview page.

3. Select **Browse** in the import dialog box to add the MIG/MAG file you previously exported from another tenant.

4. Select **Check** to verify the uploaded file.

5. The MIGs/MAGs from the list you want to import should have a tick next to them.

6. Select **Import**, as shown in Figure 3-57.

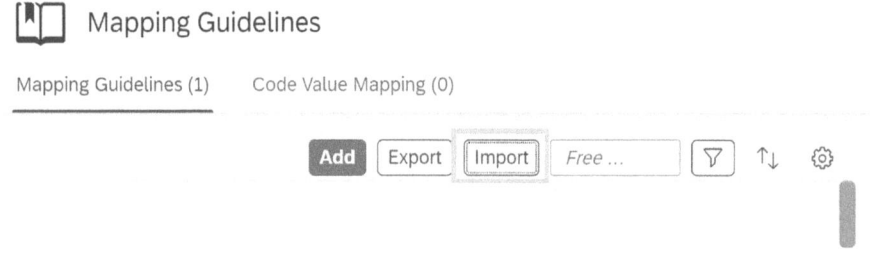

Figure 3-57. Import Mapping Guidelines

The next section discusses exporting a MAG as a document.

3.1.8.3 Export Documentation

Your MAG or MIG can be exported in the documentation format.
The following approved documentation formats are available for exporting your MIG or MAG.

- PDF

- RTF

- Excel

You can use the document file as a guide while implementing the messages and mappings in the integration solution and as reference material for bookkeeping needs.

1. Access the MAG or the Message Implementation Guideline (MIG) sections.

2. To export as documentation, choose MIG or MAG.

3. Select, then select **Export**, as shown in Figure 3-58.

4. Select the suitable choice per your needs.

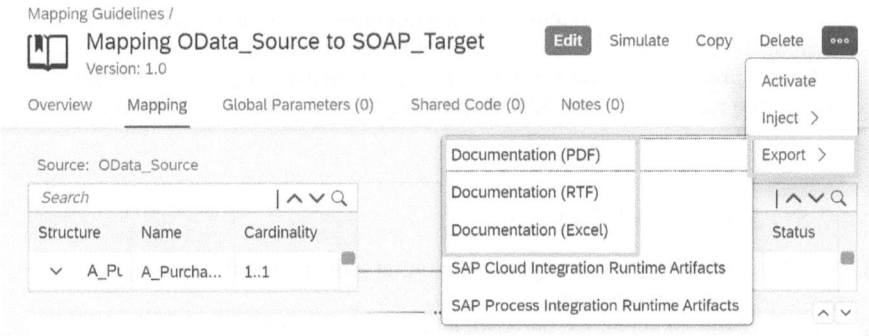

Figure 3-58. *Export MIG as document*

3.1.8.4 Export Runtime Artifacts

Export Runtime Artifacts is a feature in SAP Integration Advisor that allows you to export the runtime artifacts used in your integration scenarios. This includes the XSD schema files, WSDL files, and endpoint URLs required to execute the integration scenario.

The Export Runtime Artifacts feature is useful when you need to move your integration scenarios between different systems or when you need to share your integration scenarios with other developers or stakeholders.

The following explains how to export runtime artifacts.

1. Access the MAG or the Message Implementation Guideline (MIG) sections.

2. Choose the MIG or MAG containing the runtime artifacts you want to export.

3. Select Export, as shown in Figure 3-59.

4. Two alternatives are available for exporting the artifacts. Depending on your demands, select the best solution.

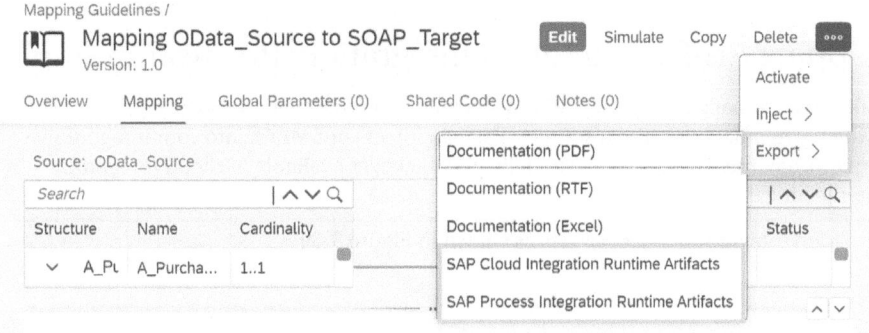

Figure 3-59. *Export runtime artifacts*

3.1.8.4.1 Consuming Runtime Artifacts

The automatically generated artifacts produced by the Integration Advisor can be utilized in various runtimes or implementations and are compatible with any interface/API format. Every runtime has particular methods, formats, prerequisites, and implementation guidelines. These factors must be considered when implementing the generated runtime artifacts into these runtimes and connecting the related apps. The goal of SAP is to provide tools and templates that speed up the integration of runtime artifacts into the various supported runtime systems: SAP Cloud Integration Runtime and SAP Process Integration Runtime.

3.1.8.4.2 Change XSD File to EANCOM

B2B flow stages used in SAP Cloud Integration have a specified naming scheme, and the Message XSD output by Integration Advisor uses the same scheme.

These Message XSDs now use the same name scheme for all EDIFACT and EDIFACT subset messages. As a result, the file name for Message XSDs for an EANCOM MIG changed to conform to this naming convention. The specifics of this naming scheme are shown in Table 3-8.

Table 3-8. *Naming Convention*

Field	Value
New Naming Convention	UN_EDIFACT <MessageType> <Version> <AssociationCode>. xsd
MessageType	The type of message that the MIG is built around in accordance with UNH > S009 > 0065
Version	This is the version in EDIFACT. When dealing with subsets, the parent EDIFACT's version is considered in accordance with UNH > S009 > 0052 and 0054.
AssociationCode	This is a code unique to the subset that designates the Message Guideline in accordance with UNH > S009 > 0057.

You have seen the creation and enhancement of MIG. You have also seen the same for the MAG in SAP Integration Advisor. But after creating the MAG, what is the next step? Is your work finished? What was the result? The next section uses this MAG in SAP Cloud Integration.

3.1.9 Push Mapping Artifacts in Cloud Integration: Practical Example

In the Destinations tab of your subaccount in the SAP BTP cockpit, you must keep the SAP Cloud Integration tenancy details into which you must inject the resources. You can inject your MAGs into your integration flow according to the requirements. There are also predelivered packages available in Cloud Integration which you can use for injecting the MAGs.

1. Open the Cloud Integration Discover Tab, as shown in Figure 3-60.

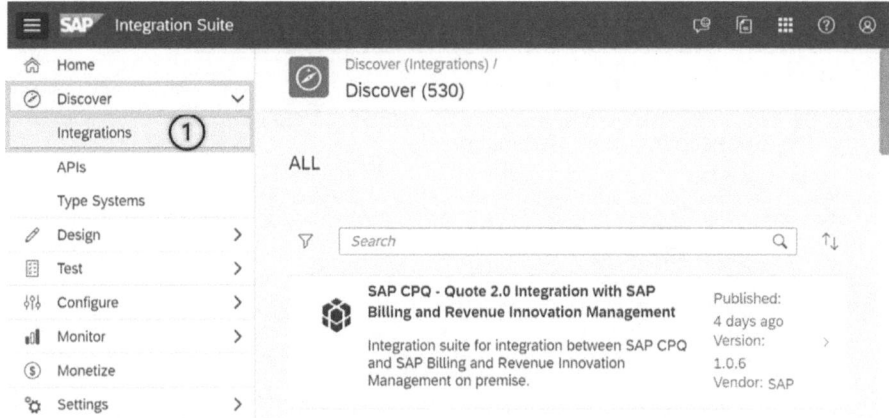

Figure 3-60. Discover (Integrations)

2. You can also use the predefined integration packages. Let's suppose you copy **EDI Integration Templates for the SAP Integration Advisor** package to your design artifact, as shown in Figure 3-61.

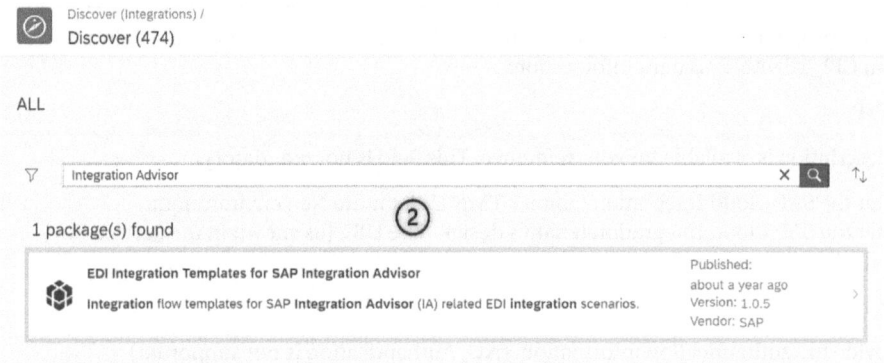

Figure 3-61. Copy integration package to design artifact

3. Navigate to your subaccount after logging in to the SAP BTP cockpit, as shown in Figure 3-62.

4. Select a new destination from the list, as shown in Figure 3-62.

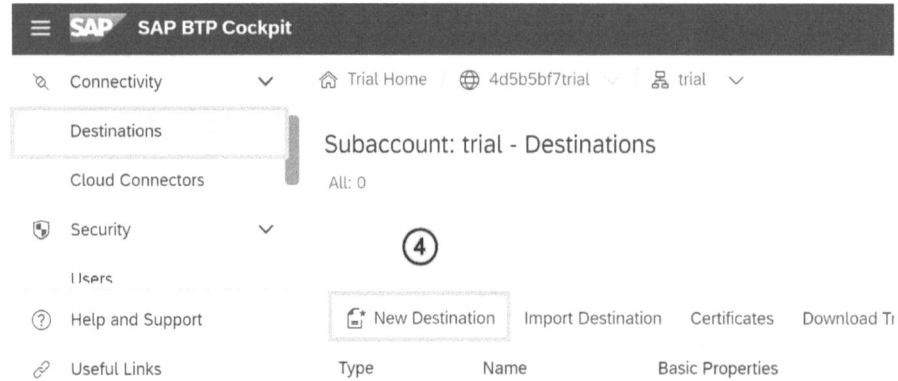

Figure 3-62. *New Destination*

5. Select Destination Configuration from the left pane and enter the information for your SAP Cloud Integration tenant, as described in Table 3-9.

Table 3-9. *Configure Destination*

Field	Description
Name	Enter a proper name for your destination from the left pane. The name should always start from CPI_TENANT<Tenant Information>.
Type	HTTP
Description	A description is available for your reference. This field is not mandatory.
URL	Enter the SAP Cloud Integration tenant's TMN URL for the Neo environment. Enter the SAP Cloud Integration tenant's design time URL (as shown in the Service Key) for the Cloud Foundry field.
Proxy Type	Internet
Authentication	Provide the authentication information. (NO_Authentication is not supported)

6. Enter the information in Table 3-9 in the Destination Configuration tab, as shown in Figure 3-63. Click **Save**.

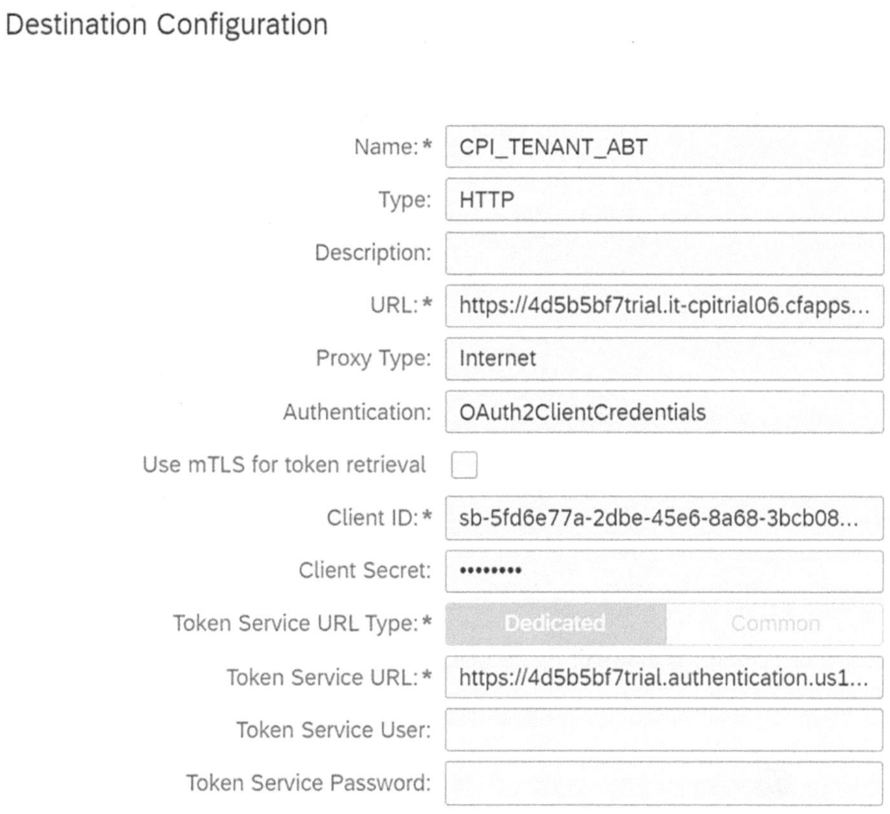

Destination Configuration
Name: *
CPI_TENANT_ABT
Type:
HTTP
Description:
URL: *
https://4d5b5bf7trial.it-cpitrial06.cfapps...
Proxy Type:
Internet
Authentication:
OAuth2ClientCredentials
Use mTLS for token retrieval
Client ID: *
sb-5fd6e77a-2dbe-45e6-8a68-3bcb08...
Client Secret:
••••••••
Token Service URL Type: *
Dedicated
Common
Token Service URL: *
https://4d5b5bf7trial.authentication.us1...
Token Service User:
Token Service Password:

***Figure 3-63.** Destination Configuration*

7. In Integration Suite, open your MAG section, and select the MAG. After selecting ooo, select **Inject ➤ SAP Cloud Integration Flow Resources**, as shown in Figure 3-64.

Mapping Guidelines /
Mapping OData_Source to SOAP_Target
Version: 1.0
Edit
Simulate
Copy
Delete
Activate
Inject >
Export >
SAP Cloud Integration Flow Resources
7
Overview
Mapping
Global Parameters (0)
Shared
Source: OData_Source
Target
Search
Structure
Name
Cardinality
A_Purcha...
1..1
Search
Structure
Name
Car...
Status
Order...
1..1

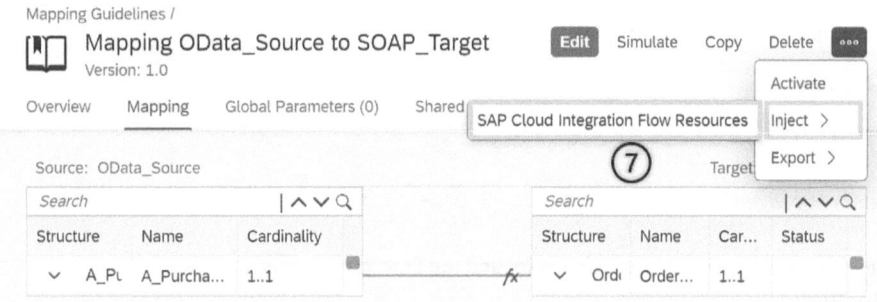

***Figure 3-64.** SAP Cloud Integration Flow Resources*

8. Select the destination you created in step 5.

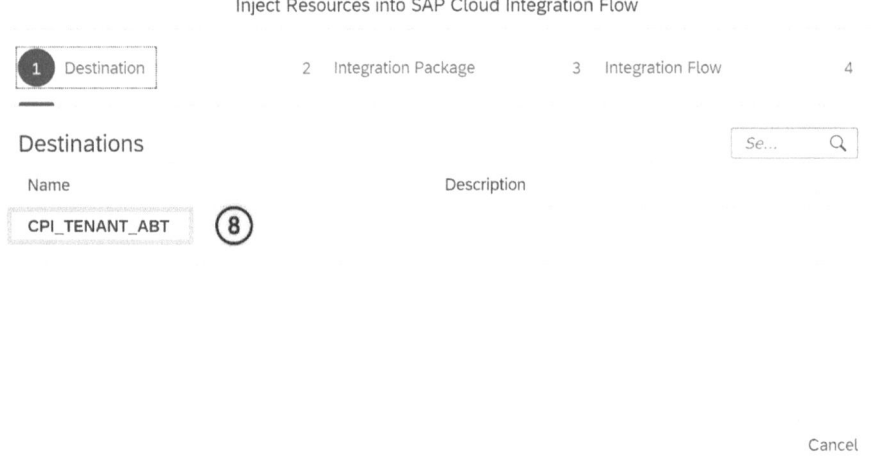

Figure 3-65. *Destination*

9. Select Integration Package, as shown in Figure 3-66.

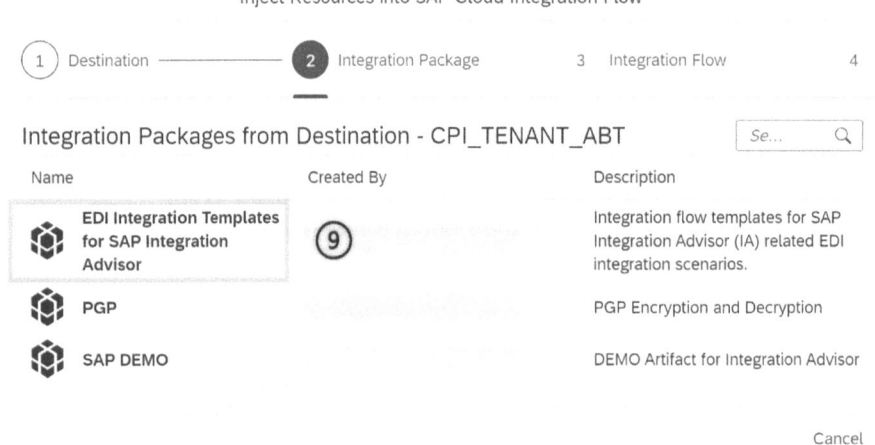

Figure 3-66. *Integration Package*

10. Select the desired integration flow. After selecting the flow, click **Inject**, as shown in Figure 3-67.

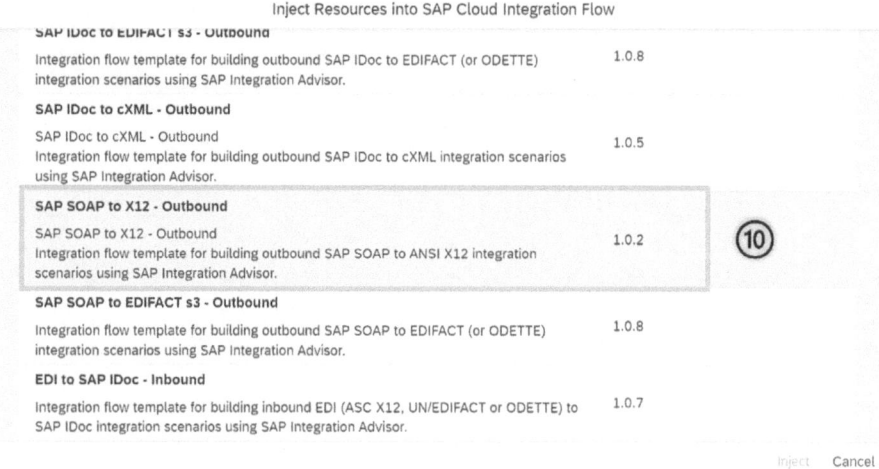

Figure 3-67. *Select integration flow*

11. The Result tab shows the information that you have selected so far. After reviewing all the details, click Close, as shown in Figure 3-68.

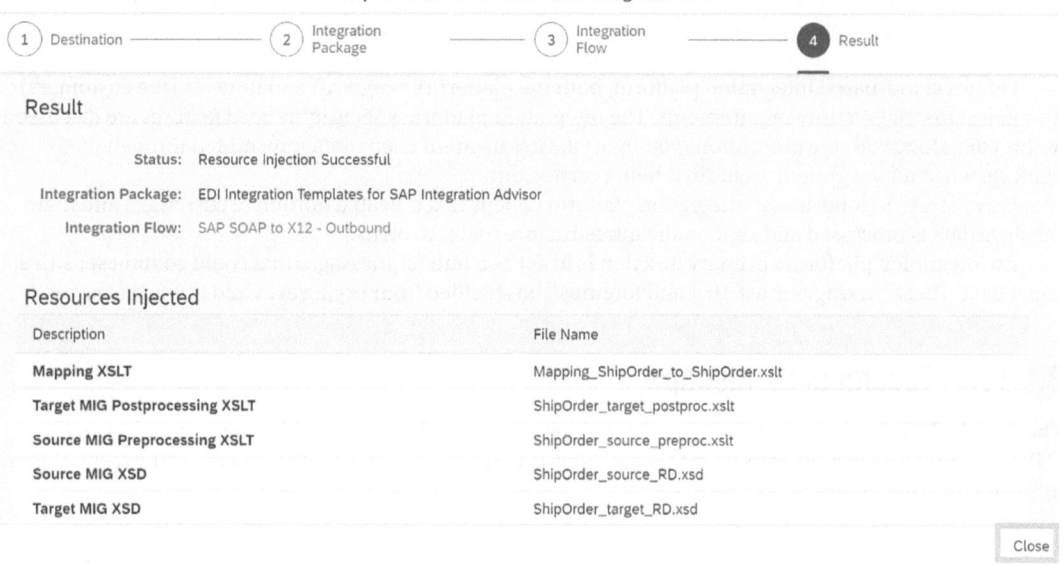

Figure 3-68. *Review details*

12. You can see the injected artifacts from the integration flow view. You can edit the integration flow according to your requirements using the Mapping artifacts, as shown in Figure 3-69.

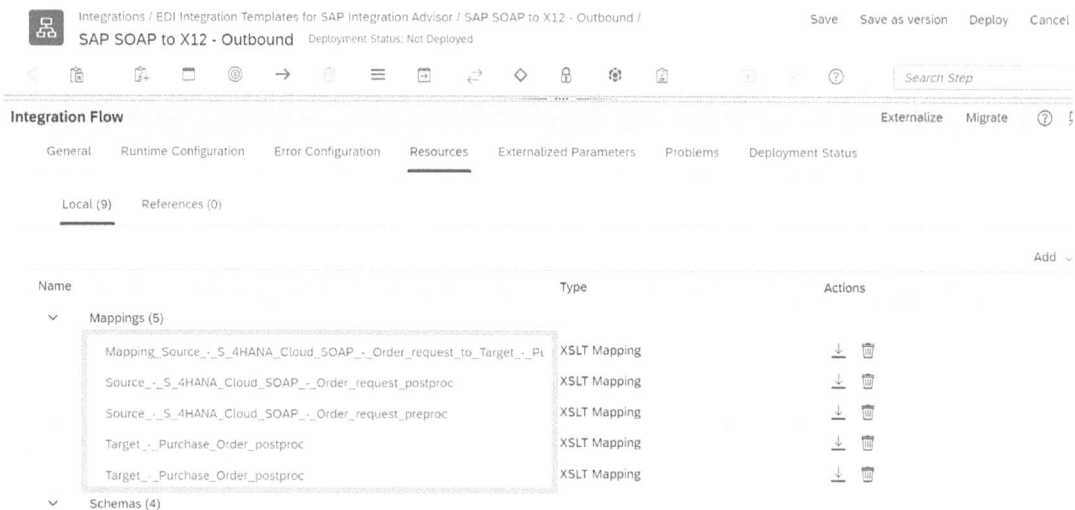

Figure 3-69. *Injected mapping artifacts in SAP Cloud Integration*

3.1.10 Security

The SAP Integration Advisor security guide gives you an overview of all the security-related features of the application, including process-related security information, identity and access management, data storage, protection, and privacy, as well as what SAP does to maintain product security standards.

Using a cloud-based integration platform, both the platform's host, SAP, and its users (the customers) are subject to strict security requirements. The integration platform's security-related features are discussed in this part, along with the precautions you may take to safeguard client data transmitted through the platform when an integration scenario is being carried out.

Users of SAP's cloud-based integration platform concur that a sizable portion of customers and their sensitive data is processed and kept on the infrastructure that SAP owns.

An integration platform's primary function is to act as a hub for messages that could contain sensitive client data. These messages must, first and foremost, be shielded from prying eyes and unlawful access.

3.1.10.1 Technical Landscape

The technical landscape comprises several parts that may securely connect with remote parts using the HTTPS protocol. Also, only users with specific rights can access the various components of the technological infrastructure due to user access design.

3.1.10.1.1 Components and Communication Path

Technically speaking, the SAP cloud-based integration platform is created as a cluster of virtual machines. Although sharing the same physical infrastructure when connecting to the platform through the Internet, users can only access one or more tenant or account-specific areas.

A virtual integration runtime that is clustered is installed on each tenant. In terms of resources and data processing on them, tenants are strictly separated from one another. Each tenant technically stores data in a different database structure.

The integration platform handles data transferred between the participating parties in real time on several virtual computers housed on the SAP cloud. Tenant clusters are clusters that belong to a particular tenant.

According to the integration platform's design, the relevant virtual machines are completely isolated from one another in terms of the associated customers. In other words, each client is allotted distinct resources (memory, CPU, and file system) via the cloud-based integration platform. Additionally, each tenant uses a unique database structure, ensuring that the data of the various clients are kept segregated. Tenant isolation is another name for this division.

3.1.10.1.2 User Access

Other components are involved when a dialog user accesses the infrastructure, such as when an administrator view monitoring data or when an integration developer installs an integration artifact, in addition to the previously described components that interact with one another when messages are processed and sent between the relevant systems.

Individuals in various positions can access the infrastructure, which is available to customers, and SAP, which provides the integration infrastructure. These are the human entry points (for dialog users).

- Devoted experts at SAP gain access to the infrastructure to offer the customer a tenant cluster.

- To create and distribute integration content and to keep track of an integration scenario in progress, experts on the customer side access the infrastructure.

3.1.10.2 Process-Related Security Aspects

The cloud-based integration platform's deployment, updating, and use processes adhere to the greatest security standards.

The SAP Integration Advisor complies with several technical SAP internal policies, procedures, recommendations, and product standards. The SAP code of business conduct and other behavioral security norms, such as communication and keeping a clean desk, are binding on both employees and operators.

For example, the SAP Secure Development Lifecycle (SDLC), which supports the implementation of measures like test-driven development and threat modeling, is used to create SAP software.

3.1.10.3 Identity and Access Management

Users of Dialog who access the platform must be verified by an identity provider. By default, SAP Identity Service is employed. SAP ID Service is the central service for handling IDs and their life cycles.

Inbound calls to the platform must be authenticated depending on the authentication mechanism selected. The load balancer performs the authentication if the client delivers a client certificate. The load balancer compares the calling component's client certificate to a list of trustworthy certification authorities before it ends the TLS connection. A user has been assigned to this certificate. The associated identity provider checks the calling entity if a basic authentication is specified. The platform supports basic authentication, OAuth, Security Assertion Markup Language, and client certificate authentication.

Access to the platform's dedicated features is restricted and secured by authorization checks. To handle the authorizations of dialog users, various authorization groups are accessible. Based on a persona, an authorization group establishes a set of specific permissions on the activities that take place throughout an integration project.

Tenant administrator and content developer are two personas connected to Integration Advisor. The duties of each role and the roles necessary for those personalities are briefly described in Table 3-10.

Table 3-10. Persona Types

Persona	Responsibilities	Roles Required	Roles Required in SAP BTP Account
Tenant Admin	Account and subscription management for customers Make IA subscription available Control users and roles in a customer account	None	Administrator
Content Developer	Read-and-consume systems Establish and update message guidelines Develop and update MAGs Update B2B license privileges.	Typesystem.Read Guidelines. ReadWrite	None
IA Administrator	Read-and-consume systems Read-and-consume systems Develop and keep up with MIGs Develop and update Mapping Guidelines Update B2B license privileges Unblock the message guidelines and MAGs that other users have locked.	Guidelines. Administrator Guidelines. ReadWrite Typesystem.Read	None

3.1.10.4 Data Protection and Privacy

The integration platform processes and stores different kinds of client data at various times. This data is protected to the highest standard, and SAP takes special steps to ensure this degree of security.

A compilation of information about a data subject makes up an information report. Such a report may need to be provided by a data privacy specialist, or an application may give a self-service option. Integration Advisor assumes that software developers, including SAP clients, can offer this data.

Consider the laws in the many jurisdictions where your company works while handling personal data. Regulations may force you to destroy the data after the end of the purpose. However, other laws can impose a lengthier retention requirement on you. Before the data is completely removed after the retention term, you must prevent unauthorized individuals from accessing it during this time.

The Integration Advisor platform only keeps data for a finite amount of time (referred to as retention time).

3.1.10.5 Logging and Tracing About Security

Data read accesses or modifications to the system configuration can be tracked using audit logs by SAP administrators or the tenant administrator. This makes it possible for administrators to take the necessary precautions to stop system abuse.

With the help of a Monitor application UI, a tenant administrator can ask for audit logs for his or her tenant. You may keep track of modifications made to tenant data in the Audit Log area. For instance, removing the mapping and message implementation instructions.

An audit log keeps track of the following information for each logged event.

- Event's nature

- Time and date of the incident

- Creator of the event (user)

- Origin of the situation

3.1.11 Summary

This chapter overviewed SAP Integration Advisor and introduced the key terminology used throughout the book. It also delved into the initial setup of SAP Integration Advisor and the creation of custom type systems. It also described how to add and delete custom messages, highlighting their disadvantages.

The chapter then moved on to MIGs, which provide detailed guidelines for creating messages that conform to specific standards. You saw a practical example of creating MIGs and explained how to work with nodes and codelists.

MAGs were introduced, and you saw a practical example of creating MAGs and learned how to work with source and target nodes, mapping cardinality, and conditional mapping. And you learned about date and time conversion, string processing, code value mapping, and functions.

The chapter concluded with sections on importing and exporting in SAP Integration Advisor, pushing mapping artifacts in Cloud Integration, and security considerations. In terms of security, the chapter briefly discussed the technical landscape, identity and access management, data protection and privacy, and logging and tracing about security.

The next chapter enters the world of APIs and discusses SAP Integration Suite's most crucial capability: API Management.

CHAPTER 4

■ ■ ■

SAP API Management

In today's digital era, integrating various systems and applications has become crucial for businesses to streamline operations and stay competitive. That's where SAP API Management comes in. Businesses require agile and scalable solutions to meet the ever-evolving customer demands. With SAP API Management, enterprises can unlock the full potential of their SAP systems by securely exposing their APIs to developers, partners, and customers. However, managing SAP APIs can be challenging, especially for those new to the game.

APIs, or application programming interfaces, allow different systems to communicate with each other seamlessly, making it possible for companies to exchange data and information with external partners and customers.

SAP API Management provides a comprehensive solution for managing and monitoring APIs, including creating and publishing APIs, securing them with authentication and authorization, and analyzing their usage.

With SAP API Management, businesses can unlock the full potential of their digital ecosystem, enabling faster innovation and growth. This chapter delves into the key features and benefits of SAP API Management and how it can help businesses achieve their integration goals.

Let's start things off by reviewing what SAP API Management is and what features it offers.

4.1 API Integration and Management with SAP API Management

Different software programs can connect via an API, which is a collection of protocols and tools. They make it possible for businesses to communicate information and features with customers, partners, and other external systems.

A complete set of tools are available through SAP API Management for developing, publishing, and maintaining APIs and tracking and evaluating their usage. It has several features, such as the following:

- **API Lifecycle Management**: SAP API Management offers a complete set of tools for managing the entire life cycle of an API, from design and development to deployment and retirement.

- **Security and access control**: To make sure that only authorized users and systems can use APIs, SAP API Management provides robust security features, including OAuth 2.0 authentication and authorization, rate restriction, and IP whitelisting.

© Jaspreet Bagga 2023
J. Bagga, *Introduction to Integration Suite Capabilities*, https://doi.org/10.1007/978-1-4842-9630-1_4

- **API analytics**: SAP API Management offers real-time visibility into API usage, including traffic, usage trends, and performance indicators, enabling businesses to optimize their API strategies and ensure they accomplish their objectives.

- **Developer portal**: SAP API Management features a developer portal where businesses can offer instructions, model code, and other materials to help developers in making better use of their APIs.

SAP API Management helps businesses accelerate their digital transformation by offering a secure and effective means of developing, publishing, and managing APIs. This simplifies communicating with partners, clients, and other stakeholders. Figure 4-1 shows the Discover APIs tab in SAP Integration Suite.

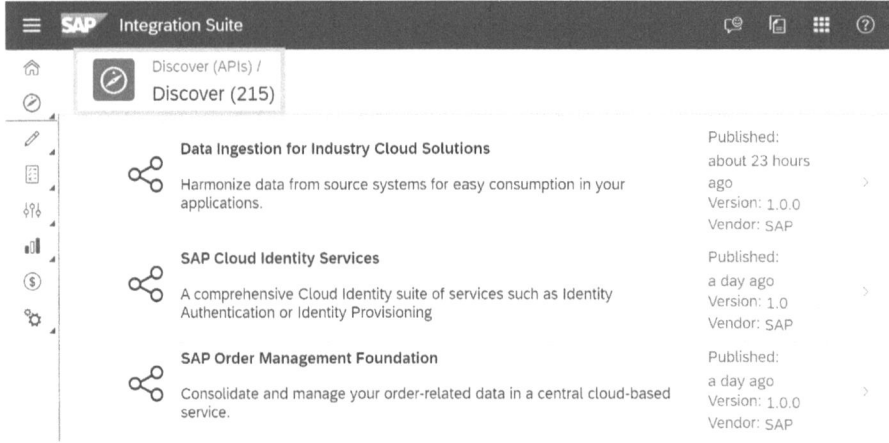

Figure 4-1. *SAP API Management Discover tab*

You are now aware of what SAP API Management is and the features it provides to the user. Let's move forward and look deeper into the components provided by SAP API Management.

4.1.1 API Management Components

SAP API Management comprises several components that work together to streamline the integration and management of the APIs. The following are the key components.

- **API Runtime**: You can successfully deploy and utilize your APIs using the API runtime. Authentication and API access requests are made by apps using the API runtime.

- **API Portal**: The API portal is the only place to design, safeguard, and publish API proxies. Simple API discovery may be made here, and as the API administrator, you can manage, monitor, secure, create, and publish the rate plans for the APIs.

- **API Business Hub Enterprise**: The API business hub enterprise allows API discovery, browsing, exploration, rate plan subscription, and app development self-service for application developers.

- **API Analytics**: It offers potent analytical tools to monitor your use of APIs. Use API Analytics to gather data about the API call's user ID, latency, and other factors.

- **API Designer**: API designers and developers can define, use, and document APIs. A range of outputs can be produced, and it supports open APIs.

- **Client SDK**: Developers can obtain client software development kits (SDKs) using open-source websites with a non-commercial license.

- **Metering**: To obtain metering information for APIs, products, and applications, you may now explore the API package.

4.1.2 Concepts of SAP API Management

It is important to understand the SAP API Management portal's organizational structure before building APIs. Its organizational structure outlines the relationships between APIs, products, apps, users, developers, and accounts within API Management.

The structure (see Figure 4-2) comprises elements like API Management Account, User, Product, Developer, API Provider, API, Application Key, and Application. While moving through the chapter, you see each word's use and meaning in the structure.

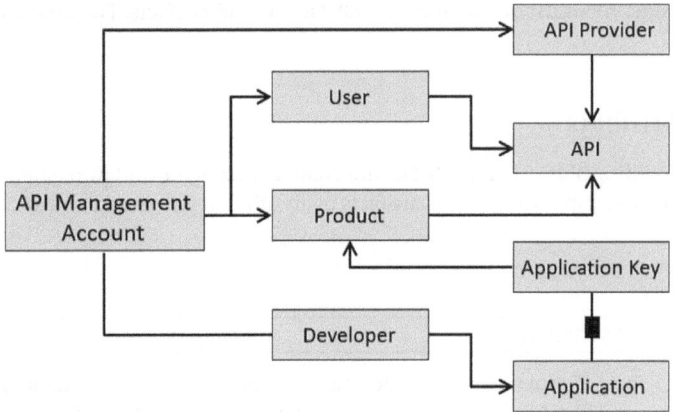

Figure 4-2. *API portal structure*

4.1.3 Configure SAP API Management

Setting up SAP API Management in SAP Integration Suite was discussed in Chapter 2. Make sure you have followed those steps, starting with SAP API Management development.

At this point, you should be aware of API Management and its related terms. Your SAP API Management portal has been set up in SAP Integration Suite. Next, let's explore the SAP API Management world with some development. The next section discusses API development.

4.1.4 API Development

API Management, which provides a single interface for managing and keeping track of the various APIs platforms, now includes real-time statistics.

API Management makes it possible for users to securely access pertinent data immediately. Although lowering the chance of security breaches, selective data can be exposed. Instead of directly accessing services, app developers use APIs built using API Management. A publicly accessible HTTP endpoint is connected to your back-end services via API Management, which also handles the security and authorization requirements for safeguarding, assessing, and monitoring your services.

The first and major step to start with SAP API Management is building an API. The next section explains how to build the API and what are the artifacts and policies related to it.

4.1.4.1 Build APIs

API portals offer a standardized framework for the definition and publication of APIs. Every API Management customer is given access to their own cloud-based API portal application. The API portal provides tools for configuring systems, creating and publishing APIs, and testing and analyzing APIs.

When building APIs, you must be familiar with SAP terms and artifacts. The next section discusses API artifacts.

4.1.4.1.1 API Artifacts

API artifacts are the building blocks of an API that are created, managed, and published using SAP API Management. Table 4-1 describes the various artifacts of an API that you need to be aware of before building APIs.

Table 4-1. *API Artifacts*

Name	Description
API Proxy	An API proxy acts as an enterprise intermediary between a client application and an API. It functions as a smart server or gateway, skillfully managing and regulating requests and responses, ensuring smooth communication between the two. API proxies can perform several tasks: rate limitation, monitoring, transformation, and security.
Proxy Endpoint	The URL endpoint that an API proxy exposes to receive inbound requests from client applications is called a proxy endpoint.
Target Endpoint	The back-end services that implement an API's functionality are target endpoints.
API Resource	Several individual entities comprise an API proxy. For instance, the API administrator wants to use an API proxy object to expose BusinessPartnerCollection as an API resource.
Operations	Methods like GET, PUT, DELETE, and POST are made.
Policy	Policies regulate the SAP API Management runtime engine. The implication is that rules are distinct from how services are described. They can be dynamically linked to these APIs or services to specify the minimum or maximum levels of various operations and service quality.
API Documentation	This concisely describes each API resource.

The next section discusses the policies presented in SAP API Management.

4.1.4.1.2 Policies

API Management can specify an API's behavior using policies. A program called a policy runs a certain function at runtime. They allow you to expand an API with common features without developing them from scratch every time. Policies offer tools for securing APIs, managing API traffic, and changing message formats. An API's behavior can also be altered by attaching scripts to policies.

The following policies defined in SAP API Management are divided into four categories.

- Security policies

- Traffic Management policies

- Mediation policies

- Extension policies

The next section goes through each policy type.

4.1.4.1.2.1 Security Policies

In SAP API Management, security policies are a collection of guidelines applied to APIs to ensure they are safe, dependable, and accessible. These guidelines safeguard APIs from various security risks, such as illegal access, data breaches, and cyberattacks. The following are some of the security policies.

- **Basic Authentication**: This Base64 policy encodes a username and password before writing the result to a variable.

- **Decode JWT**: This policy provides information about the JSON Web Token (JWT) decoding policy.

- **Generate JWT**: This policy covers creating a JWT policy.

- **JSON Threat Protection**: This feature disables specific restrictions for specific JSON structures, like arrays and strings, reducing the threat posed by content-level attacks.

- **OAuth v2.0**: This is an authorization procedure for resources protected by APIs is defined by OAuth 2.0.

- **OAuth v2.0 GET**: API Management for apps creates and manages a collection of OAuth resources.

- **OAuth v2.0 SET**: API Management generates and distributes OAuth access tokens to apps. With API Management, you can add or remove the custom attributes related to an access token.

- **Regular Expression Protection**: By allowing developers to set up regular expressions that can be tested against API traffic at runtime, API Management assists in identifying typical content-level threats that exhibit patterns.

- **Verify API Key**: Implementing this policy is one technique to prevent unauthorized access to online-accessible APIs. The application key must be validated per the Verify API Key policy to utilize your APIs.

- **Verify JWT**: This policy validates a signed JWT using a set of claims that can be customized.

- **XML Threat Protection**: Developers can address XML flaws and reduce API assaults by using API Management.

4.1.4.1.2.2 Traffic Management Policies

In SAP API Management, traffic management policies are guidelines used to regulate and manage the flow of API traffic. These rules enhance user experience, increase API performance, and guarantee that users can always access the API.

The following are some of the SAP API Management's traffic control rules.

- **Access Control**: Based on certain Identity Provisioning (IP) addresses, this policy limits access to your APIs.

- **Invalidate Cache**: The cache can be specifically invalidated by supplying an HTTP header. The cache is flushed whenever a request with the specified HTTP header is obtained.

- **Lookup Cache**: This policy writes an OAuth access token to the cache. Using a Lookup Cache policy, the OAuth token is fetched to handle following requests.

- **Populate Cache**: This policy writes an OAuth access token to the cache. A Lookup Cache policy is used to get the OAuth token and manage subsequent requests.

- **Quota**: This policy is a collection of guidelines and restrictions established by a company or service provider to cap the number of resources a user can utilize during a given time frame.

- **Reset Quota**: This policy lets you reset a specific Quota policy's limit.

- **Response Cache**: This is a set of guidelines and settings that control how API replies are cached. Response caching can enhance API performance by easing the burden on the back-end systems and lowering the latency of repeated queries.

- **Spike Arrest**: This policy restricts how many requests can be forwarded from the processing flow step where the policy is connected.

4.1.4.1.2.3 Mediation Policies

Mediation policies in SAP API Management are regulations applied to API traffic as it moves through the API gateway. These principles are applied to adapt messages to the needs of API consumers and back-end systems.

SAP API Management mediation policies can be utilized for various functions, including message transformation, content-based routing, message filtering, authentication, and authorization. You can alter the way the API gateway behaves to suit the requirements of your business by utilizing these policies.

The following are a few SAP API Management mediation policies.

- **Access Entity**: refers to guidelines and settings that specify API request access control and authorization guidelines. Your APIs are protected using the Access Entity policy, which limits who has access to them and what operations they may carry out.

- **Assign Message**: This policy enables the creation of fresh HTTP requests and responses and their modification.

- **Extract Variables**: This policy pulls data from the request message and stores it as variables that can be utilized in other policies or the API response. The Extract Variables policy makes specified data from a request message available for usage in later stages of the API flow.

- **JSON to XML**: Developers can use this policy type to translate messages from JSON to XML format.

- **KeyValueMap Operations**: You can manipulate key/value pairs in the request or response message with this policy. Key/value pairs can be changed or added to message headers or bodies using this policy, enabling customization of the API message flow. This policy allows you to add, remove, or change key/value pairs based on guidelines in the message body or headers.

- **RaiseFault**: When there are errors, you can write your own messages using this policy. If an error occurs, this policy responds to the requesting application with a FaultResponse.

- **XML to JSON**: This policy transforms XML data in the request or response message to JSON format. The message payload is converted from one format to another using the XML to JSON strategy, enabling compatibility between various systems and streamlining the handling of XML data in contemporary applications.

- **XSL Transform**: Documents can be converted from one XML format to another using the language known as extensible stylesheet language transforms (XSLT). Applications that use XML and want a different format for the same data employ this policy.

4.1.4.1.2.4 Extension Policies

You can add custom code to the API gateway to carry out additional processing and functionality that is not possible through the built-in policies by using extension policies in SAP API Management.

The API gateway's extension policies are written in JavaScript and can be used in various ways to extend its capabilities. The following are examples of extension policies in SAP API Management.

- **JavaScript**: This policy enables you to run unique JavaScript code as part of the API request and response flow. The API message flow can have customized logic and functionality added using the JavaScript policy.

- **Message Logging**: You can log API messages to a central location for monitoring and analysis using this policy in SAP API Management. This policy can record information about API calls and answers, such as headers, payloads, and status codes for requests and responses.

- **Message Validation**: This policy verifies a message and rejects it if it doesn't meet the conditions.

- **Open Connectors**: This policy is provided by API Management and can be added to an Open Connector type API.

- **Python Scripts**: This policy sets up Python script code such that it runs through an API proxy.

- **Service Callout**: This policy calls an external service within an API flow. This policy allows for additional data retrieval or processing using the API or back-end services.

- **Statistics Collector**: Back-end services are discovered and exposed as APIs using this policy. With the help of this policy, you can connect to back-end systems and instantly create APIs based on the services they expose.

4.1.4.1.3 API Providers

An API provider is a company that develops, distributes, and manages APIs. An API provider is an organization, division within an organization, or person in charge of administering and making available APIs for other users or systems. Use an API provider to specify any additional information required to create the connection, such as proxy settings and the specifics of the host you want an application to connect. Via the API portal, connections to OData-hosted systems can be configured.

Let's try to expose the APIs from the demo API provider (i.e. SAP Gateway Demo ES5).

You must first register to create an account at SAP demo ES5.

The following steps create the API provider in SAP API Management.

1. Open the API portal, and choose Configure from the navigation icon on the left in the top-left navigation bar, as shown in Figure 4-3.

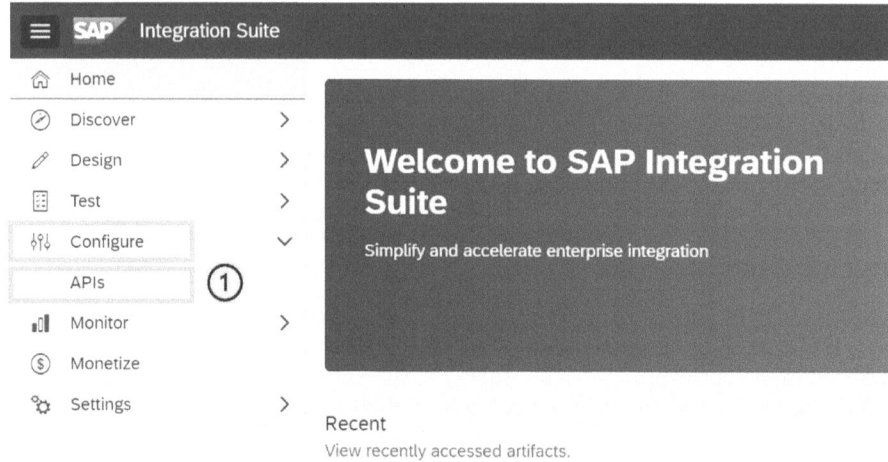

Figure 4-3. *Configure APIs*

2. Select APIs to open the Configure dashboard. Choose Create, as shown in Figure 4-4.

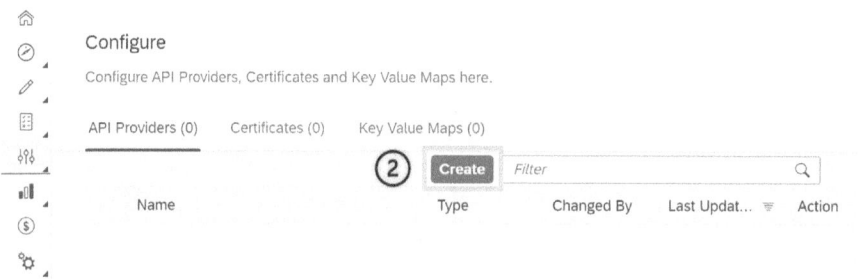

Figure 4-4. *Create API Providers*

3. In the Overview tab, provide the name and description of the API provider, as shown in Figure 4-5.

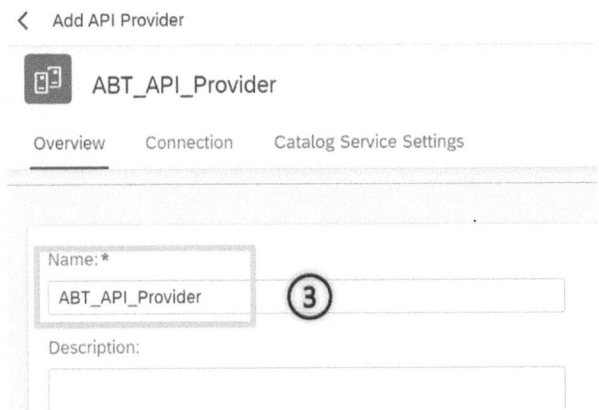

Figure 4-5. *API Provider details*

4. Choose the Connection tab's connection type by navigating there.

- Internet: Connect to the cloud-based system.

- On-Premises: Choose On-Premises if you connect the Cloud Connector to the other On-Premises system.

- Open Connectors: Open Connectors link independent RESTful APIs with other APIs from third parties.

- Cloud Integration: With cloud integration, you may reach every service endpoint.

Fill out the details in the connection tab with the following values, as shown in Figure 4-6.

- Type: Choose any connection type from the Internet, Open Connectors, On-Premises, or Cloud Integration. Choose the Internet.

- Host: Provide the API provider's host name. SAP dev center is a good example that acts as an API provider. The host name is sapes5.sapdevcenter.com.

- Port: Provide the valid port number of the API provider. In this case, it is 443.

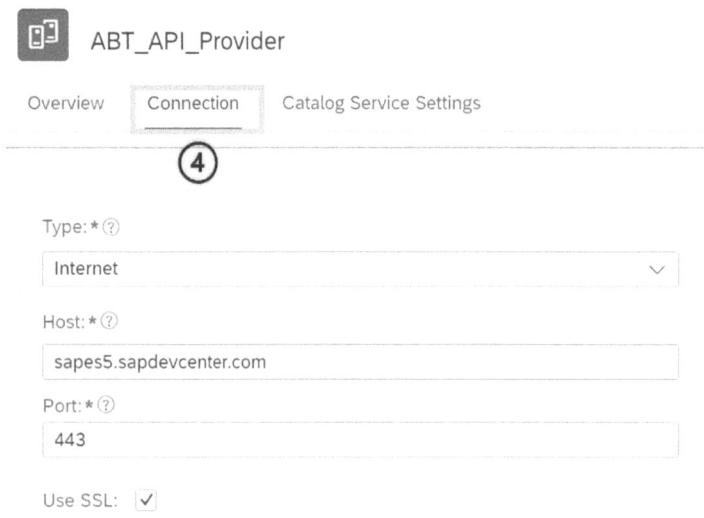

Figure 4-6. Provide details in Connection tab

5. Open the **Catalog Service Settings** tab and provide the required details, as shown in Figure 4-7.

- Path prefix: /sap/opu/pdata/IWFND

- Service Collection URL: /CATALOGSERVICE/ServiceCollection

- Authentication type: Basic

- Username: User ID from SAP Demo Server

- Password: Password from Demo Server

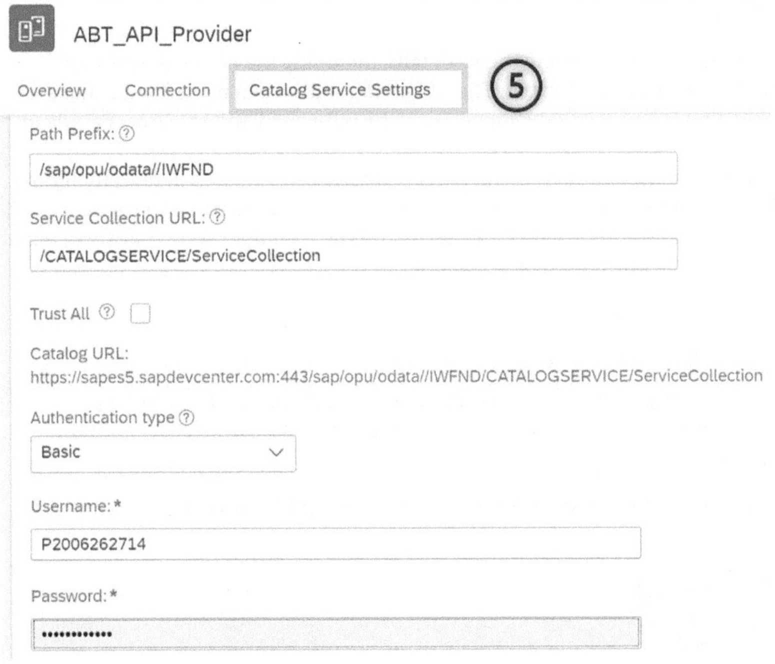

ABT_API_Provider

Overview   Connection   Catalog Service Settings   ⑤

Path Prefix: ⑦

/sap/opu/odata//IWFND

Service Collection URL: ⑦

/CATALOGSERVICE/ServiceCollection

Trust All ⑦ ☐

Catalog URL:
https://sapes5.sapdevcenter.com:443/sap/opu/odata//IWFND/CATALOGSERVICE/ServiceCollection

Authentication type ⑦

Basic   ⌄

Username: *

P2006262714

Password: *

••••••••••••

Figure 4-7. *Catalog Service Settings*

6. Save the API provider and Test the Connection. You will get the 200 Message: OK, as shown in Figure 4-8. If you do not get the 200 Message that means you have error in your Connection. To avoid getting an error while doing the Test Connection, make sure you have provided the correct details in the Connection tab.

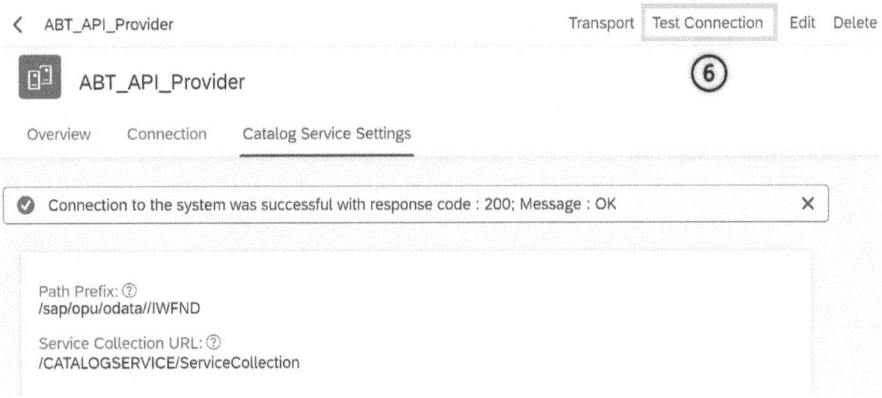

‹ ABT_API_Provider   Transport │ Test Connection │ Edit   Delete

ABT_API_Provider   ⑥

Overview   Connection   Catalog Service Settings

✓ Connection to the system was successful with response code : 200; Message : OK   ✕

Path Prefix: ⑦
/sap/opu/odata//IWFND
Service Collection URL: ⑦
/CATALOGSERVICE/ServiceCollection

Figure 4-8. *Test Connection*

You have configured the API provider and know the SAP API Management policies. The next step in building the API is to create the API proxy using the API provider.

4.1.4.1.4 Create API Proxy

You can build an API in SAP API Management by specifying its details, including the name, description, endpoint URL, parameters, headers, and response format. The API configuration can be defined using a graphical user interface (GUI), and after it has been developed, it can be published and made accessible to developers.

When building an API in SAP API Management, you can specify different API-related details, including the authentication and security procedures, rate limiting regulations, and documentation. To ensure compatibility between the API and the back-end service, you can also define the target system or service that the API connects to and any necessary transformation or mediation.

4.1.4.1.4.1 Create an API

SAP API Management allows you to create and manage APIs centrally, making it easier for developers to consume and integrate with your enterprise data and services. The following is a high-level overview of creating an API in SAP API Management.

1. In the SAP Integration Suite, navigate to the left pane. Select **Design ➤ APIs**, as shown in Figure 4-9.

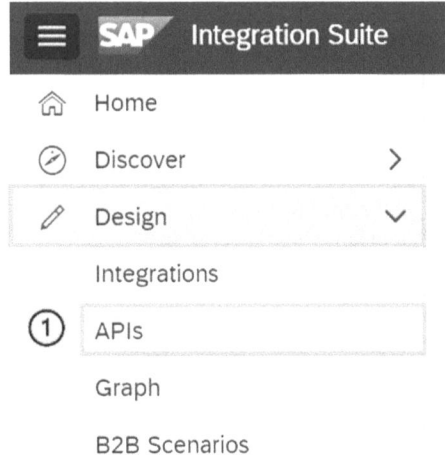

Figure 4-9. *Navigate to APIs from Design tab*

2. In the Develop dashboard click **Create**, as shown in Figure 4-10.

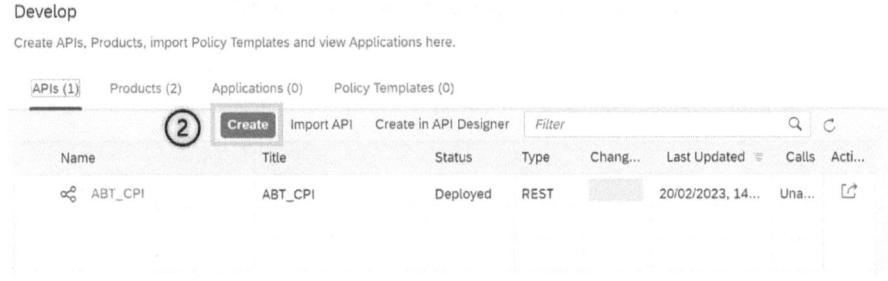

Figure 4-10. Create API

3. The wizard is started. As illustrated in Figure 4-11, you may create an API proxy by selecting the API Provider option and selecting an API of the ODATA, REST, or SOAP types. Choose the API Provider created in section 4.1.4.1.3.

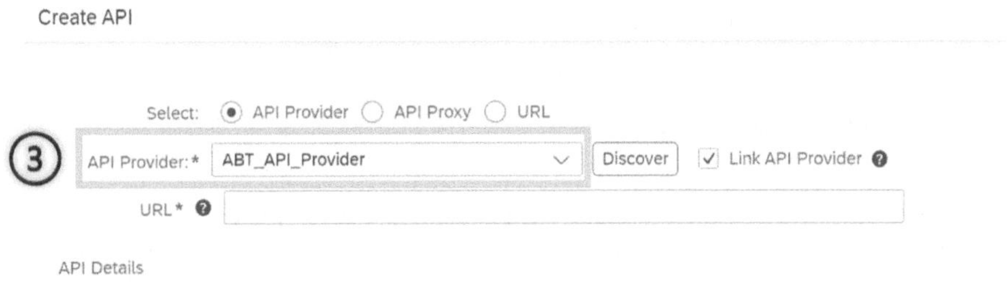

Figure 4-11. API Details

4. Navigate to the Discover tab, which leads you to the set of APIs. Select the APIs from the list to connect your API proxy with the API, as shown in Figure 4-12. These APIs are listed if you successfully developed a connection in the API provider.

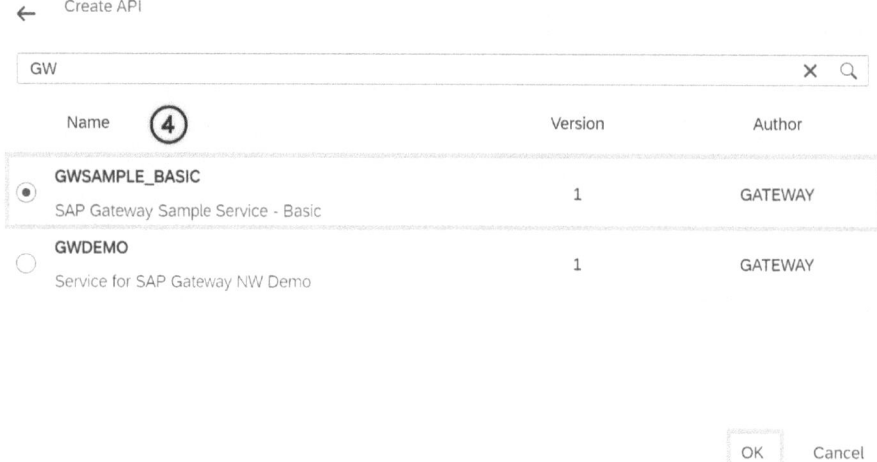

Figure 4-12. *Select API provider from Discover tab*

5. You can see the details to the wizard automatically be filled up. Check all the details and select **Create**, as shown in Figure 4-13.

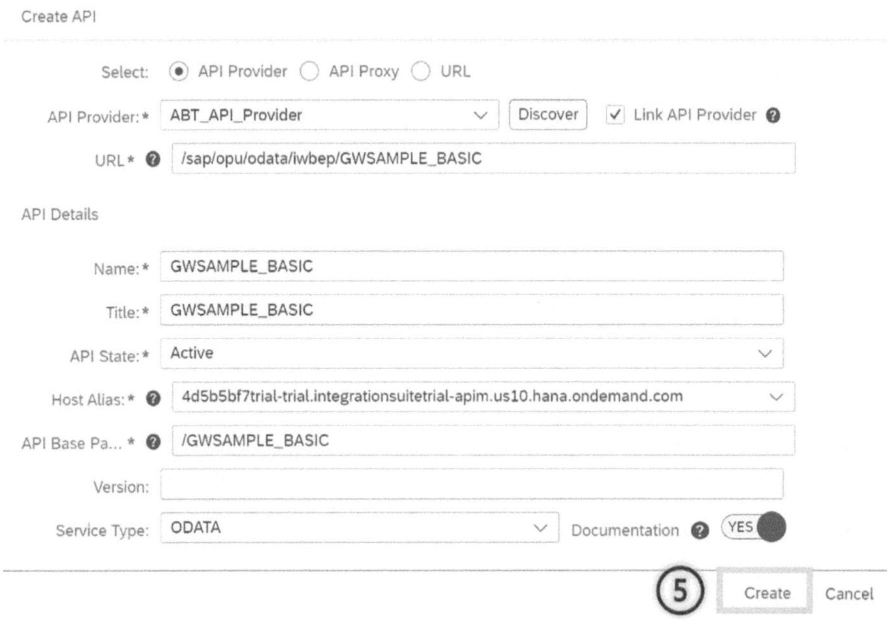

Create API

Select: ● API Provider ○ API Proxy ○ URL

API Provider: * ABT_API_Provider ⌄ Discover ✓ Link API Provider ❓

URL * ❓ /sap/opu/odata/iwbep/GWSAMPLE_BASIC

API Details

Name: * GWSAMPLE_BASIC

Title: * GWSAMPLE_BASIC

API State: * Active ⌄

Host Alias: * ❓ 4d5b5bf7trial-trial.integrationsuitetrial-apim.us10.hana.ondemand.com ⌄

API Base Pa... * ❓ /GWSAMPLE_BASIC

Version:

Service Type: ODATA ⌄ Documentation ❓ YES ●

⑤ Create Cancel

Figure 4-13. *Create API*

6. You can also navigate through the API dashboard's Overview, Target Endpoints, Proxy Endpoints, and Resources tabs.

4.1.4.1.4.2 Deploy an API

Deploying an API makes it available to clients, such as mobile apps or websites. The deployment process involves several steps, including preparing the API for deployment, selecting a hosting platform, configuring security settings, and testing the API to ensure it works properly.

Here are the general steps involved in deploying an API.

1. Navigate to the Design APIs tab.

2. Save the API.

3. Select the three dots (**●●●**) on the right. You see the Deploy option. Choose

   Deploy, as shown in Figure 4-14. Your artifact is deployed, and you can check its status to see if it has already been done so or not.

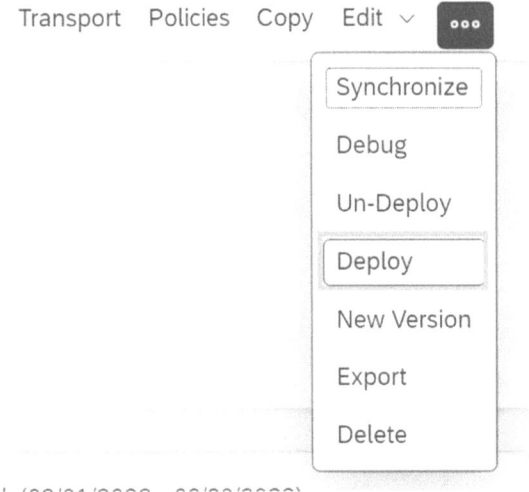

Figure 4-14. *Deploy API*

4.1.4.1.4.3 Export an API

You can export an API after creating it on the API portal. To export the API, the mandatory role required to the tenant in SAP BTP cockpit is APIPortal.Admin.

1. Log in to the API portal.

2. Navigate to the Design APIs tab.

3. In the Action column, select the **Export** button to export the specific API, as shown in Figure 4-15.

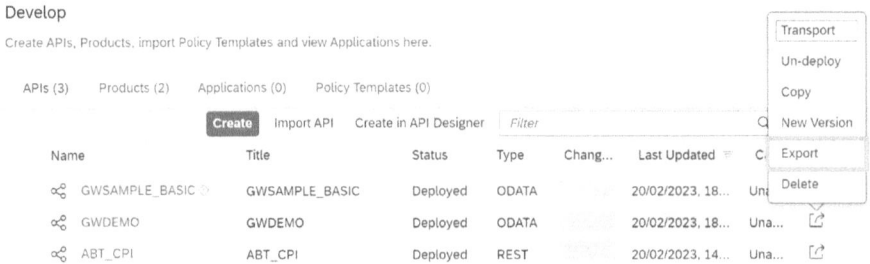

Figure 4-15. *Export*

4.1.4.1.4.4 Import an API

You can import the existing API to the existing API portal. The following explains how to import an API.

1. Open the API portal.

2. Go to the **Design ➤ APIs** window. Select **Import API** button there, as shown in Figure 4-16.

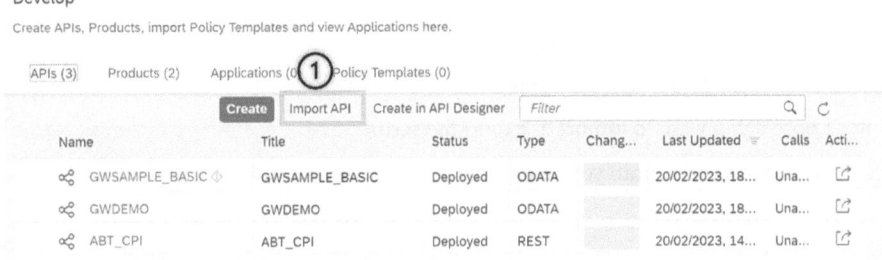

Figure 4-16. Import API

3. Browse the file name from your local computer and import the API zip file provided by your development team, once the file has been browsed click OK, as shown in Figure 4-17.

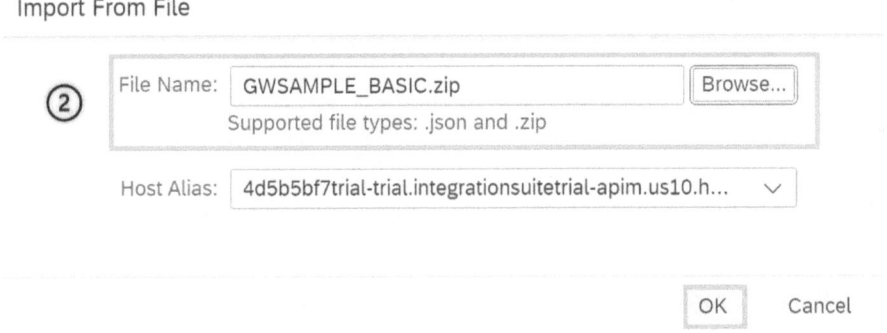

Figure 4-17. Browse ZIP file

4.1.4.1.4.5 Create a Script

A file resource or a piece of code called a *script* can be added to flows utilizing policies. The development of JavaScript, Python, and XSL scripts is supported by SAP API Management.

1. Open API Proxy and navigate to the Policies tab.

2. Click + in the Scripts section.

3. Name the script.

4. Choose one of the following from the Type field.

- JavaScript

- XSL

- Python

5. Browse the script from your device and click Add, as shown in Figure 4-18.

6. Provide the details.

- Name: Provide the name of the script collection.

- Type: Select the type of the script collection.

- Script: Choose if you want to import or export the script.

- Upload: Browse the script from your computer and click Add.

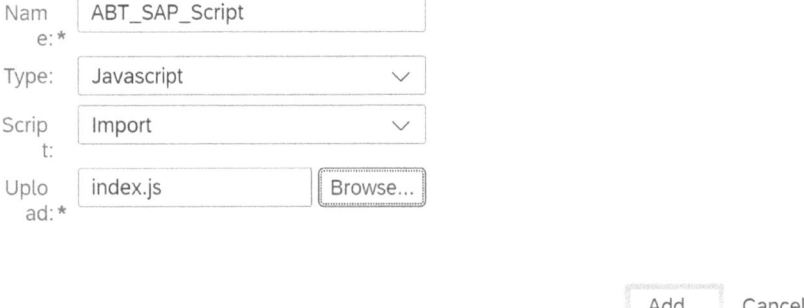

Create Script

Name:*	ABT_SAP_Script
Type:	Javascript ∨
Script:	Import ∨
Upload:*	index.js Browse...

Add   Cancel

Figure 4-18. *Create Script*

4.1.4.1.4.6 Create a Policy Template

SAP API Management provides the ability to create policy templates, which can be used to define a set of common policies that can be applied to APIs or API proxies. Policy templates can simplify applying policies to APIs by allowing you to define a standard set of policies that can be reused across multiple APIs.

The following are the general steps to create a policy template in SAP API Management.

1. Open the API portal and log in.

2. Choose **Develop** from the navigation menu on the left.

3. In the catalog, there is a list of registered APIs.

4. Choose the API from the list for which you want to generate the policy template.

5. Choose **Policies** from the Detail screen.

6. Select **Policy Template Generate** from the Policy Editor screen.

7. Follow these steps in the Create Policy Template window, as shown in Figure 4-19.

    a. In the Name field, give the template a name.

    b. In the Description tab, enter a description of the template.

    c. From the list of available policies, select the ones you need.

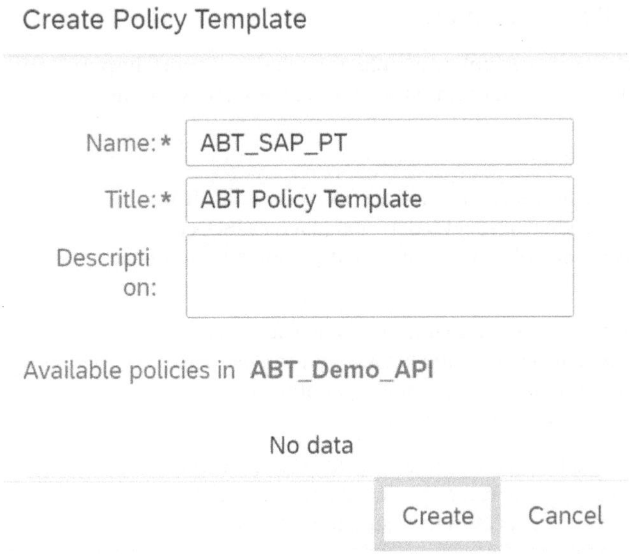

Create Policy Template

Name:* `ABT_SAP_PT`

Title:* `ABT Policy Template`

Description:

Available policies in **ABT_Demo_API**

No data

[Create] Cancel

Figure 4-19. *Create Policy Template*

In Policy Template, you can perform the additional actions as described in the next section.

4.1.4.1.4.6.1 Apply Policy Template

Add Policy Template is a feature in SAP API Management that allows you to add a predefined set of policies to an API or API proxy. These predefined sets of policies are known as policy templates, which can be created and customized according to your organization's specific needs.

1. Choose **Edit ➤ Policy Template ➤ Apply** on the Policy Editor screen.

2. Choose the desired policy templates from the Apply Policy Template window.

3. Choose **Copy** if you simply want to duplicate the policies and not the flows. Unless you want to repeat both flows and policies, choose Apply.

4.1.4.1.4.6.2 Update Policy Template

Update Policy Template is a feature in SAP API Management that allows you to modify an existing policy template. Policy templates are predefined sets of policies that can be applied to APIs or API proxies to provide consistent security, governance, and management across your organization's APIs.

The following explains how to update the Policy template in SAP API Management.

1. Choose **Policy Template Update** from the policy editing screen.

2. From the Name field in the Change Policy Template window, select the policy template that needs to be updated.

3. From the available list of policies, select the ones you need.

4.1.4.1.4.6.3 Export/Import and Delete Policy Template

SAP API Management provides several features to manage policy templates, including Export, Import, and Delete. These features allow you to efficiently manage policy templates across different environments, or remove unused or outdated policy templates.

Here's a brief overview of each of these features.

- **Export Policy Template**: This feature allows you to export a policy template from SAP API Management, and save it as a file in a specified format (such as JSON or XML). You can use this file to transfer the policy template to a different environment or share it with others.

- **Import Policy Template**: This feature allows you to import a policy template into SAP API Management from a file saved in a supported format. You can use this feature to transfer a policy template from one environment to another, or to add a new policy template to your existing set of policy templates.

- **Delete Policy Template**: This feature allows you to delete a policy template that is no longer required. You can use this feature to remove unused or outdated policy templates from your SAP API Management account to ensure that your policy templates remain well-organized and up-to-date.

The following are instructions to export, import, and delete policy templates in SAP API Management.

1. Choose **Policy Templates** from the Develop screen.

2. The catalog displays a list of possible policy template options.

3. In the Action tab, choose **Export** to export any policy template, as shown in Figure 4-20.

4. Click **Import** and browse from the to import a policy template, as shown in Figure 4-20.

5. To delete the policy template, click **Delete** option, as shown in Figure 4-20.

Develop

Create APIs, Products, import Policy Templates and view Applications here.

APIs (4)   Products (2)   Applications (0)   Policy Templates (1)

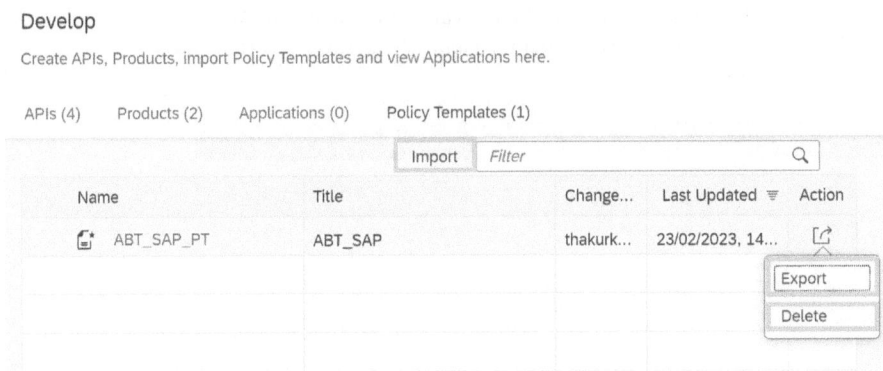

Figure 4-20. *Export/import/delete policy template*

4.1.4.1.4.7 Key Value Map

A key value map (KVM) enables the storage of key/value pairs as configuration information that may be used by various API rules. Key value maps can be accessible and used in multiple policies within the same API or across different APIs, and they are used to store and retrieve data such usernames, passwords, and API endpoint URLs.

You can perform the following functions.

- Create a key value map

- Update a key value map

- Delete a key value map

The following sections discuss each function of Key Value Map.

4.1.4.1.4.7.1 Create a Key Value Map

The following explains how to create key value maps.

1. Open the API portal and log in.

2. Choose **Configure** from the navigation menu on the left.

3. Select **Key Value Maps**, then click **Create**, as shown in Figure 4-21.

Configure

Configure API Providers, Certificates and Key Value Maps here.

API Providers (1)   Certificates (0)   Key Value Maps (0)

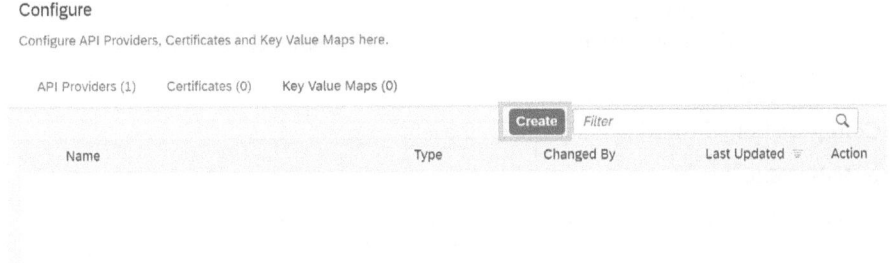

Figure 4-21. *Create key value maps*

4. Enter the mandatory information, such as the key value map name, the key, and its value. Click Save to create the key value map. Figure 4-22 shows key value map names and entry values.

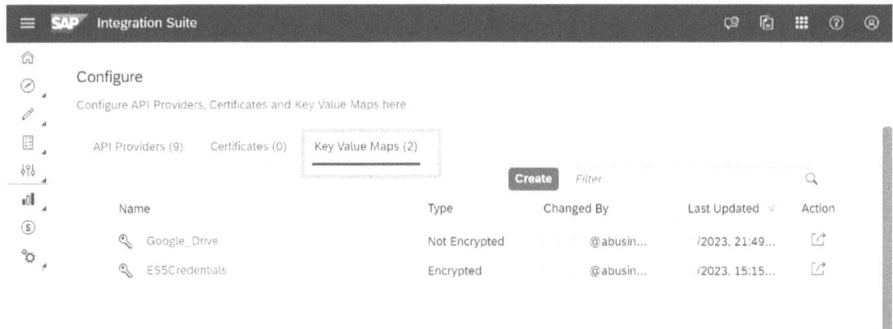

Figure 4-22. *Created key value map*

4.1.4.1.4.7.2 Update a Key Value Map

Updating a key value map in SAP API Management involves modifying the existing key/value pairs stored in the key value map or adding new ones. This can be useful when you need to update the configuration data or other information that is stored in the key value map.

The following steps explain how update a key value map in SAP API Management.

1. Choose **Key Value Map**.

2. Choose **Edit** to carry out the following actions.

    a. To add a new entry, choose Add and type the key and value in the Key Value Map. To delete an entry, select delete icon in the Action column. Restore the modifications.

    b. Choose the entry, then change the Value field if you wish to update an entry.

3. Click Save.

4.1.4.1.4.7.3 Delete a Key Value Map

Deleting a key value map in SAP API Management involves permanently removing the key value map and all the key/value pairs stored within it. This can be useful when you no longer need the configuration data or other information stored in the key value map.

The following explains how to delete a Key Value Map in SAP API Management.

1. Open the API portal and log in.

2. Choose **Configure** from the navigation menu on the left.

3. Select **Key Value Map**.

4. Choose **Delete** in the Action column, as shown in Figure 4-23, to remove an entry.

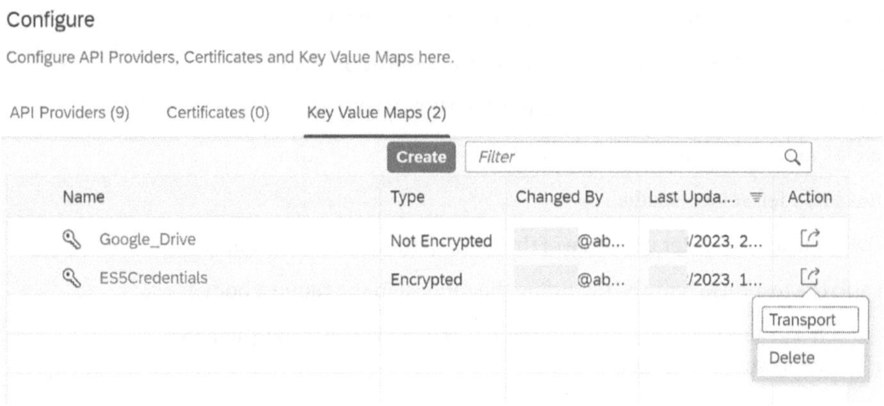

Configure

Configure API Providers, Certificates and Key Value Maps here.

API Providers (9)   Certificates (0)   Key Value Maps (2)

Name	Type	Changed By	Last Upda... ≡	Action
🔑 Google_Drive	Not Encrypted	@ab...	/2023, 2...	⬀
🔑 ES5Credentials	Encrypted	@ab...	/2023, 1...	⬀

Transport

Delete

Figure 4-23. *Delete key value map*

4.1.4.1.4.8 Test APIs

Use the API Test Console to check an API's functioning while it is being used.

With the API Test Console provided by SAP API Management, you can test your APIs. Testing is required to understand an API's behavior at runtime. You can examine the resources linked to an API's resources and carry out the actions using the test console.

OData and REST-based services can be tested using the API Test Console.

Following is the procedure for using the API Test Console.

1. Open the API portal and log in.

2. From the left pane, navigate to the **Test ➤ APIs** option.

3. You see a list of APIs when the API Test Console opens. Choose the API you want to test, as shown in Figure 4-24.

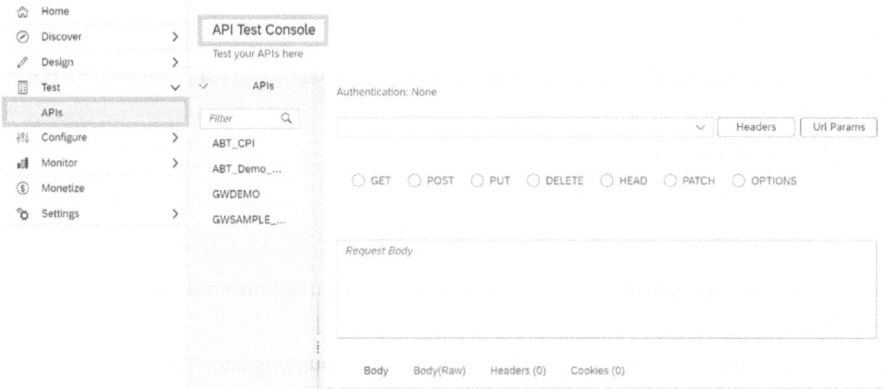

Figure 4-24. *API Test Console*

4. Enter the service URL if you are aware of it and it contains the API.

5. To choose the kind of authentication, select the Authentication type.

6. Select any required method, GET, POST, DELETE, HEAD, PATCH.

7. With the PUT and POST methods, enter the request body.

8. To include a header, select Header.

9. Choose **URL Params** to enter the value of the query parameter. Select **Send**.

10. You can add the response body by providing the request in the request body.

11. Select **Launch API Viewer** to see the transactions related to the testing activity you carried out, as shown in Figure 4-25.

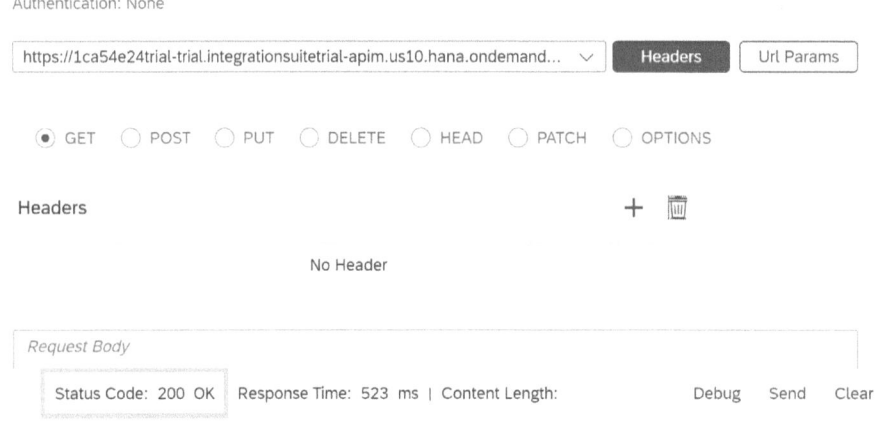

Figure 4-25. Successful test connectivity

4.1.4.1.4.9 Debug API

In SAP API Management, you may debug an API proxy by looking into the specifics of each stage in the flow to troubleshoot and monitor it.

1. Open the API portal and log in.

2. Open the Developer portal from the left menu.

3. Select the API which needs to be debug.

4. Input the body, header, and parameters. Select the **Debug** button in bottom right of the screen.

5. It is advised to duplicate the tab you are working on: one for debugging and one for sending the request through the endpoint URL.

6. On the Debug Viewer page, click Start Debug.

7. Select **Refresh** to view the transaction information when a debugging session is active.

8. Select **Stop Debug** once you've recorded enough queries, as shown in Figure 4-26.

Figure 4-26. API Debug Viewer

9. Use the tips shown in Table 4-2 to debug the API.

Table 4-2. Debug API Icons

Icons	Description
	Shows a condition determined by the API
	Identifies a change in the execution flow's state
	Provides details on the current flow
	Shows the outcome of a condition's execution
	Indicates the presence of an error during the execution of the policy

4.1.4.1.5 Custom Attributes

Custom Attributes are user-defined metadata fields available for API proxies, API products, and API developers in SAP API Management. With the help of custom attributes, you may provide these entities with more details or traits that can be used for tracking or reporting.

Custom Attributes can be of many sorts, such as text, numeric, or date, and can be defined and managed through the SAP API Management user interface or through APIs. Moreover, custom attributes can be given default values and be designated as mandatory or optional.

4.1.4.1.5.1 Add Custom Attributes to Products

Custom attributes are additional metadata that can be added to products in SAP API Management to provide more information about the product. These attributes can be used to filter, search, or categorize products based on specific criteria.

The following explains how to add custom attributes to products in SAP API Management.

1. Open the API portal and log in.

2. Open the Developer portal from the left menu of SAP Integration Suite.

3. Select **Products**, followed by the product to which you want to add the custom attribute.

4. Select **Custom Attributes** from the page with the product details.

5. Select **Add** in the Custom Attributes section. Provide the custom attribute a name and value. Click **Add** to add more attributes.

6. Click the Delete icon in the Actions column to remove a custom attribute.

7. Save the changes.

4.1.4.1.5.2 Add Custom Attributes to an Application

You may manage your applications, add custom attributes to them, and create applications on behalf of other application developers.

At the application level, values have been assigned to the custom attributes. These values facilitate simple configuration and runtime access. In this case, the administrator may assign custom attributes while creating an application at the application level on the developer's behalf. As a result, you may use this to improve the functionality of your API and carry out attribute-specific runtime enforcements.

There are the following Role of the Enterprise administrator for an API Business Hub which are as follows:

- Create an application on behalf of a user and provide the user with the application key and value.

- You can build new applications in different settings by using the same application key and value (for instance, production and non-production).

- At the application level, specify unique properties and control the logic of API calls. (Section 4.1.4.4 covers how to create an application in SAP API Management.)

4.1.4.1.6 Configure Load Balancing

An API proxy's load-balancing features can be set up using the API Administration API portal.

1. Open the API portal and log in.

2. Choose **Develop** from the navigation menu on the left. In the catalog, there is a list of registered APIs.

3. Find an API proxy that has previously been connected to an API provider.

4. To enable the load-balancing feature, select the slide button.

5. From the API Provider selection list, pick more API providers.

   • The Fallback Server field appears when one of the additional API providers is selected.

   From the list of available API providers, pick one to serve as the fallback server. The fallback server receives all traffic if the load balancer determines that the other API providers are not accessible.

6. Enable Retry by selecting the slide button. If the retry option is chosen, a request is reissued each time a response failure, such as HTTP timeout, occurs. A request is also sent out each time a response is obtained, and it matches a value listed by the Response Code.

7. The algorithm used by default is Round Robin. However, you may also select the weighted and least-connection algorithms. Each API provider receives a request from the Round Robin algorithm in the order they are mentioned in the target endpoint HTTP connection.

8. Maximum Failure is activated by choosing the slide button. When enabled, it keeps track of how many unsuccessful API proxy requests result in the request being routed to a different API provider.

9. Select **Save**, as shown in Figure 4-27.

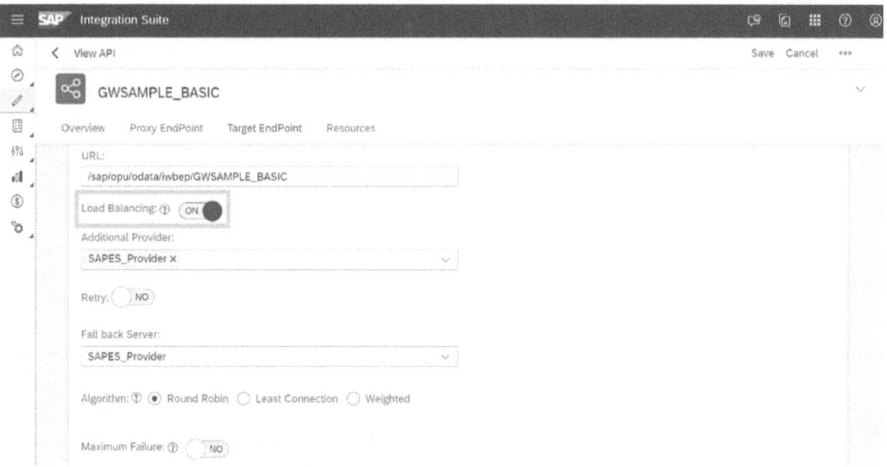

Figure 4-27. *Load Balancing*

You have completed the first step of API development that is creating the API. The second step is publishing the API, which is discussed in the next section.

4.1.4.2 Publish APIs

Publish APIs is the process of making your APIs accessible to other developers and users through a developer portal or marketplace platform. With the help of the Publish APIs feature, you can design a unique developer portal where you can promote your APIs, offer API documentation, and let developers sign up, subscribe, and utilize your APIs on their own terms.

A collection of APIs is a product. It has metadata for monitoring and analytics that is particular to your company. For instance, all CRM-related APIs may be combined into a single CRM solution. It is easier to combine relevant APIs into a single product and publish it than to publish APIs separately. A product is published to the catalog after the APIs have been added, making it accessible to application developers for browsing.

A product is connected to one or more APIs when it is created. Moreover, different products can be connected to the same API. After you link an API to a product, all of its features, including its resources and documentation, are implicitly incorporated into the product.

An application developer can learn which APIs are offered on the API portal by using a product. You choose the product to add in the application when you create it. SAP API Management produces an application key and secret for each application you build. To acquire access to numerous products, use this key.

4.1.4.2.1 Develop an API Product

Developing an API product in SAP API Management involves creating a package of related APIs that can be used by developers to build applications and services. By developing an API product, organizations can provide a standardized and consistent set of APIs that can be used across different applications and platforms.

The following are instructions to develop an API product in SAP API Management.

1. Open the API portal. Open the API in the Design tab. Navigate to the Products tab and click Create, as shown in Figure 4-28.

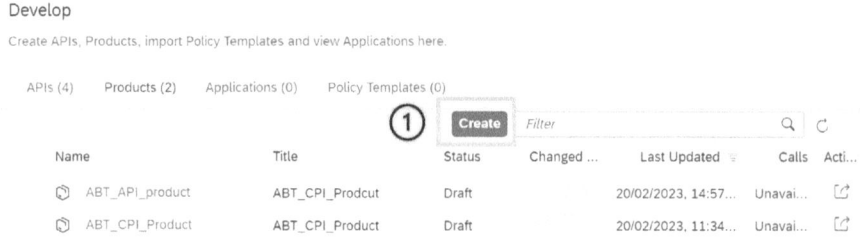

Figure 4-28. *Create an API product*

2. In the **Overview** tab, specify the product name and title to the product. In Quota, specify the number of times the API proxy is called. Request Every means (per) minute, hour, or day, as shown in Figure 4-29.

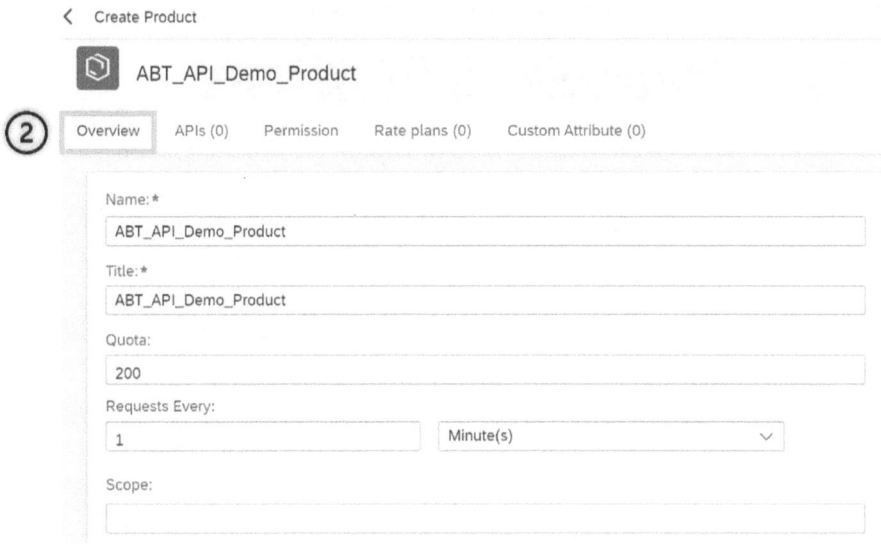

Figure 4-29. *Specify details*

3. In the APIs tab, select the APIs to be added in the Products. Click **Ok**, as shown in Figure 4-30.

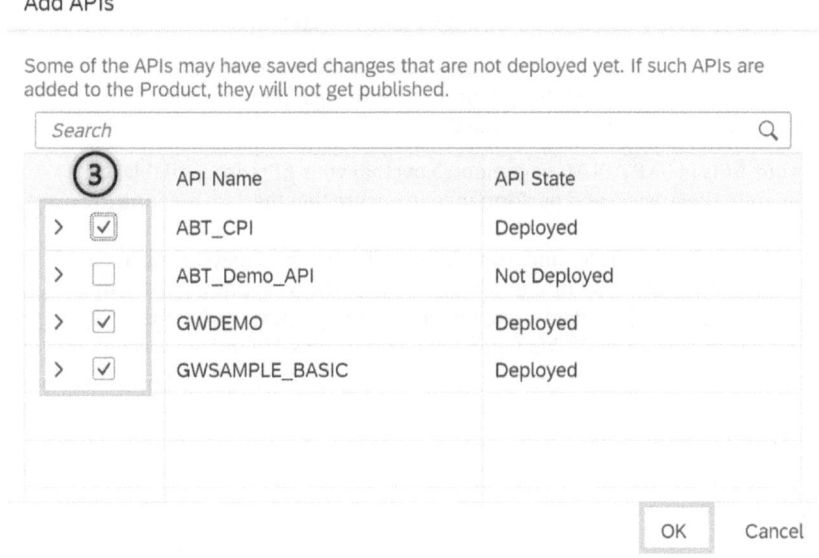

Figure 4-30. *Add APIs*

4. Click **Publish** to publish the API product, as shown in Figure 4-31.

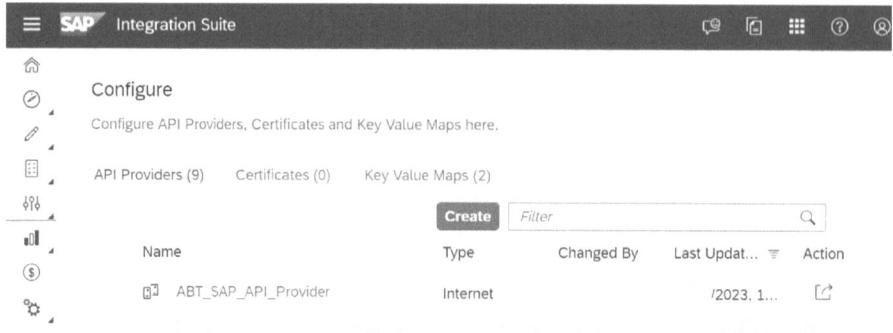

Figure 4-31. *API product*

4.1.4.2.2 View Applications

As an administrator, you may now use the API portal to access the subscribing applications.

1. Open the API portal and log in.

2. Choose Develop from the navigation menu on the left.

3. Go to the Apps tab page by navigating.

    • A list of applications is provided. If you have the role APIPortal.Admin and are logged into the API portal, you can view the application key and secret.

    • During the current month, you can see how many calls each API in an application has received. Both the information screen for each program and the Calls column for each application display the data.

Great job on publishing your APIs in SAP API Management! Now that your APIs are available to consumers, it's important to analyze their usage and performance to ensure that they are meeting the needs of your users.

SAP API Management provides several tools for analyzing APIs, including the Analytics dashboard, which provides detailed metrics on API usage, response times, and error rates. By analyzing these metrics, you can identify areas for improvement and make data-driven decisions to optimize your APIs.

In addition to the Analytics dashboard, SAP API Management also provides API tracing and debugging tools, which allow you to monitor and diagnose issues in real-time. This can be especially useful when dealing with complex APIs or when troubleshooting issues reported by users. The next section walks through how to analyze your APIs in SAP API Management.

4.1.4.3 Analyze APIs

Analyzing an API in SAP API Management involves examining the API to understand its functionality, behavior, and usage patterns. This analysis is typically done to help developers and administrators understand how the API works and how it can be used to build applications and services.

Full analytics capabilities are offered by API Management to assist customers in comprehending the various trends in API performance and consumption. Using runtime data from the APIs, the API Analytics server analyses the data. The runtime data is collected and analyzed, and then presented using charts, headers, and key performance indicators (KPIs).

Make use of the analytics dashboard to see the results. The key to improving business decisions is managing APIs, attracting the right application developers, fixing problems, and utilizing the detailed analytics perspective.

Two types of analytics dashboards are accessible for examining API reports.

- **API Analytics**: The analytics dashboard provides a thorough view of analytical graphs and KPIs pertinent to how the APIs are being used, as shown in Figure 4-32. The dashboard also shows graphs and KPIs for errors, such as the total number of API problems, system failures, and policy errors.

- **Advanced API Analytics**: You may analyze your API usage and performance with the help of the brand-new analytics dashboard from Advanced API Analytics, which offers strong tools and in-depth reports. Each report page contains details on important API metrics that are important to both business customers and API developers. The reports are classified across report pages.

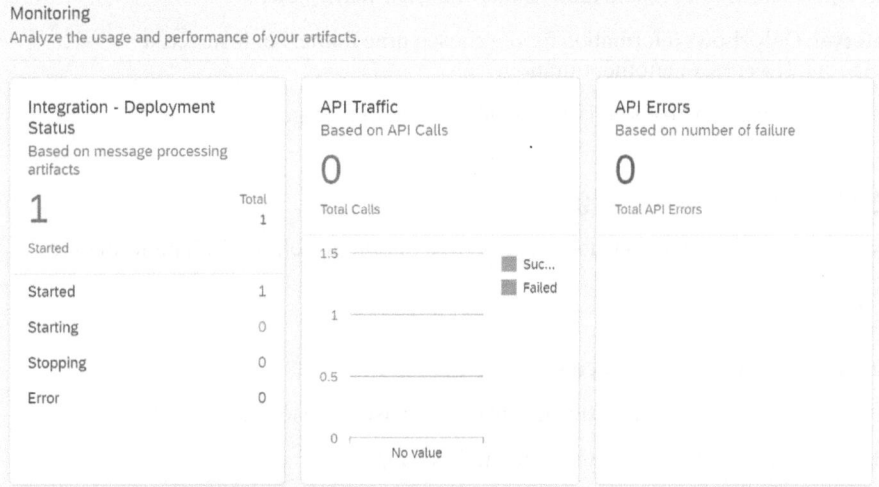

Figure 4-32. *API analytics dashboard*

4.1.4.3.1 API Analytics

Some analytical graphs and KPIs are provided by API Analytics (KPIs). The dashboard comes preconfigured with these graphs and metrics.

4.1.4.3.1.1 Analytics Dashboard

A web-based interface called the Analytics dashboard in SAP API Management offers real-time insights and visualizations on the performance, usage, and behavior of your APIs and API customers. You may track and examine many aspects of your APIs, including traffic volume, response times, problems, and trends, as well as the actions of your API users, including usage trends, demographics, and feedback, using the Analytics dashboard.

The following are features of the Analytics dashboard.

- **Views**: The dashboard provides us the three views.

    - **Performance**: This view displays charts and KPIs for performance. Total API hits and average API response times are two examples of performance-related KPIs. You can also look at an API traffic graph.

    - **Error**: This view presents graphs and KPIs pertaining to errors. For instance, you may track error-related KPIs (e.g., the overall system error count, the overall API policy error count, etc.), or you can show a chart of the top 5 API policy error rates.

    - **Custom**: This view displays the unique charts you've created by selecting the chart type, measurements, and dimensions in the left-hand pane.

- **Time Interval**: Only shows information for the chosen time frame. For instance, a period of months, weeks, or another duration.

- **Resize Charts**: Charts can be resized from small to medium to large.

4.1.4.3.1.2 Analytics Dashboard: Working

In the form of charts and KPIs, the analytics dashboard presents a thorough picture of API performance and faults.

1. Open the API portal and log in.

2. Open Analyze portal from the left menu.

3. Choose one of the following views using the drop-down list in the top-left corner.

4. For each chart on the dashboard, you can do the following things.

    - Details: Choose Details from the top-right menu of any chart to see the details page. You can select criteria and switch between different chart kinds on the details page.

    - Resize: Adjust the chart's size by resizing it (small, medium, or large).

Choose the desired time at the dashboard's top to review data for a specified period.

4.1.4.3.2 Advanced API Analytics

The majority of the reports displayed graphically on the analytics dashboard were created utilizing eye-catching charts. To display data in the way that best suits your needs, you can select from a variety of chart formats. You can easily scan and analyze key API metrics thanks to the chart-oriented layout, which aids in your ability to make smarter business decisions. The report's Overview, Health, and Usage pages are where the analytical data is dispersed, with each page containing details on important API metrics, as shown in Figure 4-33.

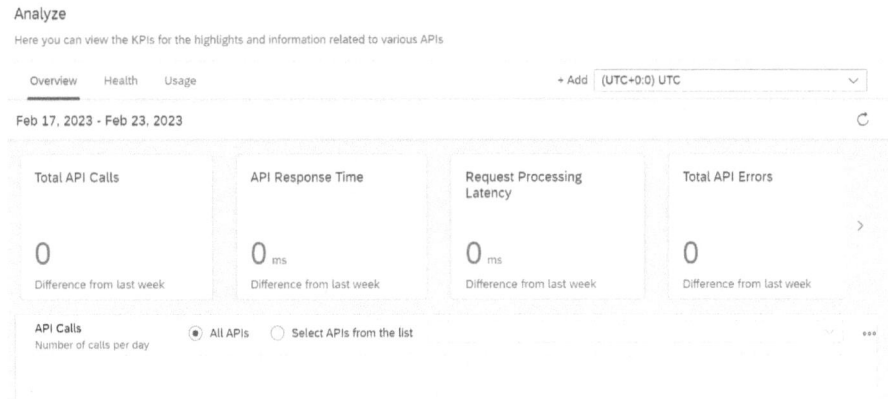

Figure 4-33. *Analyze dashboard*

- **Overview**: The most significant API metrics are enumerated on the Overview page. Developers as well as users can rapidly check reports and follow trends in the API for the previous seven days on the Overview page. Business customers have access to the most used APIs, the overall number of API requests, and the top application developers. Developers can learn further about underperforming APIs as well as the variables that affect them, such as latency and response time.

- **Health**: The Health page provides information on your APIs' KPIs. On the Health tab, where they can also examine patterns in API problems over the previous seven days, API developers may instantly track the metrics for APIs that have an impact on their performance. The information on typical API error types and the normal API response time are the most crucial monitoring metrics. Moreover, API developers can discover the most recent error responses and the total number of incorrect queries performed for an API.

- **Usage:** KPIs for your APIs are detailed on the health page. The metrics for APIs that affect their performance may be tracked instantaneously by API developers on the health page, where they can also look for trends in API issues over the last seven days. The most important monitoring metrics are those related to usual API error kinds and average API response times. Moreover, API developers can learn about the total number of improper API queries made as well as the most recent error answers.

4.1.4.3.2.1 Create and work with Custom Reports

You can design unique representations for API metrics that are essential to your company's operations. You may group all these API metrics together and analyze them in a single window using the Custom view option.

1. Select the +Add button to create a custom view in the analytics dashboard, as shown in Figure 4-34.

Analyze

Here you can view the KPIs for the highlights and information related to various APIs

Figure 4-34. *Add custom view*

2. Give your new custom report a name in the Create Custom View dialog and hit OK, as shown in Figure 4-35.

Create Custom View

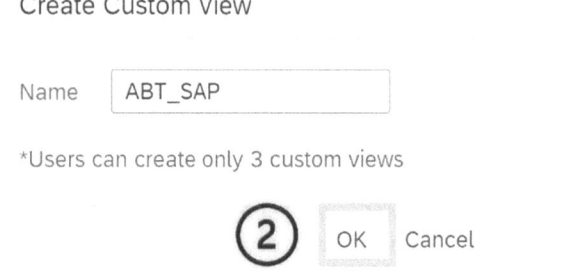

Figure 4-35. *Create Custom View*

3. Give the chart a title and a description, as shown in Figure 4-36.

Figure 4-36. *Title and Description in custom report*

4. In the Dimensions drop-down menu, select the API measure you wish to employ. Choose the desired API metric from the Measurements menu, as shown in Figure 4-37.

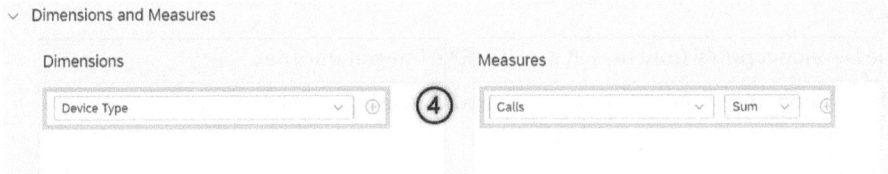

Figure 4-37. *Dimensions and Measures*

5. Click **Save** after selecting OK to save the graph, as shown in Figure 4-38.

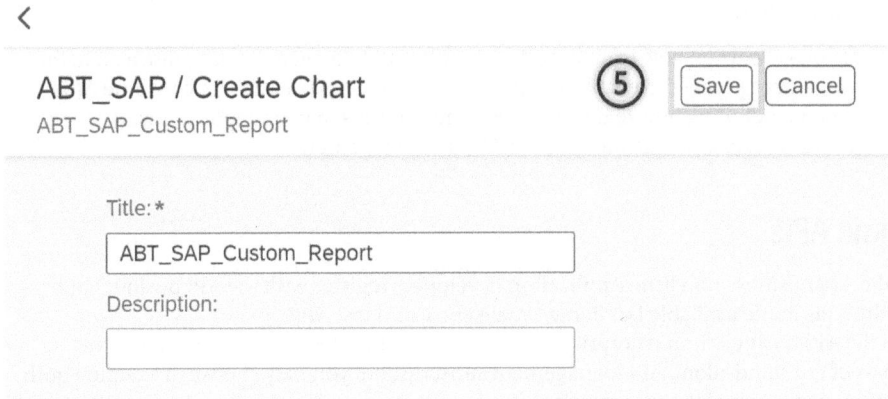

Figure 4-38. *Save custom chart*

6. Choose the custom view, then click New to add new charts.

7. Choose the API metric you want to track and measure in Dimensions and Measurements in the Create Chart window. Give the chart a name and a succinct description in the Title and Description areas.

8. There are buttons to conceal the date-range picker, examine a grid of all your custom charts, and refresh the reports in the upper-right corner of the date-range picker.

9. Each custom chart you include in your custom view has an Action column with Edit and Delete buttons.

4.1.4.3.2.2 Create Custom Dimensions and Measures

For tracking analytics data, Advanced API Analytics offers a variety of standard dimensions and metrics. You can make your own measures and dimensions if you require ones that aren't in the list, though.

1. Choose Custom Measure in the analytics dashboard's +Add drop-down menu.

2. In the Add Custom Metric window, enter the name you want to add for tracking data. The only data entered during this stage is the names of the custom dimensions and measures.

3. Hit **OK**.

4. Open the Developer portal from the left menu in SAP Integration Suite.

5. Select the API from the list of APIs for which you wish to use the custom measure to gather data.

6. Choose **Edit Policies**.

7. Add the Statistics Collector policy to the proxy endpoint's preflow.

8. Open the Statistics Collector policy payload attached to the API proxy.

9. Go to the analytics dashboard after creating the custom dimension or measure. Add a custom view and make custom charts with the unique dimensions or measures you created.

Great job! Now that you've learned how to analyze APIs in SAP API Management, let's move on to the next step in the API Management process: consuming APIs. This chapter explores how to consume APIs within SAP API Management, including how to discover, test, and use APIs in your applications.

The next section covers how to consume the APIs in SAP API Management.

4.1.4.4 Consume APIs

Use APIs by way of the API business hub firm. Application developers register with the API business hub company, research the APIs made available by clients, create apps, and test APIs.

If you've added the API business hub enterprise as a capability with Integration suite or if you've subscribed to it as part of the stand-alone API Management subscription, you may choose to examine both the new and the traditional designs of the user interface.

You still see the API business hub enterprise's traditional appearance when you log in for the first time. To display the new design, however, utilize the toggle switch. The site administrator is the only person who may use the toggle switch.

4.1.4.4.1 Onboard Application Developer

Onboarding an application developer in SAP API Management involves providing the credentials and access rights to consume the APIs published in the API portal. This process enables application developers to discover, explore, and consume APIs in a secure and controlled manner.

Onboarding users into the API business hub enterprise can only be done using the Add User flow or self-registration processes. Before granting an application developer access to the API business hub enterprise, the API administrator must onboard them.

4.1.4.4.1.1 Create an Account with API Business Hub Enterprise

The API business hub enterprise is a cloud-based service offered by SAP that provides a central repository for enterprise APIs, enabling organizations to easily discover, consume, and manage APIs across their IT landscape. Creating an account in the API business hub enterprise is an essential step for organizations that want to take advantage of the features and benefits of the service. Here are the general steps to create an account with API business hub enterprise in SAP API Management.

1. To access the API business hub enterprise page, select the option in the top-right corner of the API portal home screen, as illustrated in Figure 4-39.

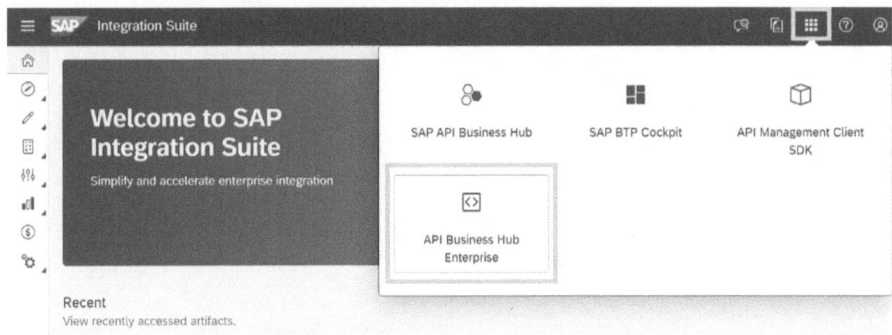

Figure 4-39. *API Business Hub Enterprise*

2. Figure 4-40 illustrates how to access the Manage menu.

Figure 4-40. *Manage page*

3. Choose **Registered Users** from the Manage Users tile, as shown in Figure 4-41.

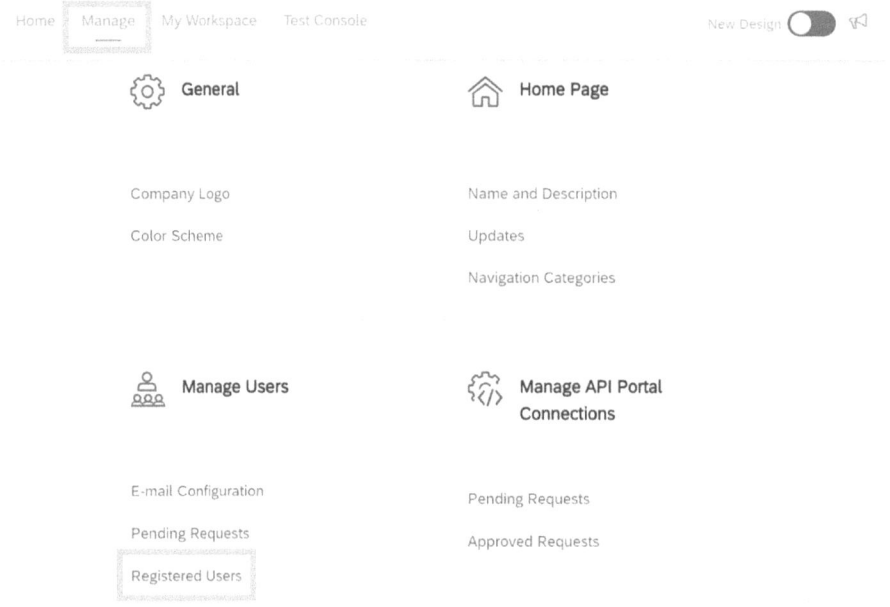

Figure 4-41. *Registered Users*

4. Type your email address in the Administrator section, then click the acknowledgment hook to confirm. Click the + icon in the section for registered users, as shown in Figure 4-42.

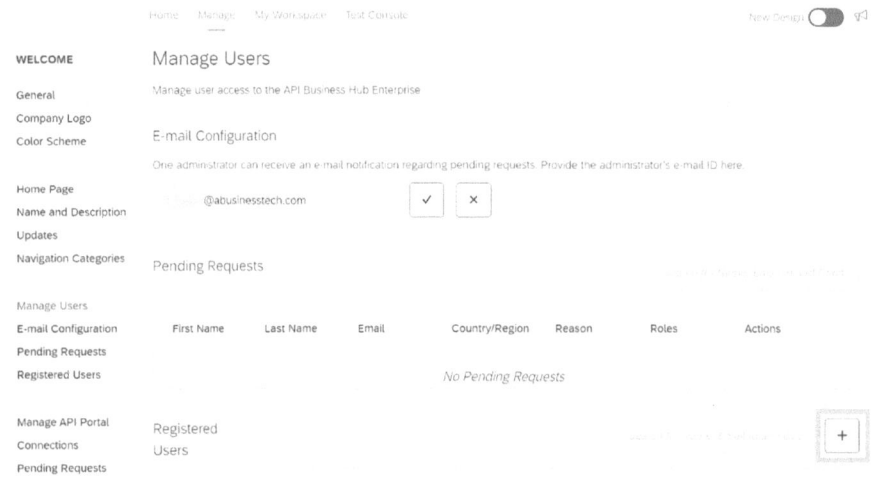

Figure 4-42. *Manage Users*

5. Enter information displayed in Figure 4-43. Your trial account's email address; user ID, the account's user ID; first and last name, whatever name you choose; email ID, your trial account's email address; assigned roles, all the open roles; reason, any legitimate reason.

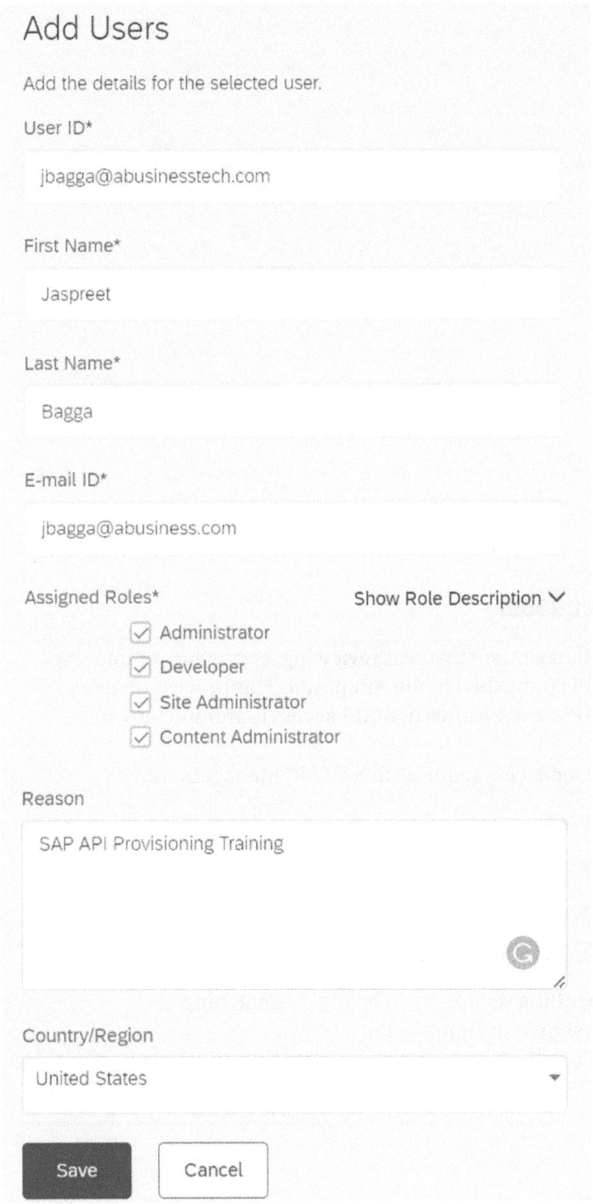

Add Users

Add the details for the selected user.

User ID*

jbagga@abusinesstech.com

First Name*

Jaspreet

Last Name*

Bagga

E-mail ID*

jbagga@abusiness.com

Assigned Roles*   Show Role Description ⌄

☑ Administrator
☑ Developer
☑ Site Administrator
☑ Content Administrator

Reason

SAP API Provisioning Training

Country/Region

United States ▾

[Save]   [Cancel]

Figure 4-43. *Add Users*

6. The end product should look like Figure 4-44.

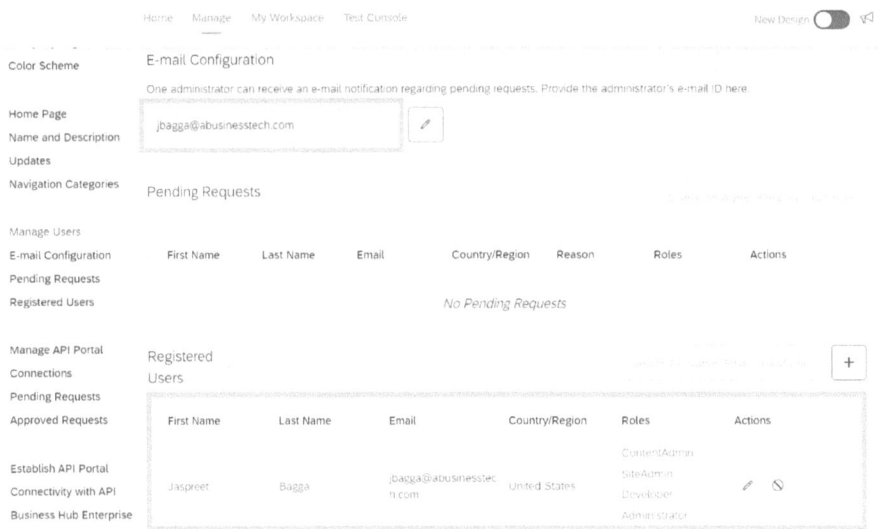

Figure 4-44. *Registered Users*

4.1.4.4.1.2 Manage the Users' Access Requests

Managing the users' access requests in SAP API Management involves reviewing, approving, or denying requests from users seeking access to consume APIs published in the API portal. This process enables organizations to control who can consume their APIs and ensures that API access is granted only to authorized users.

The following steps explain how to manage user access requests in SAP API Management.

1. Log in to the API business hub enterprise.

2. Navigate to **Manage ➤ Manage Users**.

3. Provide the email address in the textbox.

4. Go to the Pending section to see the open requests.

5. Approve or reject the request in the pending section by selecting the matching action item in the Actions column, as shown in Figure 4-45.

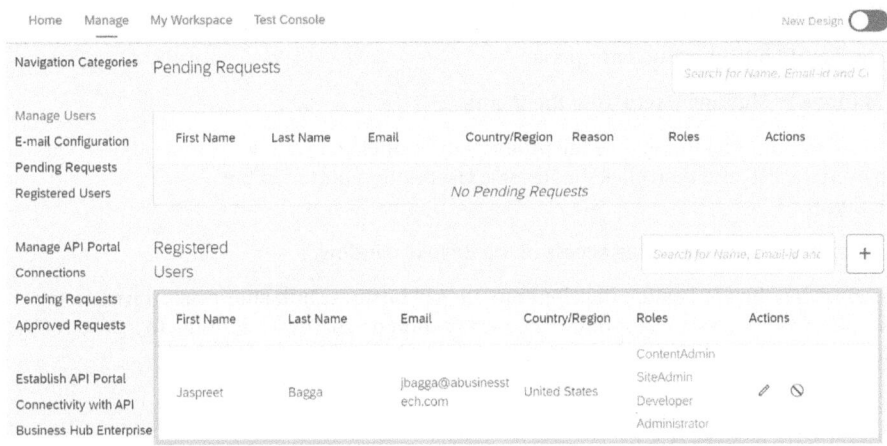

Figure 4-45. *Manage Registered Users*

6. Once a consumer registers for API Enterprise Hub and the administrator approves the access, you get a Welcome email from SAP API Management to begin exploring the published enterprise APIs and subscribe applications to the products/API to generate API keys which is used for Verify API Key policy, as shown in Figure 4-46.

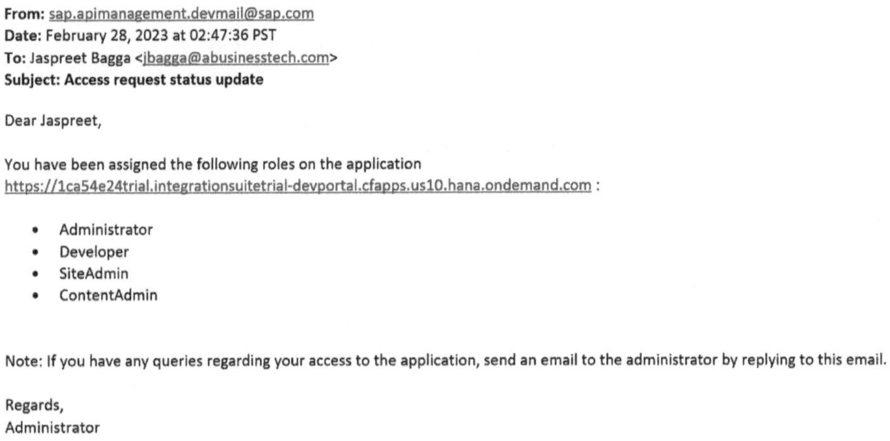

From: sap.apimanagement.devmail@sap.com
Date: February 28, 2023 at 02:47:36 PST
To: Jaspreet Bagga <jbagga@abusinesstech.com>
Subject: Access request status update

Dear Jaspreet,

You have been assigned the following roles on the application
https://1ca54e24trial.integrationsuitetrial-devportal.cfapps.us10.hana.ondemand.com :

- Administrator
- Developer
- SiteAdmin
- ContentAdmin

Note: If you have any queries regarding your access to the application, send an email to the administrator by replying to this email.

Regards,
Administrator

Figure 4-46. *Welcome message from SAP API Management*

4.1.4.4.1.3 Revoke Access

Revoking access in SAP API Management involves revoking the API access permissions of users who are no longer authorized or who no longer require access to the APIs. This process enables organizations to control who can access their APIs and ensures that API access is granted only to authorized users.

The following are the steps to revoke access in SAP API Management.

1. Connect to the API business hub enterprise.

2. Choose **Manage** ➤ **Manage Users** from the menu.

3. Visit the Registration section. Choose the program developer you want to restrict access to from the list, and then click the Reverse User Action button in the Actions column.

4. Provide a justification for canceling access in the Revoke window.

Once you have onboarded the application developer to the API business hub enterprise, the next step is to create the new application and assign a product. The next section discusses subscribing to a product by creating an application.

4.1.4.4.2 Subscribe to a Product

The product list can be found on the dashboard. Also, you can view the different resources that each product provides.

1. Log in to the API business hub enterprise.

2. You can find the item on the dashboard or by using the search bar to enter its name.

3. Select **Subscribe** in the Product screen.

4. You can sign up for the following.

   • Add to Existing Application: The list of applicants. Choose the required application.

   • Create New Application: Entering the name and title, create an application. By default, the chosen Product is added to the application.

5. Select **Subscribe**, as shown in Figure 4-47.

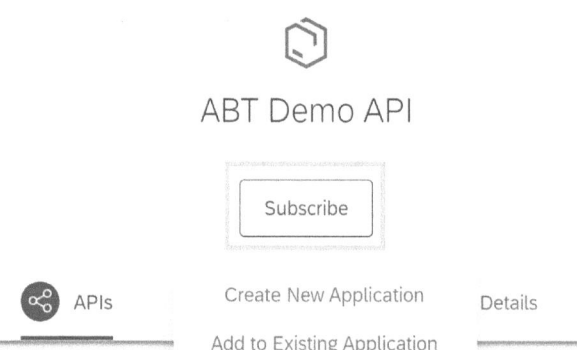

Figure 4-47. *Subscribe to a product*

4.1.4.4.3 Create an Application

Creating an application in SAP API Management is a process of creating a new application that can consume and interact with the APIs published in SAP API Management.

SAP API Management provides a developer portal that allows developers to discover, explore, and consume APIs. To access the APIs, developers must register an application in the developer portal. When an application is registered, it is assigned a unique client ID and secret, which is used to authenticate and authorize the application to access the APIs.

Similar APIs are combined in API Management to create products, which are then published in the catalog. An application developer registers with the API business hub enterprise by completing the required registration. After successfully enrolling, the application developer can look at the products and APIs needed to build an application. Upon successful application creation, the system creates an application key and secret.

You must pass the generated application key to access any APIs in the application you created that are protected by the Verify API Key policy. But, you must submit an OAuth token to access APIs protected by the OAuth policy. This token can be generated by fusing the newly established application key and application secret.

The following explains how to create an application in SAP API Management.

1. Log in to the API business hub enterprise.

2. Open the My Workspace page.

3. The Apps section displays any applications you've already generated. For a produced application, you may check how many calls were made overall in the most recent month.

4. Choose ⊞ the option to create an application in the Application section, as shown in Figure 4-48.

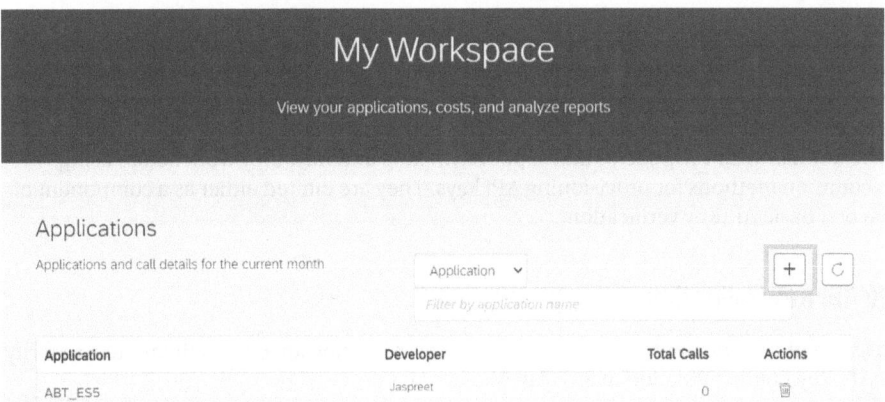

Figure 4-48. *Add Application*

5. Provide the application's title, description, and call back URL in the Create an application dialog (all optional).

6. Choose ⊞ products to include in this application.

7. Save the application.

8. After providing the products to the application, you can see the information of the application with the secret key, as shown in Figure 4-49.

Application Info

Callback URL:

Created by: ▓▓▓▓▓ @abusinesstech.com   Version: **1**   Last Modified: **Mar 4, 2023**

Application Secret: **s6QK3fyqzRAjGyRI** ⎘   Calls This Month: **0**

Application Key: **quvw21G7aYEETpnLMnRleM8yKhQMYvqY** ⎘   Regenerate Key

Figure 4-49. *Application information*

4.1.4.4.4 Consume an Application

After creating an application, you can use the APIs per your corporate needs.

The application developer obtains an API key upon subscribing, which the application must use to authenticate itself each time it makes an API call. API keys offer a straightforward method for app authentication.

Applications obtain API keys through API Management, which also provides you the choice to give your APIs policy-based API key-based authentication. However, the API proxy, not the API product itself, is responsible for the key enforcement. As a result, you must make sure that all API proxies, as well as the resources those API proxies define, implement key validation in some way.

Make sure you are familiar with the policies that support API keys and how they work before using them. There are two common methods for provisioning API keys. They are offered either as a component of an OAuth verification or a Basic APIKey verification.

4.1.4.4.4.1 Verify API Key Validation

An API provider can authenticate and approve requests made to an API using an API key thanks to a security feature called Verify API Key policy validation in SAP API Management.

4.1.4.4.4.2 OAuth 2.0 Validation

API Management supports standard OAuth flow. Only the client credentials grant type is currently supported by SAP for OAuth.

The application key and secret should be written down before you begin.

Request by using the following.

- OAuth token URL

- Method: POST

- Custom Header: Application key> Application Secret> Authorization Value

In the payload, the grant type is client credentials.

The result from the OAuth validation call is returned as a JSON payload. The access token is included after successful validation. Follow the access token as it descends.

The default setting for the expiry time is 3600 seconds (1 hour). You can customize the information in the answer and the expiration date.

4.1.4.4.5 Analyze Application

Use analytics tools to examine application performance, error rates, and usage.

API Management provides in-depth analytics capabilities to comprehend application utilization. Headers, charts, and key performance indicators display the runtime data (KPIs).

Go to My Workspace as an application developer to access the analytics data. All the data for the applications you have subscribed to are shown by default in the analytics area. Based on the context of the application developer, all charts are displayed.

There are two ways to look at the analytics information.

- **Performance Analytics**: Shows the performance-related charts and KPIs for the selected time window. Table 4-3 lists the graphs used to assess each app's performance.

Table 4-3. *Performance Analytics*

Chart Name	Description
Traffic Across all APIs	This graph shows the total number of API requests that all applications have made.
Slowest APIs	Based on response time, this graph shows which APIs are the slowest.
Top APIs	This graph displays the most often used APIs.
Top Products	Based on the quantity of API queries made for each unique product, this graph shows the most popular items.
Top Applications	Based on the volume of API requests made to each application, this graph shows which applications are the most often utilized.

- **Error Analytics**: shows the error-related KPIs and charts for the selected time window. The graphs used to view error analytics across all applications are described in Table 4-4.

Table 4-4. *Error Analytics*

Chart Name	Description
Total Errors	This graph shows the overall mistakes.
Error Prone APIs	This graph shows the total amount of API failures.
Error Prone Applications	This graph shows the number of API problems related to the application.

4.1.4.4.6 Test API Behavior During Runtime

To test how APIs behave during runtime, use the API Test Environment.

You can test your APIs with the Test Environment. To comprehend an API's behavior at runtime, testing is necessary. It enables you to act and explore the resources connected to an API. You may test REST-based and OData-based services with it as well.

1. Log in to the API business hub enterprise.

2. Navigate to the test console.

3. Select the desired API from the list shown on the left side. Choose GWSAMPLE_BASIC API from SAP demo EZ5.

4. Select the Method type.

   - GET

   - POST

   - PUT

   - DELETE

5. Enter the service URL if you know it, and it contains the API.

6. To choose the kind of authentication, choose Authentication. You have the following selections.

   - None: No verification is necessary.

   - A basic authentication username and password should be provided.

7. With the PUT and POST methods, enter the request body.

8. To include a header, select Header.

9. Enter the query parameter value in URL Parameters and then select Send.

10. Provide the Response Body in the Request Body if required for the API.

11. Choose Launch API Viewer to see the transactions related to your testing activity, as shown in Figure 4-50.

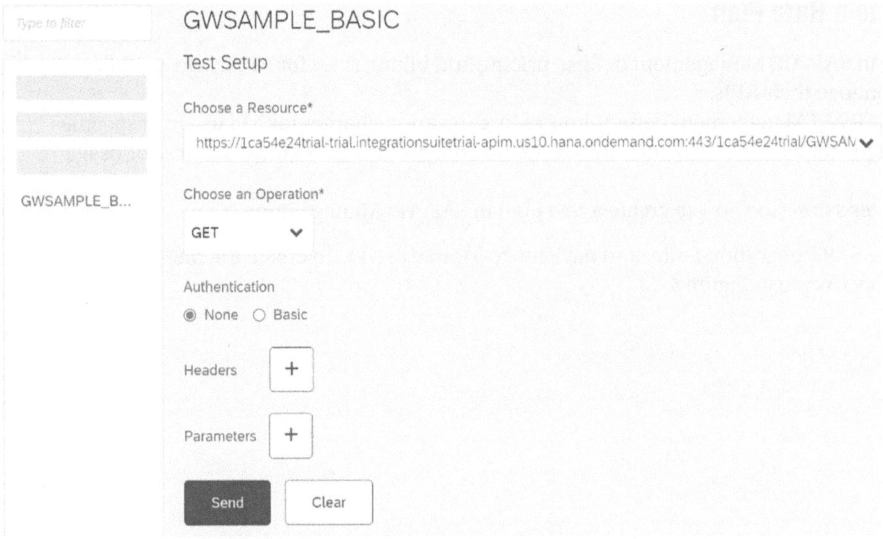

GWSAMPLE_BASIC

Test Setup

Choose a Resource*

https://1ca54e24trial-trial.integrationsuitetrial-apim.us10.hana.ondemand.com:443/1ca54e24trial/GWSAN ⌄

Choose an Operation*

GET ⌄

Authentication

◉ None ○ Basic

Headers +

Parameters +

Send   Clear

Figure 4-50. *Test API*

The next section discusses how to monetize APIs.

4.1.4.5 Monetize APIs

Every API provider can create money by employing the monetization capability tool that API Management offers.

The Monetize API feature in SAP API Management is designed to help organizations monetize their APIs by creating a revenue stream from API usage. With this feature, you can set up pricing plans for your APIs, define usage limits, and generate revenue reports based on actual API usage.

The Monetize API feature allows you to define pricing plans for your APIs based on usage limits such as the number of requests or the volume of data transmitted. You can also set up different pricing tiers for users or customer segments. This enables you to create a revenue stream based on actual usage from your APIs.

The Monetize API feature provides detailed reports on API usage and revenue generation. It lets you monitor your API usage and revenue in real time, identify usage patterns, and make data-driven decisions to optimize your API monetization strategy.

As an application developer in the API business hub company, you can create applications and incorporate items in them.

API Management provides two features.

- Rate Plan Service

- Billing Service

4.1.4.5.1 Rate Plan Service

Users can establish pricing plans and associate them with objects using API Management. By establishing a tariff plan, you can charge the people who make the applications that utilize your APIs.

4.1.4.5.1.1 Create a Rate Plan

Creating a rate plan in SAP API Management defines pricing and billing rules for API usage, enabling organizations to monetize their APIs.

A rate plan in SAP API Management defines how an organization charges for API usage. It specifies the rate (price) charged for each API call and any additional charges for value-added services or usage thresholds.

The following steps describe how to create a rate plan in SAP API Management.

1. Log in to SAP Integration Suite, and navigate to Monetize API. To create the rate plan, click **Create** in Figure 4-51.

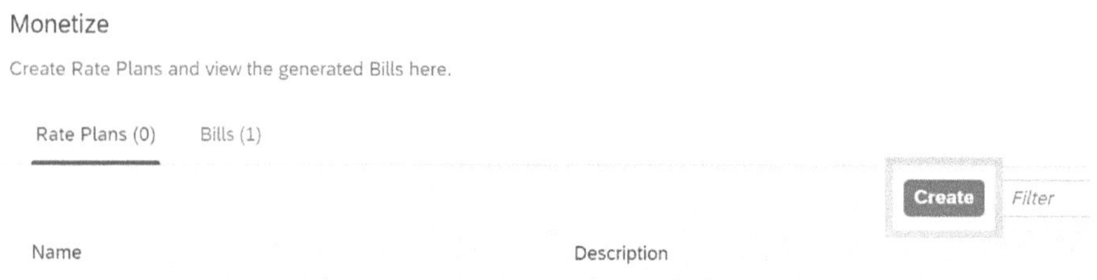

Monetize

Create Rate Plans and view the generated Bills here.

Rate Plans (0)   Bills (1)

Create   Filter

Name   Description

Figure 4-51. *Create Rate Plans*

2. Provide the following details on the Create Rate Plan screen.

   - Name: The pricing plan's name (ABT_Plan)

   - Description: A sketch of the strategy (ABT rate plan)

   - Frequency: Monthly

   - Currency: Euro

   - Basic Charge: The user's initial payment made after signing up for the service covered by this pricing plan (20)

   - Rate per API Call: The cost of one API call in euros (0.1, 0.2, 0.3)

   - Plan Type: Basic or Tier

3. After providing all details, as shown in Figure 4-52, click Save.

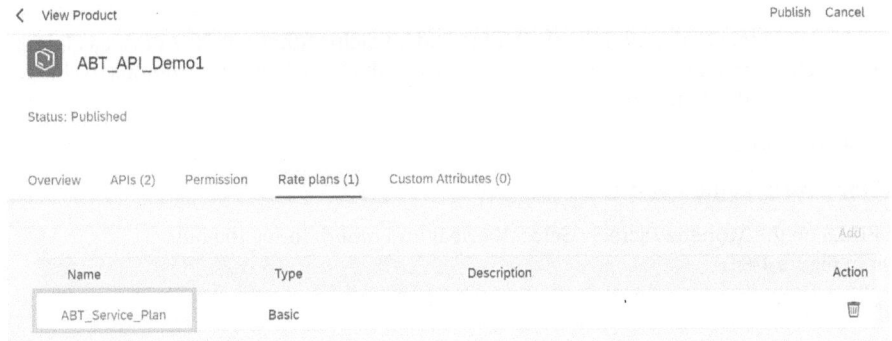

< View Rate Plan

◨ ABT_Plan   Save   Cancel

Name: *

ABT_Plan

Description:

Frequency: *

Monthly

Currency: *

EUR   ⌄

Plan Type:

Tier   ⌄

Figure 4-52. *View Rate Plan*

4.1.4.5.1.2 Link the Rate Plan to an API Product

Linking a rate plan to an API product in SAP API Management associates pricing and billing rules with an API product. By linking a rate plan to an API product, organizations can monetize the usage of the API product and charge developers based on their API usage.

The following steps explain how to link a rate plan to an API product in SAP API Management.

1. Log in to the API portal, and navigate to Design ➤ APIs.

2. Choose the product to which you tend to attach the rate plan from the list in the Products tab. Select the product created in section 4.1.4.4.2. Select the product as ABT_API_Demo that you have already created.

3. Select the product.

4. Select and save the rate plan, as shown in Figure 4-53.

< View Product   Publish   Cancel

▣ ABT_API_Demo1

Status: Published

Overview   APIs (2)   Permission   Rate plans (1)   Custom Attributes (0)

                                                                    Add

Name	Type	Description	Action
ABT_Service_Plan	Basic		🗑

Figure 4-53. *Add rate plan to product*

4.1.4.5.1.3 Update the Rate Plan

Updating a rate plan in SAP API Management enables organizations to adjust their pricing strategy, change their billing rules, and add or remove value-added services. It allows organizations to adapt to changing market conditions, customer needs, and business objectives and optimize their revenue from API usage.

The following explains how to update a rate plan in SAP API Management.

1. Log in to the API portal.

2. Navigate to the Monetize tab.

3. Choose the rate plan that needs to be updated. Make any necessary adjustments by clicking Edit.

4. Click **Save**.

4.1.4.5.1.4 Delete a Rate Plan

Deleting a rate plan in SAP API Management removes a rate plan created and associated with API products or APIs. Deleting a rate plan in SAP API Management removes all associated pricing and billing rules, and any existing subscriptions or usage data associated with the rate plan is affected.

The following explains how to delete a rate plan in SAP API Management.

1. Open the API portal and log in.

2. Select **Monetize** from the menu bar.

3. RATE PLANS can be selected from the Monetize screen.

4. To remove a rate plan, choose it in the Actions column, and select Yes.

4.1.4.5.2 Billing Service

The API business hub enterprise and the API portal offer billing options.

The billing service allows you to examine and download the bill details for a certain developer and month.

4.1.4.5.2.1 Billing Service in the API Portal

The Billing Service in SAP API Management handles the billing and invoicing functions for API usage. It is a key component of the monetization feature in SAP API Management, which enables organizations to charge developers for using their APIs and generate revenue.

1. Log in to the API portal.

2. Choose **Monetize** from the menu bar.

3. Choose BILLS on the Monetize screen. Select the year and month to see the bill, as shown in Figure 4-54.

Figure 4-54. Bills in API portal

4.1.4.5.2.2 Billing Service in SAP API Business Hub

The Billing Service in SAP API Business Hub is a feature that enables API providers to monetize their APIs by charging developers for their API usage based on the pricing and billing rules defined in the rate plans. It is a part of the monetization functionality offered by SAP API Management that enables organizations to generate revenue from their API programs.

1. Log in to the API business hub enterprise. Get to My Workspace by navigating.

2. The Cost section includes two charts showing the billing information for each application the developer has subscribed to, as shown in Figure 4-55.

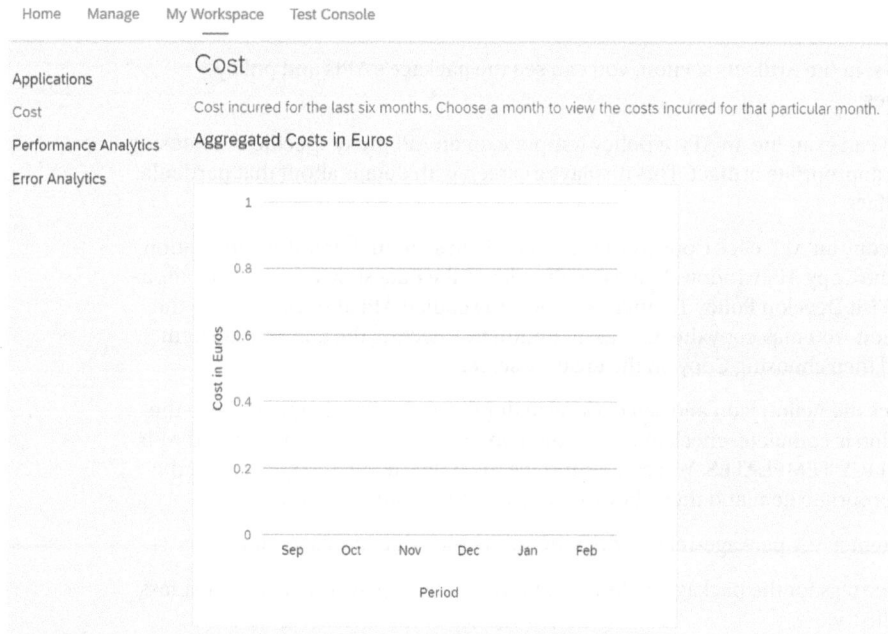

Figure 4-55. Billing cycle in SAP API Business Hub Enterprise

The next section covers the predelivered API packages provided by SAP.

4.1.4.6 Discover API Packages

The SAP API Business Hub hosts the API packages offered by API Management. These API packages may be found at the API portal.

Log in to the API gateway and choose Discover to browse the packages listed in the following categories.

- **Highlights**: The Featured section lists packages that have been featured. The Newest section shows the most recent packages that have been released.

- **All**: All of the packages in the portal are listed in this section.

Each package has its tile that displays the name of the product, its rating, and a summary of the bundle. A Partner label is displayed on the package if a partner created the content.

4.1.4.6.1 Package Details

A package is a container that can accommodate many kinds of content. It often includes policy templates and APIs. Links and documents may also be included.

Each package includes the following information.

- **Overview**: These details regarding the package are provided in the overview.

  - Description

  - Supported Platform

  - Categories

  - Created On

- **Artifacts**: In the Artifacts section, you can see the package's APIs and policy templates.

  - You can examine an API, a policy template, or an API proxy specific by choosing the appropriate artifact. This displays a screen with details about that particular artifact.

  - To copy an API, click Copy from the Action icon's menu. Enter the information in the Copy API window. You can edit or leave the data shown for API Details as is. Visit Develop Policy Templates to view the copied API after completing the action. You may copy the APIs as an option by selecting the appropriate item and then choosing Copy on the ensuing screen.

  - Click the Action icon and select Copy to duplicate a policy template. Once the action is complete, check out the copied policy template by going to Build APIs POLICY TEMPLATES. You can reproduce the policy template by selecting the appropriate item and then choosing Copy on the ensuing screen.

- **Documents**: Any package-related documents are included in this section.

- **Tags**: The tags for the package include country, product, keyword, lines of business, and industry.

- **Ratings**: User reviews and comments on the bundle are included in this section.

- **View in SAP API Business Hub**: APIs can be examined and copied to the API site at the package level. If you want to perform tasks like testing the API or writing code, go to SAP API Business Hub by choosing View in API Business Hub. You may get the same package on the SAP API Business Hub by clicking the link.

Figure 4-56 shows the SAP Order Management Foundation API package, which is a sample API package showing the different details of the API package.

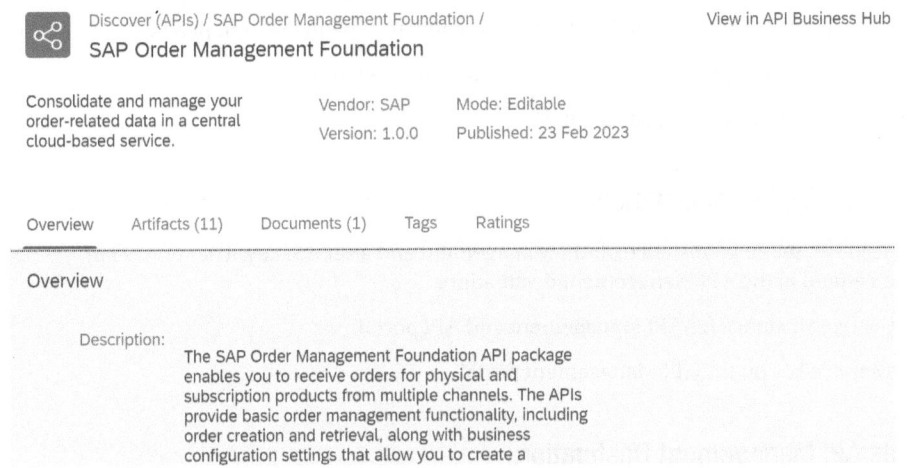

Discover (APIs) / SAP Order Management Foundation /   View in API Business Hub
SAP Order Management Foundation

Consolidate and manage your   Vendor: SAP   Mode: Editable
order-related data in a central   Version: 1.0.0   Published: 23 Feb 2023
cloud-based service.

Overview   Artifacts (11)   Documents (1)   Tags   Ratings

Overview

Description:
             The SAP Order Management Foundation API package
             enables you to receive orders for physical and
             subscription products from multiple channels. The APIs
             provide basic order management functionality, including
             order creation and retrieval, along with business
             configuration settings that allow you to create a

Figure 4-56. *Discover tab in SAP API Management*

The next section explains how to transport APIs.

4.1.4.7 Transport APIs

In SAP API Management, *transport APIs* means transferring API artifacts, such as policies, configurations, and APIs, from one environment to another in a secure and regulated manner. This is frequently necessary when creating, testing, and deploying APIs in various settings, such as development, staging, and production environments.

4.1.4.7.1 Enable Content Transport

Configure the service instances, destinations, and a route between the source and destination nodes to enable the transit of APIs and associated artifacts.

4.1.4.7.1.1 Creating a Content Agent Instance

Create a service instance and service key for the content agent in your source subaccount. The service key information is required while constructing the content assembly service destination.

1. Create a service instance in the SAP BTP cockpit.

2. Create a service key for the content agent.

3. Note the URL, ClientID, and client secret after the service key has been produced since this information is required when the content agent's HTTP destination content assembly service is built. Use these steps to reproduce the data.

    a. Select the content agent service instance you created to expand the right pane.

    b. Select the content agent service key name to view the credentials.

    c. Choose the JSON tab, then copy the link.

4.1.4.7.1.2 Create an API Portal Instance

On your source subaccount, set up an instance of API Management and a service key. The service key information must be defined in the API Management destination.

1. Create a service instance for API Management and API portal.

2. Create a service key on the API Management portal.

4.1.4.7.1.3 Create API Management Destination

To enable you to send API requests and get API material from the API portal workspace, create a destination in your source subaccount named API Management.

1. Go to the SAP BTP cockpit in your web browser and go to your source subaccount.

2. In the left-hand window, select the Destinations tab.

3. Choose a new destination.

4. Configure all the mandatory details in the Destination Configuration tab.

5. Select Save.

To confirm that the destination was added properly, you may make a check connection. The "APIManagement was connected to" pop-up notice displays when you run a connection check.

4.1.4.7.1.4 Add Source Node in Transport Management Applications

The Transport Management Applications feature in SAP API Management allows organizations to manage the life cycle of their APIs by creating and deploying API proxies to different environments. The Add Source Node functionality is part of this feature that enables organizations to add source nodes to their API proxies during deployment.

1. Log in to the SAP BTP cockpit using your web browser to access your Transport subaccount.

2. Instances and subscriptions can be selected in the left pane.

3. Choose **Cloud Transport Management** in Subscriptions > Browse to Application.

4. Move to the left pane and select Transport Nodes.

5. To add a source node, click +.

6. Give the node a memorable name in the New Node dialog box, such as source node1, and turn on the Allow Upload to Node option.

7. Click **OK**.

4.1.4.7.1.5 Create Transport Management Destination

API providers can define and configure the transport destinations for their APIs by using the Create Transport Management Destination feature in SAP API Management. This involves specifying the destination type, transport protocol, server host, port, and other relevant details required to access the destination.

1. Go into SAP BTP cockpit in your web browser and go to your source subaccount.

2. In the left-hand window, select the Destinations tab.

3. Choose a new Destination.

4. Configure all the mandatory details in the Destination Configuration tab.

5. Opting for Edit > New Property allows you to add a new property to the SAP Cloud Transport Management service. The first text field should contain the source System ID. The second text field should be filled out with the source node that you designated in Transport Management Apps.

6. Select **Save**.

You may also run a check connection to make sure you added the destination properly. When you perform a connection check, the following pop-up notice appears: The connection to TransportManagementService has been made.

4.1.4.7.1.6 Add Destination Node in Cloud Transport Management Applications

The Add Destination node in the Cloud Transport Management Applications of SAP API Management is a feature that allows you to create and configure destinations for your API proxies. Destinations are virtual representations of back-end systems that route requests from API proxies to the actual back-end systems.

1. Log in to SAP BTP cockpit using your web browser to access your Transport subaccount.

2. Instances and subscriptions can be selected in the left pane.

3. Choose **Cloud Transport Management** in Subscriptions > Browse To Application.

4. Go to the left pane and select Transport Nodes.

5. To add a source node, click +.

6. Choose the Allow Upload to Node option and give the node a meaningful name, such as destination node1, in the New Node dialog box.

7. Choose Multi-Target Application from the Content Type drop-down menu.

8. Select the TMS Deploy destination you created in the Transport subaccount from the Destination selection.

9. Select **OK**.

4.1.4.7.1.7 Connect Source and Destination Node

The Connect Source and Destination Node feature in SAP API Management is a part of the transport APIs functionality that enables you to move APIs between different environments, such as development, testing, and production. This feature allows you to connect the source and destination nodes in SAP API Management, which enables the seamless transfer of APIs and their configurations.

1. From the left pane, select Transport Routes.

2. Decide to add a route for transport.

3. Provide the transport route's name and a brief description.

4. Pick the source and destination nodes from the list of current transport nodes.

5. Select **OK**.

4.1.4.7.2 Trigger Content Transport

After the system has been configured for transport, the API proxy and API artifact can be moved from the source to the target API site.

The API proxy is sent to the destination when transport is initiated, together with any related artifacts, such as the API provider, key store certificate, trust store, and key value maps. But you can also control the movement of each object separately artifact.

4.1.4.7.2.1 API Proxy Transport: Source to Destination

The destination API portal always imports the deployed state of API proxies.

1. Open the API portal and log in.

2. Choose Develop from the navigation menu on the left.

3. Choose the API (GWSAMPLE_BASIC) you wish to transport on the APIs tab page.

4. Select the Transfer option after clicking the Action button next to the API, as shown in Figure 4-57. The Transport option is available on the API's Details page, which you can open as an alternative.

Figure 4-57. *Transport API Proxy*

5. Describe the Transport pop-up and select Yes.

6. To get to the transport nodes, follow these instructions to the Transport subaccount.

   a. Go to your Transport subaccount by logging in to the SAP BTP cockpit through your web browser.

   b. Instances and subscriptions can be selected in the left pane.

   c. Navigate to Application in Subscriptions Cloud Transport Management.

   d. Go to the left pane and select Transport Nodes.

7. Choose the destination node that directs you to the desired API gateway.

8. Look for the API you used to initiate the transport in the destination node's Transport Description column.

9. To import the chosen API from the queue to the destination node, select Import Selected.

4.1.4.7.2.2 API Provider Transport: Source to Destination

In SAP API Management, API Provider Transport: Source to Destination is a feature that allows API providers to move APIs and their associated artifacts, such as API proxies, from one environment to another, such as from a development environment to a production environment. This feature is part of the Transport APIs functionality, which enables organizations to manage their APIs and related assets across different environments, such as testing, staging, and production.

1. Open the API portal and log in.

2. Select Configure from the navigation menu on the left.

3. Choose the API provider (ABT_SAP_API_Prover) you wish to send from the page in the API Providers tab.

4. Select the **Transport** option after selecting the Action icon next to the API provider. The Transport option is available on the information page of the required API provider, which you can visit as an alternative, as shown in Figure 4-58.

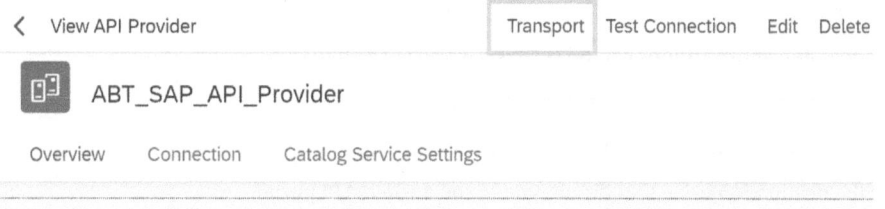

Figure 4-58. *Transport API Provider*

5. Describe the Transport pop-up and select Yes.

6. To get to the transport nodes, go to the Transport subaccount and follow these instructions.

    a. Go to your Transport subaccount by logging in to the SAP BTP cockpit through your web browser.

    b. Instances and subscriptions can be selected in the left pane.

    c. Navigate to Application in Subscriptions Cloud Transport Management.

    d. Go to the left pane and select Transport Nodes.

7. Choose the destination node that directs you to the desired API portal.

8. Search for the API provider you used to start the transport in the destination node's Transport Description column.

9. Select **Import Selected** to import the selected API provider to the target node from the queue.

4.1.4.7.2.3 Certificate Transport: Source to Destination

In SAP API Management, Certificate Transport: Source to Destination is a feature that enables the secure transportation of certificates for authentication and encryption between the source and destination environments when transporting APIs between them. This feature ensures that the certificates required to access APIs in the destination environment are securely transported from the source environment to maintain the integrity and security of the API programs.

1. Open the API portal and log in.

2. Choose **Configure** from the navigation menu on the left.

3. On the Certificates tab page, select the certificate you wish to transport.

4. Choose the Transport option after selecting the Action icon next to the certificate. Alternatively, you can open the certificate and choose the Transport option on the information page.

5. Provide a description in the Transport pop-up and select Yes.

6. To get to the transport nodes, go to the Transport subaccount and follow these instructions.

    a. Go to your Transport subaccount by logging in to the SAP BTP cockpit through your web browser.

    b. Instances and subscriptions can be selected in the left pane.

    c. Navigate to Application in Subscriptions Cloud Transport Management.

    d. Go to the left pane and select Transport Nodes.

7. Choose the destination node that directs you to the desired API portal.

8. Look for the certificate for which you initiated the transport in the destination node's Transport Description column.

9. Select **Import** to import the chosen certificate to the destination node from the queue.

4.1.4.7.2.4 Key Value Maps Transport: Source to Destination

Key Value Maps Transport: Source to Destination is a feature in SAP API Management that allows for the transport of key value maps between environments or systems, which is useful in the context of Transport APIs. Key value maps often store configuration data, such as endpoints, passwords, or other values varying between environments or systems.

1. Open the API portal and log in.

2. Choose **Configure** from the navigation menu on the left.

3. On the Key Value Maps page, in the Tabs menu, select the key value map you want to send.

4. After clicking the Action button next to the key value map, choose Transport. The Transport option is available on the details page of the key value map; you can also open it instead, as shown in Figure 4-59.

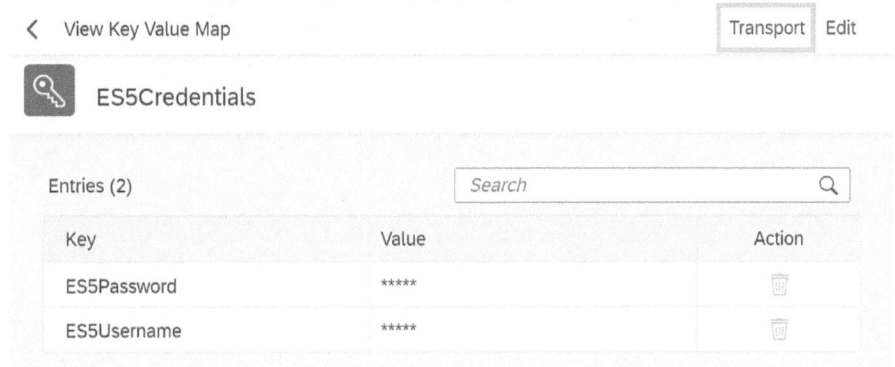

Figure 4-59. *Transport key value maps*

5. Provide a description in the Transport pop-up and select Yes.

6. To get to the transport nodes, go to the Transport subaccount and follow these instructions.

   a. Go to your Transport subaccount by logging in to the SAP BTP cockpit through your web browser.

   b. Instances and subscriptions can be selected in the left pane.

   c. Navigate to the application in Subscriptions Cloud Transport Management.

   d. Go to the left pane and select Transport Nodes.

7. Choose the destination node that directs you to the desired API gateway.

8. Search the destination node's Transport Description column for the key value map for which you initiated the transfer.

9. Select Import to import the key value map to the destination node from the queue.

4.1.4.7.2.5 Product Transport: Source to Destination

The Product Transport: Source to Destination feature works by creating a transport package that contains all the relevant artifacts for the API product, such as the API definition, policies, security configurations, and other related artifacts. The package is then moved from the source environment, where the API product was developed or modified, to the destination environment, where it is deployed and made available to consumers.

1. Open the API portal and log in.

2. Choose Design from the navigation menu on the left.

3. On the Products page, select the product you wish to move.

4. Choose the **Transport** option after selecting the Action icon next to the product. Alternatively, open the desired product and choose the Transport option on the details page, as shown in Figure 4-60.

Figure 4-60. *Transport product*

5. Provide a description in the Transport pop-up and select Yes.

6. To get to the Transport Nodes, go to the Transport subaccount and follow these instructions.

   a. Go to your Transport subaccount by logging in to the SAP BTP cockpit through your web browser.

   b. Instances and subscriptions can be selected in the left pane.

   c. Navigate to the application in Subscriptions Cloud Transport Management.

   d. Select **Transport Nodes** from the list.

7. Choose the destination node that directs you to the desired API portal.

8. Look for the item for which you initiated the transfer in the destination node's Transport Description column.

Select Import to bring the product out of the queue and into the target node.

4.1.4.8 Working with Policies in API Proxies: Practical Example

You have learned all the operations performed in SAP API Management so far. Next, let's create the API proxy with certain policies in it and test it.

4.1.4.8.1 Prerequisites

First, you must create some things in advance.

- Refer to section 4.1.4.1.3 to create the API provider.

- Refer to section 4.1.4.1.4 to create the API proxy.

- Use GWSAMPLE_BASIC API from SAP Gateway Demo ES5.

4.1.4.8.2 Create Key Value Maps

The Key Value Maps feature in SAP API Management allows API providers to store and retrieve key/value pairs that can be used in API policies and other configurations. It provides a way to define reusable variables that can be referenced across different API products and policies, making it easier to manage and maintain API configurations.

The following explains how to create key value maps in SAP API Management.

1. Choose **Key Value Maps** in Configure APIs.

2. Create the new key value map and specify the ES5 username and password, as shown in Figure 4-61.

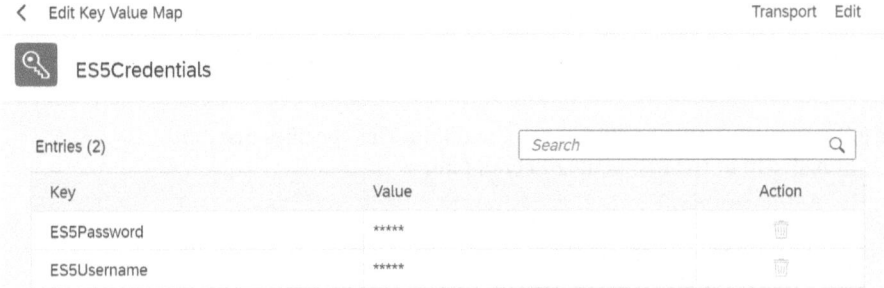

Figure 4-61. *Edit Key Value Map*

4.1.4.8.3 Add Policy to the API Proxy

Create a policy to provide guidelines for the API, such as security measures or traffic management.

1. Select **Policies** after selecting the API proxy, as shown in Figure 4-62.

Figure 4-62. *Policies*

2. Select the flow in Policy Editor to specify the policy; there are proxy endpoints and target endpoints.

3. In the proxy endpoint preflow, select Verify API Key policy on the right.

4. Select the policy and add it to the flow, as shown in Figure 4-63.

Figure 4-63. *Verify API Key*

5. Provide the policy name, and in the stream drop-down, select the incoming request.

6. Give the following code to the Verify API Key Condition. This specifies that the API key is provided in the header of the API endpoint.

```
<VerifyAPIKey async='true' continueOnError='false' enabled='true'
xmlns='http://www.sap.com/apimgmt'>
        <APIKey ref='request.header.APIKey'/>
</VerifyAPIKey>
```

7. Open the target endpoint. In the preflow, specify the Key Value Map Operations and Basic Authentication policies, as shown in Figure 4-64.

Figure 4-64. *Add basic authentication and key value maps operation policies*

- **Key Value Maps *Operation***: From the left panel, add the Key Value Map Operation. Name it *GetCredentials* and make the stream an incoming request.

  Provide the following code in the condition string that gets generated when selecting the policy type. You may adjust the code based on your environment.

```
<KeyValueMapOperations mapIdentifier="ES5Credentials" async="true"
continueOnError="false" enabled="true" xmlns="http://www.sap.com/apimgmt">

    <Get assignTo="private.username">
  <Key>
    <Parameter>ES5Username</Parameter>
  </Key>
</Get>
<Get assignTo="private.password">
<Key>
    <Parameter>ES5Password</Parameter>
</Key>
</Get>

    <Scope>environment</Scope>
</KeyValueMapOperations>
```

- **Basic Authentication**: Place the basic authentication policy from the policies list. Name the policy *SetCredentials*.

  Provide the following code to the Basic Authentication condition.

```
<BasicAuthentication async='true' continueOnError='false' enabled='true'
xmlns='http://www.sap.com/apimgmt'>

    <Operation>Encode</Operation>
    <IgnoreUnresolvedVariables>true</IgnoreUnresolvedVariables>

    <User ref='private.username' />

    <Password ref='private.password' />

     <Source>request.header.Authorization</Source>

    <AssignTo createNew="false">request.header.Authorization</AssignTo>
</BasicAuthentication>
```

8. Save and deploy the API proxy.

4.1.4.8.4 Create an Application

The Create Application feature in SAP API Management enables API consumers to create applications that can access and consume APIs. It provides a way for API providers to manage access to their APIs and enforce security policies, such as authentication and authorization.

The following steps instruct how to create an application in SAP API Management.

1. On the right side of the dashboard, navigate to Explore our Ecosystem. Select API Business Hub Enterprise, as shown in Figure 4-65.

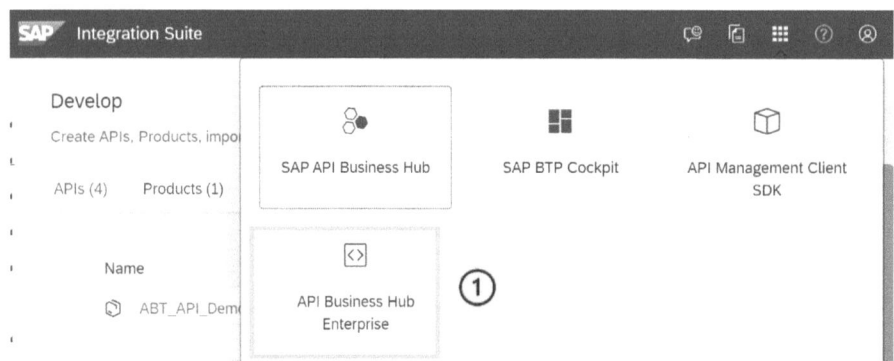

Figure 4-65. *API Business Hub Enterprise*

2. You are directed to the API business hub enterprise portal. Select your API product, as shown in Figure 4-66.

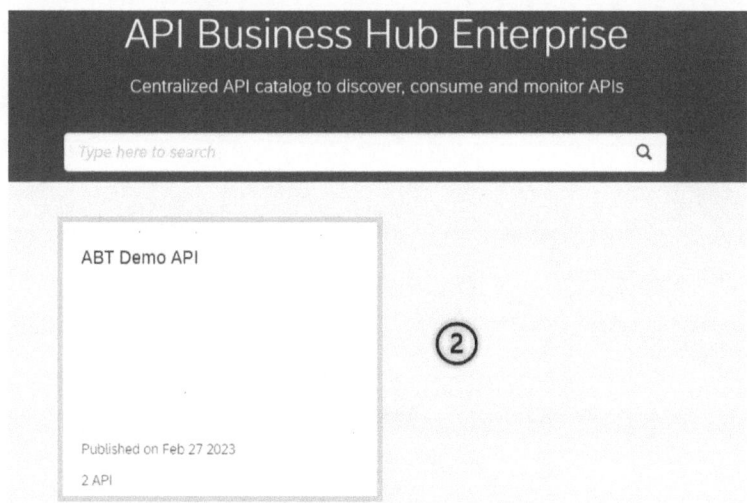

Figure 4-66. *API Business Hub Enterprise dashboard*

3. Click **Subscribe** and **Create New Application**, as shown in Figure 4-67.

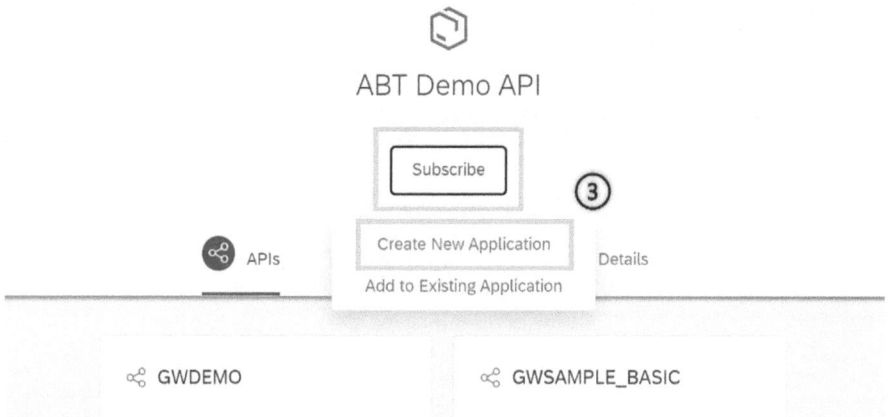

Figure 4-67. *Create New Application*

4. Provide the application title and select the API products related to the API proxies.

5. You can see the application information where the API key is generated successfully, as shown in Figure 4-68.

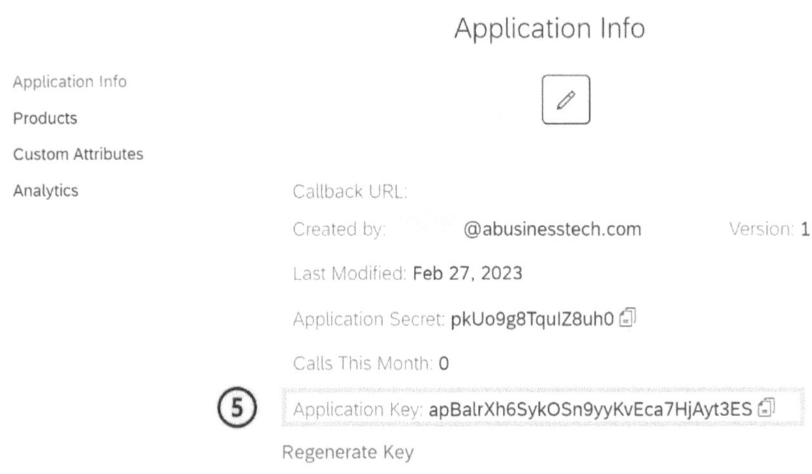

Application Info

Application Info

Products

Custom Attributes

Analytics

Callback URL:

Created by: @abusinesstech.com   Version: 1

Last Modified: Feb 27, 2023

Application Secret: pkUo9g8TquIZ8uh0

Calls This Month: 0

⑤ Application Key: apBalrXh6SykOSn9yyKvEca7HjAyt3ES

Regenerate Key

Figure 4-68. *API Key*

4.1.4.8.5 Test API Proxy

The Test API Proxy feature in SAP API Management allows API providers to test their API proxies before publishing them to API consumers. It provides a way to validate the behavior and performance of the API proxy, ensuring that it works as expected and meets the requirements of the API consumers.

The following steps explain how to test the API Proxy in SAP API Management.

1. Select **Test APIs** from the left panel of the SAP Integration Suite dashboard. Decide which API proxy you want to test, as shown in Figure 4-69.

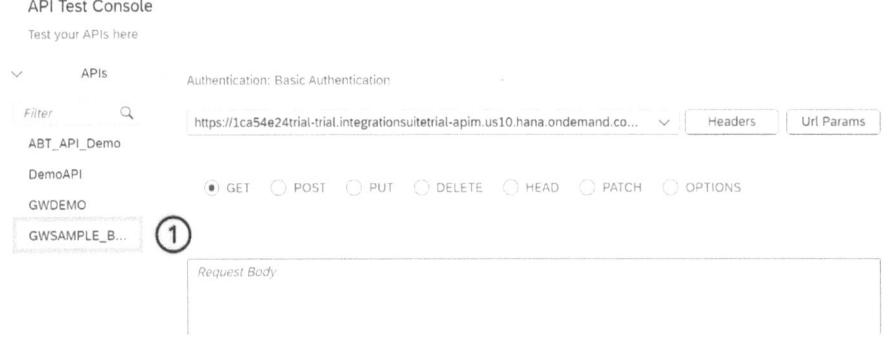

Figure 4-69. *API Test Console*

2. Send the request before adding the API key in the Headers tab. You see the *Failed to resolve API Key variable request.header.APIKey* error, as shown in Figure 4-70.

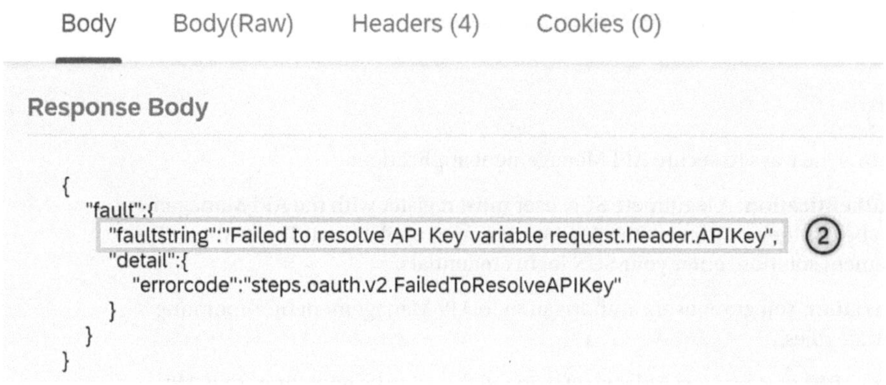

Body   Body(Raw)   Headers (4)   Cookies (0)

Response Body

```
{
    "fault":{
        "faultstring":"Failed to resolve API Key variable request.header.APIKey",  ②
        "detail":{
            "errorcode":"steps.oauth.v2.FailedToResolveAPIKey"
        }
    }
}
```

Figure 4-70. Failed API Key variable

3. Add the API key in Headers, as shown in Figure 4-71.

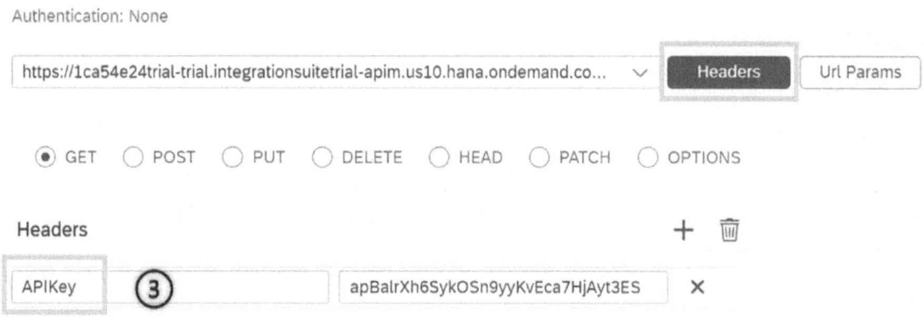

Authentication: None

| https://1ca54e24trial-trial.integrationsuitetrial-apim.us10.hana.ondemand.co... ∨ | **Headers** | Url Params |

◉ GET   ○ POST   ○ PUT   ○ DELETE   ○ HEAD   ○ PATCH   ○ OPTIONS

Headers   + 🗑

| APIKey   ③ | apBalrXh6SykOSn9yyKvEca7HjAyt3ES | ✕ |

Figure 4-71. Add APIKey to Headers

4. Send the request. The 200 Success result is shown in Figure 4-72.

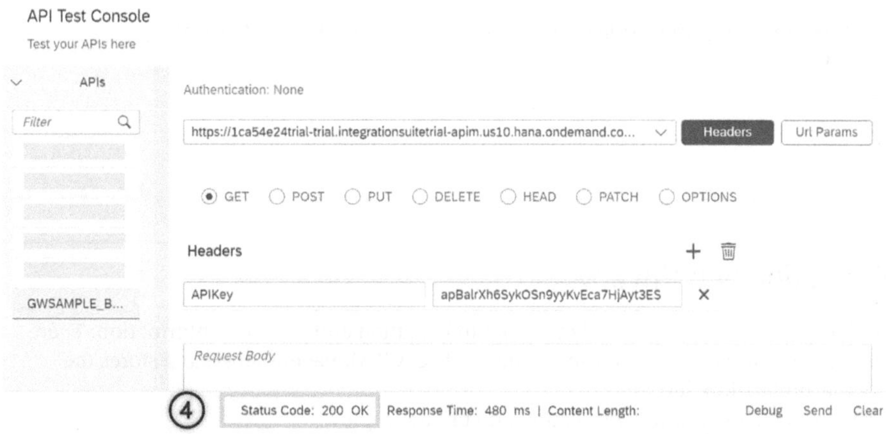

API Test Console
Test your APIs here

∨   APIs

Filter   🔍

Authentication: None

| https://1ca54e24trial-trial.integrationsuitetrial-apim.us10.hana.ondemand.co... ∨ | Headers | Url Params |

◉ GET   ○ POST   ○ PUT   ○ DELETE   ○ HEAD   ○ PATCH   ○ OPTIONS

Headers   + 🗑

| APIKey | apBalrXh6SykOSn9yyKvEca7HjAyt3ES | ✕ |

GWSAMPLE_B...

Request Body

④   Status Code: 200 OK   Response Time: 480 ms | Content Length: Debug   Send   Clear

Figure 4-72. Successful test

The next section walks through SAP API Management security.

4.1.5 Security

The following list shows the ways to secure API Management applications.

- **User Authentication**: A legitimate SCN user must register with the API Management solution before they can access the API Management application. To use the API Management solution, enter your SCN login credentials.

- **Authorization**: You grant users authorization in API Management by appointing appropriate roles.

- **Securing APIs**: Protect your APIs by referring to the security guidelines that API Management supports.

  - Basic Authentication

  - OAuth v2.0

  - OAuth v2.0 GET

  - OAuth v2.0 SET

  - Verify API Key

  - SAML Assertion Policy

- **Security Best Practice**: Information on controlling API access with OAuth, API keys, and other threat prevention measures is available in security policies. API Management supports the following policies.

  - Access Control

  - JSON Threat Protection

  - XML Threat Protection

  - Message Validation Policy

  - Regular Expression Protection

- **Traffic Management**: The following traffic control policies are supported by API Management.

  - Quota

  - Spike Arrest

  - Concurrent Rate Limit

4.1.5.1 Data Protection and Privacy

Businesses have a legal responsibility to safeguard customer information and personal information. There are tools and features to assist you in fulfilling these standards. The API Management system stores the private information of the application developer.

This section discusses the following data protection and privacy topics.

- User consent

- Read access logging

- Information report

- Erasure

- Change log

Let's start with user consent.

4.1.5.1.1 User Consent

API Management processes and stores several types of customer data. This data is protected at the greatest level of security, and SAP takes extra measures to ensure it.

SAP clients should modify the API Management to utilize their identity provider to comply with user consent. The user authorization mechanism required to permit the collection and transmission of personal data of a natural person, such as a customer, contact, or account, should be made available to clients of SAP who utilize the custom identity provider.

4.1.5.1.2 Read Access Logging

Read Access Logging (RAL) tracks and records read access to sensitive data. A law, an external company policy, or an internal company policy may designate certain data as sensitive. API Management keeps no highly sensitive personal information.

4.1.5.1.3 Information Report

An information report is a compilation of facts about a data topic. An application may offer a self-service option for such a report, or a data privacy specialist may be required to provide it.

The user ID is stored in API Management regarding personal data about data subjects. When a user creates an API artifact, such as an API proxy or an API product, this user ID is kept and can only be retrieved (read) from that artifact.

4.1.5.1.4 Erasure

Personal data can also contain referenced information. Managing referenced data first, followed by other data, including business partner data, is problematic when deleting and blocking data.

API Management records the administrator's API portal user ID. For business purposes, the user ID must be kept on file. When the resources that the administrator of the API portal established are deleted, the user ID is likewise deleted.

The first and last names, user IDs, and email addresses of users who have logged into the developer portal are saved by API Management. All personal data stored in the program is destroyed when access is denied to the linked user.

4.1.5.1.5 Change Log

Whether for auditing purposes or legal obligations, changes to personal data should be documented so that it is possible to trace who made them and when.

The corporation claims that users cannot change their personal information via the API Management application. However, the API Management application syncs changes made to the identity provider, and API Management records the sync operation. The identity provider should monitor any changes made.

4.1.6 Monitoring and Troubleshooting

With the SAP Help Portal, you can submit a report of an incident or mistake.

Use this component for your incident, as shown in Table 4-5.

Table 4-5. *Reporting Component*

Component Name	Component Description
OPU-API-OD	SAP API Management: On Demand

It is suggested to give the following details when reporting an incident.

- Landscape details (Canary, EU10, US10)

- The page's URL where the incident or problem takes place

- The actions or clicks taken to repeat the mistake

- Videos, screenshots, or the entered code

If you and SAP agree to it after subscribing, you can do application performance tests for SAP API Management as a customer. You need the approval to carry out these tests as there is a SaaS product that can impact further application performance due to high volume/penetration tests. However, it is best to monitor the performance against the suggested vendor-specified threshold values and report in case of any metrics fall outside the range.

4.1.6.1 Check the Status of Custom Domain Virtual Host Certificates

You may monitor the state of the API Management custom domain virtual host certificates and proactively identify issues by using the SAP Cloud Application Lifecycle Management (ALM) tool.

The SAP Cloud ALM platform allows you to keep track of environment backlogs and the progress of automation tasks in connection to the execution status, application status, start delay, and runtime of various SAP Cloud solutions and services to connect the health service endpoint with the Cloud ALM application.

The API *https://host>:port>/api/1.0/Health*, which has been provisioned in API Management, can be used with Cloud ALM to provide information about the specifics of the custom domain virtual host certificates expiry. With this API, you can view details about the certificates that users have supplied for their unique domain virtual hosts.

4.1.7 Summary

This chapter was a comprehensive guide to SAP API Management, covering the entire API development life cycle, from building and publishing APIs to analyzing, consuming, and monetizing them. It is a useful resource for developers, architects, and IT professionals looking to implement SAP API Management in their organizations.

Thus far, most of the things in the world of APIs have been covered. The next chapter covers another SAP Integration Suite capability: SAP Trading Partner Management.

CHAPTER 5

SAP Trading Partner Management

This chapter dives into SAP Trading Partner Management (TPM) features and components and explores how to set it up and develop B2B scenarios using its web UI. It also covers the basics of creating trading partner agreements and activating and copying them. Partner Directory data and the process of exporting data are also discussed.

The chapter looks at the integration flow configuration of sender-side communication flows, interchange processing flows, receiver communication flows, and the payload indicator in the integration flow. It also covers the importance of monitoring B2B scenarios and ensuring data protection and privacy through identity and access management.

By the end of this chapter, you should clearly understand the essential concepts of TPM and be equipped with the knowledge to develop and implement B2B scenarios effectively. So, let's get started and explore TPM.

5.1 Overview of SAP Trading Partner Management

The SAP software package includes the TPM module, which assists enterprises in managing their commercial connections with third parties like clients, vendors, and suppliers.

Through TPM's consolidated platform, organizations can manage their trade partner information, including contact information, contract terms, and transaction data. Additionally, this module enables users to build and maintain an extensive database of their business partners and track their progress against predetermined key performance metrics (KPIs).

Collaboration between businesses and their trading partners is facilitated by TPM, ensuring productive and effective cooperation. Giving companies real-time visibility into their trading partner network enables them to improve customer satisfaction, cut expenses, and streamline their supply chain management.

Let's take a closer look at TPM's features.

5.1.1 Features of SAP Trading Partner Management

TPM is an SAP Integration Suite capability designed to help manage business relationships between companies, including customers, vendors, and partners. Some of the key features of TPM include the following.

- Build and keep up trading partner profiles that meet their B2B needs. This process is part of creating a profile of one's own business with all the details for setting up B2B scenarios, such as contact information, IDs, the communication protocol and its settings, and the B2B standard.

© Jaspreet Bagga 2023
J. Bagga, *Introduction to Integration Suite Capabilities*, https://doi.org/10.1007/978-1-4842-9630-1_5

- Create trade partner agreement templates per the specifications of your B2B scenarios.

- Use the templates to create trade partner agreements that include the requirements of the trading partners.

- Publish the automatically created runtime artifacts for the B2B scenarios specified in the contracts into the SAP Cloud Integration partner directory. This guarantees that a single integration flow processes each B2B message separately during runtime.

These features are enabled by several components that work together to provide a seamless B2B integration experience; let's delve into those components and how they work.

5.1.2 Components of SAP Trading Partner Management

The following are components of TPM.

- **Company**: The leading trading partner is referred to as the *company*. The trading partner is typically in charge of the system for managing trading partners, which contains the unique data and settings required for setting up B2B scenarios with trading partners.

- **Target Partner**: A trading partner is a firm, group, or subsidiary with which the owner does business, and the interchange of electronic business data is necessary.

- **Trading Partner Profile**: Creating a B2B scenario for the electronic exchange of business data with a trading partner can be facilitated using the reusable information about a firm, organization, or subsidiary provided in a trading partner profile. The information needed to set up communication with the trade partner, identify the trading partner in the B2B message payload, and identify the type of B2B standard being used are all included. Additionally, it offers more data that paints a complete picture of the commercial situation.

- **Trading Partner Agreement Template**: A B2B scenario is described by the Trading Partner Agreement Template from the client's perspective. A B2B scenario describes the sequence of business transactions that details what kind of interchange should be sent to or received from a trade partner at each transaction stage. A payload exchanged between trading partners during a business transaction is known as an interchange. It could contain a single business document or a collection of several.

- **Trading Partner Agreement**: An agreement between two trading partners that have opted to exchange specific interchanges per a specified B2B scenario made possible by a trading partner agreement is known as a trading partner agreement. A trading partner agreement covers all aspects of the collaborative business process. Each agreement is based on a template and addresses the specifics of a chosen trade partner on the other side of a business-to-business transaction, such as the B2B standard and identifiers used in business transactions. Also, it contains the SLA (service level agreements) and QoS (quality of services) requirements needed to fulfill the contractual obligations.

- **Message Implementation Guide:** Integration Advisor collaborates closely with TPM. The message implementation guidelines (MIGs) for the business documents in an interchange of a business transaction activity can be defined using Integration Advisor. It is a specification that details how to use a type system, such as a B2B standard, de facto standard, or custom structure, to implement and process a certain message type or business document/message structure.

- **Mapping Guideline:** A mapping guideline specifies how a source MIG and target MIG should be mapped. It displays the mapping of the defined nodes at each side, describes each mapping component in depth with definitions or notes, and offers further transformational instructions. A MAG may be assigned in a trading partner agreement if both the company's and the trading partner's MIGs are assigned.

5.1.3 Development

Development in TPM involves customizing and configuring the TPM functionality to meet an organization's specific needs.

Typically, development in TPM involves defining the business partner roles, configuring the partner determination procedure, defining the partner functions, and mapping the partners to the respective partner functions. The development process also involves defining the communication channels and settings for partner communication and setting up the authorization and security roles for the trading partners.

Furthermore, development in TPM also includes creating custom programs, reports, and interfaces to integrate with external systems or automate business processes related to TPM.

The next section goes through the terminology in TPM.

5.1.3.1 Basic Concepts

TPM has some basic terminology you should be familiar with.

- A **B2B scenario** describes the sequence of business transactions between the participating trading partners within an established business environment required to complete a particular business process.

- A **business message**, also known as a *business document*, is the precise business data payload transmitted from a source system (known as the sender) or received by a target system (known as the receiver) during a business transaction phase. The source system may send one or more business messages in a single business transaction phase and may be received by the target system in a single business transaction step.

- In a B2B setting, a **business transaction** is an entity. By exchanging business information or things per the predetermined level of service quality, it creates a synchronized state in information systems that are starting and reacting (QoS).

- A **business transaction pattern** offers a business transaction's one-way or two-way pattern with the use, boundary conditions, specific business semantics, and, most importantly, the specific QoS associated with the pattern.

- A **business transaction step** is a single step that takes place during a one-way or two-way business transaction. It outlines the appropriate methods for exchanging commercial information between starting and responding to trading partners or vice versa (a two-way business transaction). It consists of all essential and involved elements for exchanging a particular piece of business information or an item between the two trading partners to complete a particular stage of the business transaction. A step in a business transaction may be from the company's perspective and may go either inward or outward.

- An **inbound direction** is the flow of business messages from the trade partner to the company during a transaction.

- An **outbound direction** is the flow of business communication from the organization to the trading partner during a business transaction.

- An **initiating trading partner** initiates a commercial transaction with a reacting trading partner.

- A **reacting trading partner** responds to a business transaction from the initiating trading partner.

- An **interchange** is a stage in a business transaction that includes all elements required to exchange B2B-related payload data, except for communication protocol data. These are exchangeable envelopes; envelopes for functional groups; and one or more business messages, each with a header and trailer containing associated information.

- The **UN/CEFACT Modeling Methodology** (UMM) is an approach that is specifically designed to capture the business needs of business processes that involve enterprises that are represented by trading partners. UMM was developed by the United Nations Center for Trade Facilitation and Electronic Commerce, a body that creates global standards. UMM's primary goal is to depict these collaborative business processes in a platform-independent way so that the rights, obligations, and agreements between the trade partners may be considered.

- While sending messages with a trading partner, an **identifier** is used to identify them specifically and to specify how to exchange data. The identifier comprises a collection of fields that aid in identifying a certain business partner during B2B transactions, as follows.

  - **Identification** is a name or value that, according to the specified identification scheme, identifies the trading partner.

  - **Type System** is a transaction system specifying how to exchange business papers. The supported types are ASC X12, SOAP, UN/EDIFAC, and IDOC.

  - **Scheme Code/Scheme Name** serves as the reference framework within which the trading partners are each specifically identified. There are numerous ways to identify a trading partner, such as by name, identification number, or employee information. But, to uniquely identify the trade partner, you must have a frame of reference that implies that the number displayed is the identification number. The scheme name and scheme code appear here.

- **Agency Code/Agency Name** identifies the trading partners to be recognized and assigns names to them.

The next section looks at the TPM Web UI and reviews the Design and Monitor sections in SAP Integration Suite.

5.1.3.2 SAP Trading Partner Management Web UI

The web UI has Design and Monitor sections in the B2B Scenarios portion of the Cloud Integration tenancy. The following topics are in the Design tab.

- Business Profile

- Trading Partners

- Agreement Templates

- Agreements

- Partner Directory Data

Each tab enables creating and maintaining the specific B2B scenario component. A corporate profile can only be present once in a TPM application. The Getting Started tutorial is written under the premise that your application does not yet have a company profile. The first step is to make a profile for your business. Figure 5-1 shows the Trading Partner Management's Design tab, where you can see the Company Profile, Trading Partners, Agreement Templates, Agreements, and Partner Directory Data tabs. Each tab is discussed in the upcoming section.

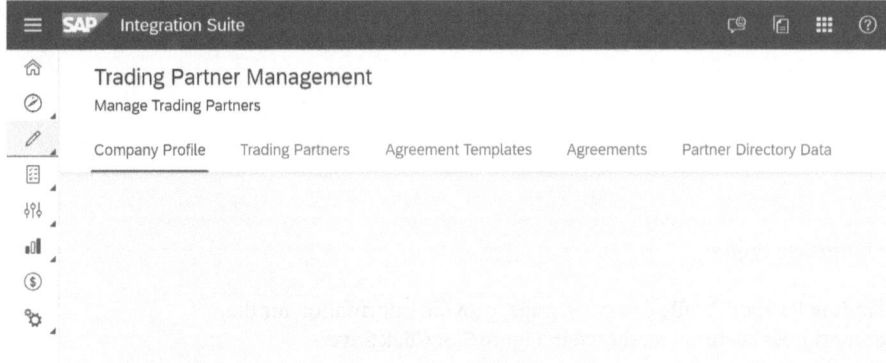

Figure 5-1. *Design tab in SAP Trading Partner Management*

You can keep track of both your successful interchanges and the interchanges that led to an error using the two tiles in the Monitor section.

- **Interchanges**: By choosing this tile, you can explore the specifics of the exchanges that result in a successful commercial transaction. The last 24 hours' worth of processed interchanges are shown by default on the tile. To view the desired exchanges, modify the filter choice.

- **Interchanges Without Assignment**: The unassigned interchanges that led to the error are shown on this tile. Each exchange's message processing log is also available for viewing.

5.1.3.3 B2B Scenarios

Let's talk about the development of B2B scenarios in TPM.

Let's start by creating the company profile, moving then to creating a trading partner profile.

5.1.3.3.1 Create a Company Profile

A certain form of the trading partner is called a *company* in this context. This trading partner represents the user (owner) of the TPM system. Here are the steps involved in setting up a new trading partner.

1. Open the Design tab in Integration Suite. Navigate to B2B Scenarios. In Company Profile, click **Create Profile**, as shown in Figure 5-2.

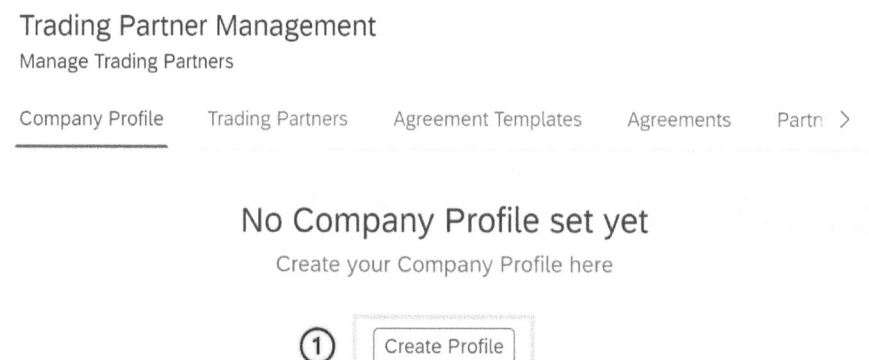

Figure 5-2. *Create a Company Profile*

2. On the Trading Partner Profile Overview page, provide information for the Details and Address sections, as shown in Figure 5-3. Click **Save**.

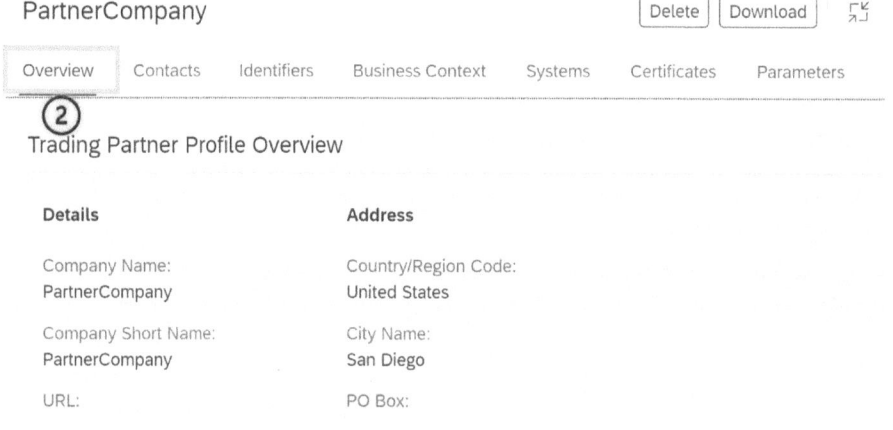

Figure 5-3. *Basic company details*

3. Navigate to the Contacts tab and click Create Contact. Provide the mandatory and basic details to the Contacts Overview page. If you want this newly established contact to be the preferred contact for communication, select the Preferred Contact checkbox. The firm profile overview provides these contact details.

4. Select **Create** in the Identifiers tab. Provide the necessary details to the Identifiers tab, as shown in Table 5-1. Provide information in Identification, Type System, Scheme, and Agency in the Details field, as shown in Figure 5-4.

Table 5-1. *Identifiers Tab*

Field Name	Description
Identification	Assign the identifier a valid ID.
Type System	From the drop-down list, select a type system.
Scheme	Choose a plan from the drop-down menu.
Agency	Enter the agency's name and code in the appropriate areas.

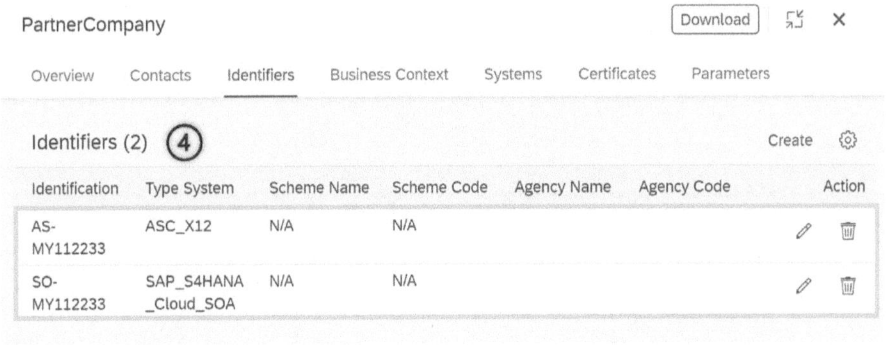

Figure 5-4. *Identifiers*

5. Navigate to the Business Context tab, and in the right corner, click Edit. Specify the details found in Table 5-2, and click **Save**.

Table 5-2. *Business Context Tab*

Field Name	Description
Business Process	From the drop-down list, choose the business process. More than one entry can be chosen.
Business Process Role	Choose your company's process role from the drop-down menu. More than one process role may be assigned.
Industry Classification	With the help of the drop-down list, select the sector to which your company belongs. You may add several entries.
Product Classification	Choose the category that your product falls within. You may add several entries.
Country/Region	Choose the country information from the drop-down menu.

6. Navigate to the Systems tab. To add the system details, select Create. Give the system a real name in the Name area, and then select the system type in the Type field. Set your system's purpose and click Save, as shown in Figure 5-5. The system is successfully constructed.

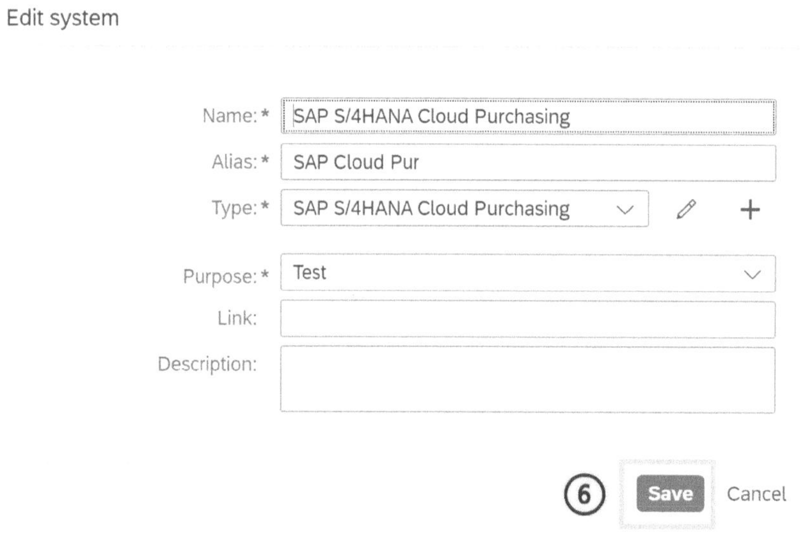

Edit system

Name: * SAP S/4HANA Cloud Purchasing

Alias: * SAP Cloud Pur

Type: * SAP S/4HANA Cloud Purchasing

Purpose: * Test

Link:

Description:

⑥ **Save** Cancel

Figure 5-5. *Edit system*

7. When your system type isn't shown, select +, fill out the form with the new system's information, and then select Apply. The Type field is filled with the newly established system, as shown in Figure 5-6.

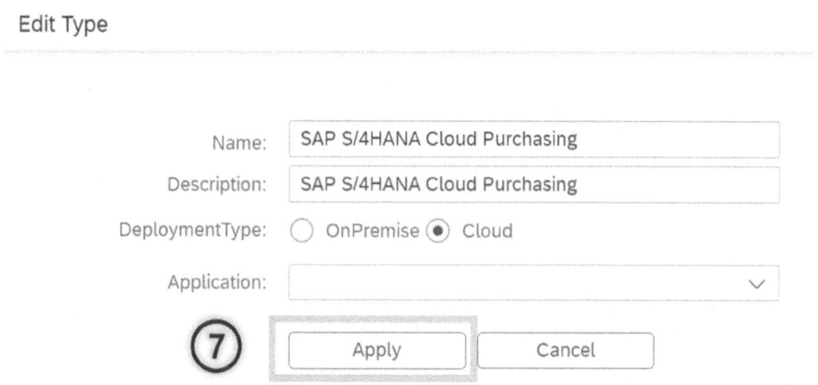

Edit Type

Name:	SAP S/4HANA Cloud Purchasing
Description:	SAP S/4HANA Cloud Purchasing
DeploymentType:	○ OnPremise ● Cloud
Application:	⌄

⑦ [Apply] [Cancel]

Figure 5-6. *Set up system type*

8. For the newly established company profile, you must also configure the type system and communication channel details to set its status to Incomplete.

    • To do this, choose Create from the Type System tab on the newly constructed system by selecting it.

    • Choose the name and version of the type system from the drop-down list in the Create Type System dialog.

    • Choose **Save**. It is successful in adding the type system. To update the system, select Save, as shown in Figure 5-7.

Create Type System

Name: *	SAP S/4HANA Cloud SOAP ⌄
Version: *	1905 × ⌄
Input version manually:	☐

⑧ [Save] Cancel

Figure 5-7. *Create Type System*

9. Select **Create** in the Communications tab. Provide all the mandatory and basic details for the Communications tab, as shown in Figure 5-8. Table 5-3 shows the details of the Communications tab that must be updated.

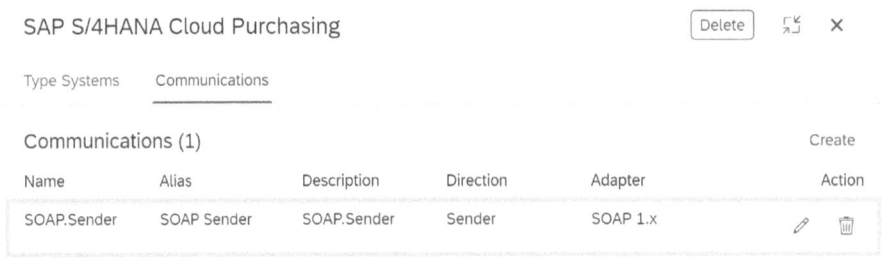

Figure 5-8. *Communications in company profile*

Table 5-3. *Communications Tab*

Field	Description
Name	Enter a proper communication name.
Alias	Give the communication an alias.
Description	Describe the communication.
Direction	Select a sender or receiver for the communication channel.
Adapter	An adapter can be chosen from the drop-down menu.

10. Different tabs are opened based on the selected sender or receiver adapter. Fill out all the mandatory fields based on the adapter.

11. Select **Create** from the Certificates tab. Provide the alias and browse the certificate.

Now that you know how to set up a company profile, I walk you through how to create a trading partner profile.

5.1.3.3.2 Create Trading Partner Profile

A *trading partner* is a corporation, organization, or subsidiary that engages in online business-to-business (B2B) transactions.

A trading partner profile offers details about a business, group, or subsidiary that are important for creating a trading partner agreement. It includes the business context, the contact, the systems involved, the communication parameters, the certificates, the active B2B scenarios, the message implementation guidelines, the functional or application acknowledgments, as well as the identifications and additional parameters required to uniquely identify the trading partner in a receiving interchange and set the necessary information in the envelopes of a sending interchange.

1. Go to TPM after signing into your application.

2. To add a new Trading Partner, select Create.

3. Update the fields in the Details section of the Overview page, as shown in Figure 5-9. Choose **Save**.

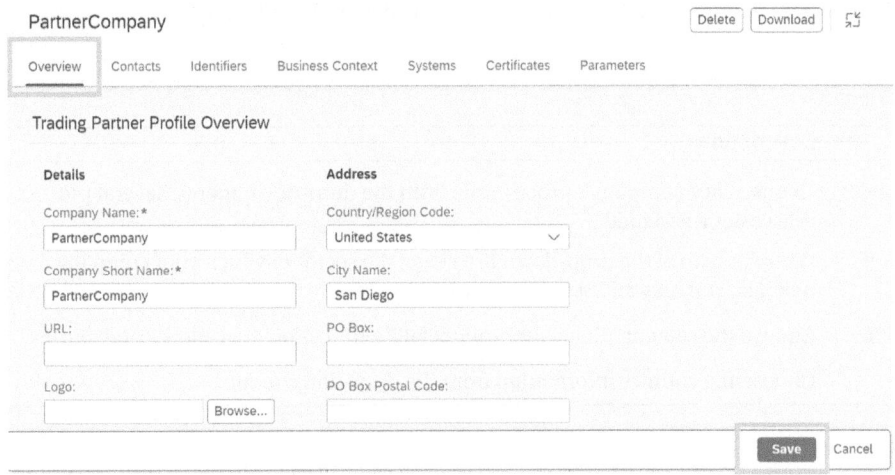

Figure 5-9. *Trading Partner Profile Overview*

4. Navigate to the Contacts tab and choose Create. Provide the mandatory details and choose Save.

5. Select **Create** in the Identifiers tab. Provide the details shown in Table 5-4, and click Save. The identifier has been created in the Identifiers tab, as shown in Figure 5-10.

Table 5-4. *Identifiers Tab*

Field Name	Description
Identification	Give the identifier a genuine ID.
Alias	Type an alias for the identifier here.
Type System	From the drop-down list, select a type system.
Scheme	Choose a plan from the drop-down menu.
Agency	Enter the agency's name and code in the appropriate areas.

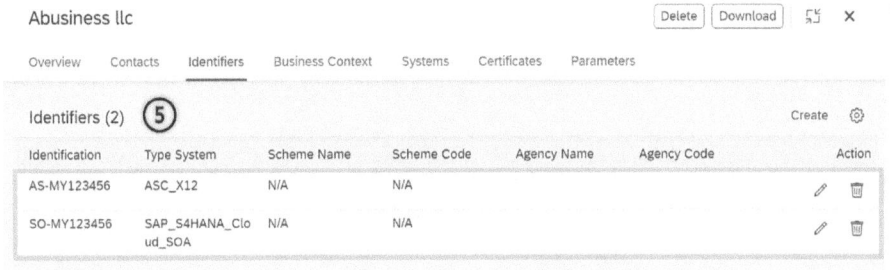

Figure 5-10. *Identifiers*

6. Navigate to Business Context and select Edit. Provide the details described in Table 5-5.

Table 5-5. *Business Context*

Field Name	Description
Business Process Role	Choose your company's process role from the drop-down menu. Several process roles can be assigned.
Industry Classification	With the help of the drop-down list, select the sector to which your company belongs. You may add several entries.
Product Classification	Choose the category that your product falls within. You may add several entries.
Country/Region	Choose the country information from the drop-down menu.

7. Navigate to the Systems tab and choose Create. Provide the details described in Table 5-6. The system is created in the Systems tab, as shown in Figure 5-11.

Table 5-6. *Systems Tab*

Field Name	Description
Name	Name the system.
Alias	Insert a system alias here.
Type	From the drop-down list, choose the system type. Select the Edit icon to change the system type information. Click Add, enter the information for the new system, and then select Apply if your system type is not listed. The Type field now be filled up with the newly established system.
Purpose	From the drop-down list, choose the type system's objective. Choose from Dev, Test, or Prod. The value you enter here indicates the transaction's purpose in the payload during the agreement's activation.
Link	Provide a system link.
Description	Provide a brief description of the system.

Figure 5-11. *Systems*

8. For the newly formed company profile, you must additionally configure the type system and communication channel data; as a result, its status is set to Incomplete.

    a. Select **Create** from the Type System tab on the newly constructed system to do this.

    b. Choose the name and version of the type system from the drop-down list in the Create Type System dialog.

    c. Click **Save**. It is successful in adding the type system. To update the system, select Save.

9. Select **Create** in the Communications tab, and fill in the mandatory information, as shown in Table 5-7.

Table 5-7. Communications Tab

Field	Description
Name	Enter a proper communication name.
Alias	Give the communication an alias.
Description	Describe the communication.
Direction	Select a sender or receiver for the communication channel.
Adapter	An adapter can be chosen from the drop-down menu.

When you select the AS2 receiver adapter, keep the fields shown in Table 5-8.

Table 5-8. Communication Details (AS2)

Field	Description
Recipient URL	Enter the receiver system's URL here.
Proxy Type	Choose the proxy type you want to employ while connecting to the receiving system.
Authentication Type	Choose a procedure for authentication.
Credential Name (Basic Authentication)	Provide the credentials needed for simple authentication.

In the Processing tab, update the fields described in Table 5-9.

Table 5-9. Processing Tab

Field	Description
File Name	Name the AS2 file specifically.
Append Timestamp	Provide a timestamp at the end of the file name.
Message ID Left Part	Indicate the AS2 message ID's left side. You may use regular expressions or '*'.
Message ID Right Part	Enter the AS2 message ID's right side. You may use regular expressions or '*'.
Own AS2 ID	Set your unique AS2 ID. You may use regular expressions or '*'.

You can specify the other fields, such as those in the Security tab and the MDN tab, as shown in Figure 5-12. Here in the connection tab we have created a receiver as AS2 in Cloud Integration and provided its endpoint as the Receipient URL.

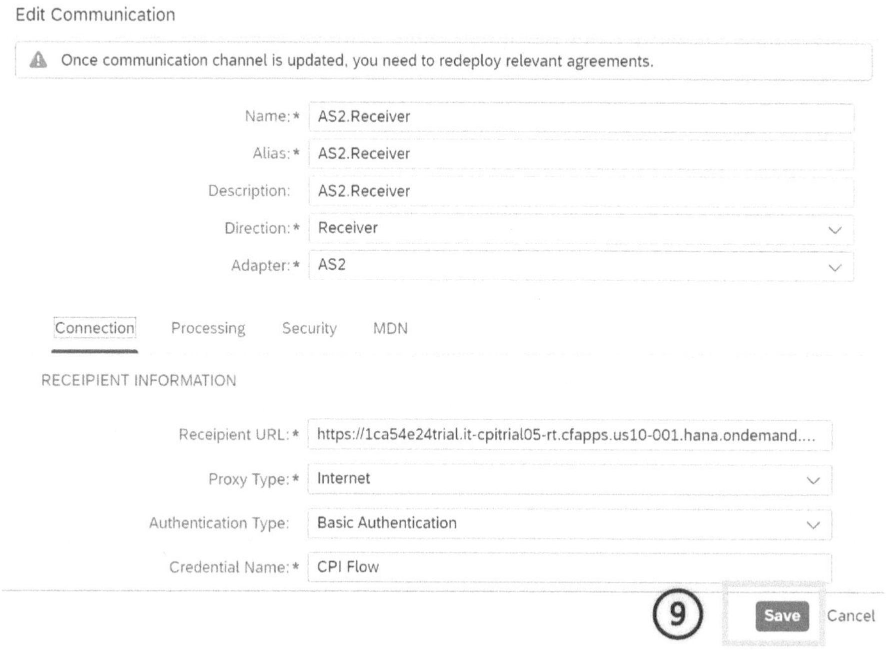

Edit Communication

> ⚠ Once communication channel is updated, you need to redeploy relevant agreements.

Name: *	AS2.Receiver
Alias: *	AS2.Receiver
Description:	AS2.Receiver
Direction: *	Receiver ⌄
Adapter: *	AS2 ⌄

Connection   Processing   Security   MDN

RECEIPIENT INFORMATION

Receipient URL: *	https://1ca54e24trial.it-cpitrial05-rt.cfapps.us10-001.hana.ondemand....
Proxy Type: *	Internet ⌄
Authentication Type:	Basic Authentication ⌄
Credential Name: *	CPI Flow

⑨   **Save**   Cancel

Figure 5-12. *Edit Communication*

10. Each Connection type has different fields. Provide the necessary details based on the chosen adapter type.

11. Select **Create** from the Certificates tab after navigating there. Provide the alias and browse the Certificate file. Click Add. The certificate has been successfully added. With the actions button, you can perform additional actions on this certificate.

Now, let's look at how to create an agreement template.

5.1.3.3.3 Create Agreement Template

A semantical choreography definition template for an agreement includes one or more business transactions. When generating the trading partner agreement, default configurations from the template level can be used and overridden.

1. Access your application after logging in, then click the Agreement Templates tab.

2. Select **Create**.

3. Enter the name, description, and version information in the Details field of the Overview tab.

4. Keep the fields listed in Table 5-10 updated in the My Company Details section.

Table 5-10. *Company Details*

Field	Description
System	From the drop-down menu, choose a system.
Type System	A type system can be chosen from the drop-down menu.
Type System Version	A type system version can be chosen from the drop-down menu.
Identifier	Choose an identifier using the value offered.

5. Navigate to the B2B Scenarios tab and select Edit. Select Choose Business Transaction. In the Details tab, provide the Name, Business Transaction Pattern, and My Company Role. Click **Save**, as shown in Figure 5-13. There is a new entry made.

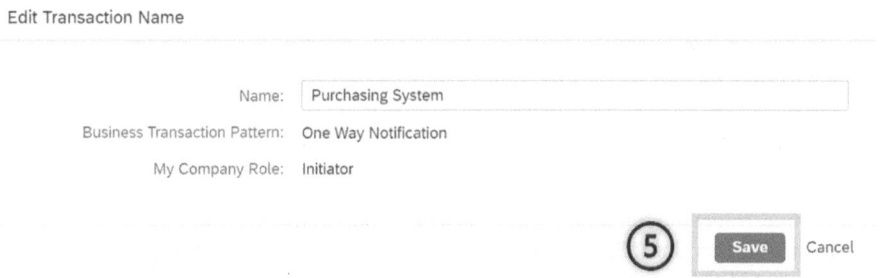

Edit Transaction Name

Name: Purchasing System

Business Transaction Pattern: One Way Notification

My Company Role: Initiator

⑤ Save Cancel

Figure 5-13. *Edit Transaction Name*

6. Choose the communication channel step on the sender side of the newly established transaction. Choose a value from the Communication field's drop-down menu, as shown in Figure 5-14.

Figure 5-14. *Communication channel*

7. On the sender side, select the Interchange step. Select a value from the list in the Message Implementation Guideline (MIG) field, as shown in Figure 5-15. For this you must create the MIG from the Integration Advisor, which we have learned in Chapter 4. Remember, this is the source MIG.

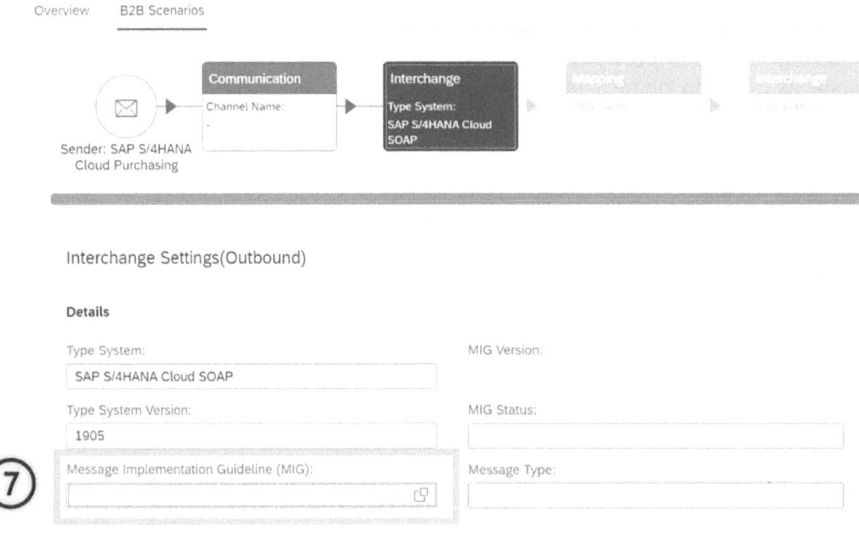

Figure 5-15. *Select the MIG*

You can change the version numbers for the type systems ASC X12 and UN/EDIFACT in the Type System Version field. Pick a MIG from the list, then click **Choose**, as shown in Figure 5-16.

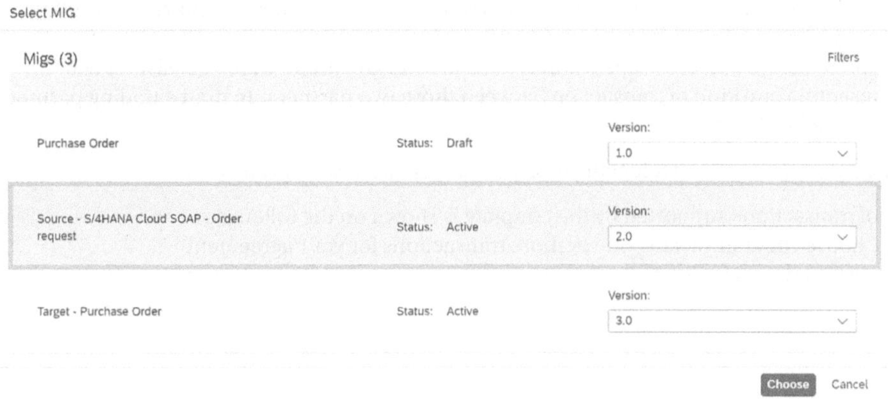

Select MIG

Migs (3)   Filters

| | | Version: |
| Purchase Order | Status: Draft | 1.0 ⌄ |

| Source - S/4HANA Cloud SOAP - Order request | Status: Active | Version: 2.0 ⌄ |

| Target - Purchase Order | Status: Active | Version: 3.0 ⌄ |

Choose Cancel

Figure 5-16. *Select the MIG version*

8. Select an option in the Custom Functional Acknowledgment drop-down menu.

9. Enable the Custom Integration Flow checkbox if you intend to utilize custom integration flows for the pre- and post-processing your exchange phase.

10. The Process Direct Address area is where you should enter the URL for your custom integration flow.

11. On the receiver side, pick the communication channel and then a value from the drop-down list for the field Communication.

12. Choose a value from the drop-down menu for the field Receiver Functional Acknowledgment Channel.

13. On the receiver side, choose the Interchange form.

14. Select a MIG from the list in Value Help for the Message Implementation Guideline (MIG) field and click Choose.

15. Choose a selection for the Number Range field from the drop-down menu.

16. Enable the Custom Integration Flow checkbox if you wish to use custom integration flows for your pre/post-processing of the interchange phase, much like the sender.

17. The Process Direct Address area is where you should enter the URL for your custom integration flow.

18. If you want to enable the receiver's functional acknowledgment, select the Enable checkbox next to Receiver Functional Acknowledgment.

19. If you want to modify the template's parameters, select Parameters.

20. Choose **Add Parameters** from the Parameters tab's drop-down menu.

21. Choose **Save**.

The next section demonstrates how to create a trading partner agreement.

5.1.4 Create a Trading Partner Agreement

Using a B2B scenario, two trading parties have opted to exchange specific business data or items to complete the agreed trading/business process in a particular business setting. This agreement is known as a trading partner agreement.

The final phase in the design of a B2B transaction is to create an agreement, which consists of two trading partners representing one kind of transaction between those two partners. To draft a trading partner agreement, adhere to the following steps.

1. Go to the Agreements tab after logging in to your application. Select Create.

2. The list of transactions supported by the template is shown on the following screen. Choose **OK** after picking one or more transactions for your agreement from the list, as shown in Figure 5-17.

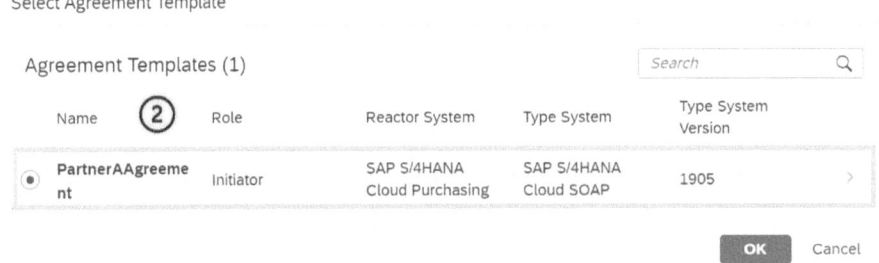

Figure 5-17. *Select Agreement Template*

3. Enter the information shown in Table 5-11 in the Details and My Company Details section of the Overview tab.

Table 5-11. *My Company*

Field	Description
Name	The agreement's name
Description	Describes the agreement in detail
Version	A copy of your agreement

The fields in the My Company section are listed in Table 5-12.

Table 5-12. *My Company Fields*

Field	Description
System	From the drop-down menu, choose a system.
Type System	A type system can be chosen from the drop-down menu.
Type System Version	A type system version can be chosen from the drop-down menu.
Identifier	With the offered value assistance, choose an identification. With the button, you may also generate an identification.
Identifier as Trading Partner	Choose a corporate identification that serves as the trading partner. With the button, you may also generate an identification.

Keep the fields listed in Table 5-13 in the section for trading partner information.

Table 5-13. *Trading Partner*

Field	Description
Name	Enter the trading partner's name.
System	From the drop-down menu, choose a system.
Type System	A type system can be chosen from the drop-down menu.
Type System Version	A type system version can be chosen from the drop-down menu.
Identifier	With the offered value assistance, choose an identification. With the button, you may also generate an identification.
Identifier as Company	Choose a trading partner identification for the company. With the button, you may also generate an identification.

After providing all the details in the Agreement section, you see the Overview page, as shown in Figure 5-18.

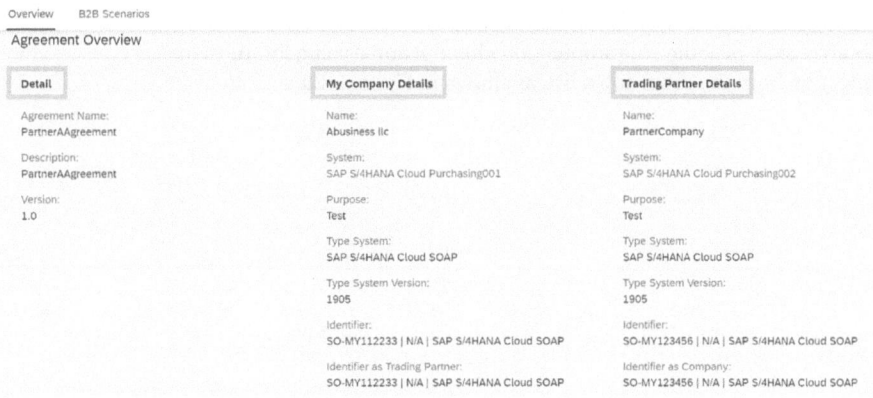

Figure 5-18. *Agreement Overview*

4. Find the B2B Scenarios tab by navigating. The transactions created using the agreement template are shown on this tab. Select the communication channel step on the sender side, as shown in Figure 5-19.

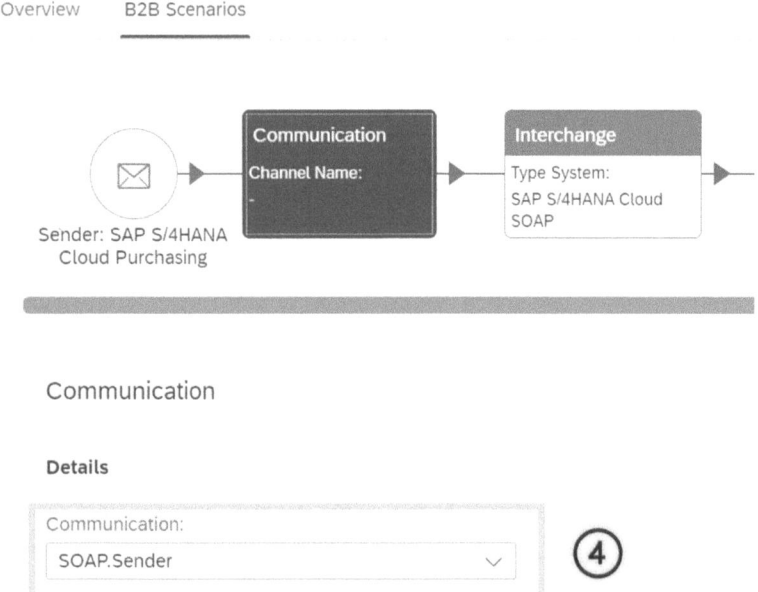

Figure 5-19. *Communication on sender side*

5. On the sender side, select the Interchange step. Select a MIG from the list in the Value Help for the Message Implementation Guideline (MIG) field and click Choose.

6. On the sender side, select the Mapping step. With the value assistance offered for the Mapping Guideline field, choose a MAG.

7. On the receiver side, pick the communication channel and a value for the field Communication from the drop-down list, as shown in Figure 5-20.

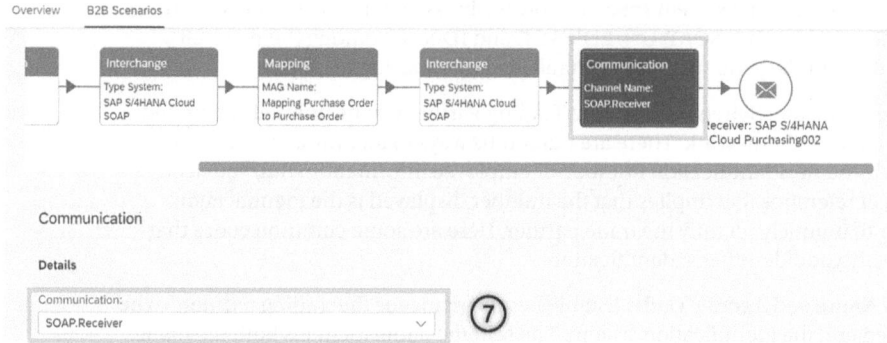

Figure 5-20. Select the receiver communication

8. On the receiver side, choose the Interchange form. Select a MIG from the list in the Value Help for the Message Implementation Guideline (MIG) field and click Choose.

9. Choose a value for the field Number Range from the drop-down menu.

10. Enable the Custom Integration Flow checkbox if you wish to use custom integration flows for your pre/post-processing of the interchange phase, much like the sender. Enter your integration flow's address path in the Process Direct Address field. If you want to turn on functional acknowledgment for the receiver, select the Enable checkbox next to Receiver Functional Acknowledgment, as shown in Figure 5-21.

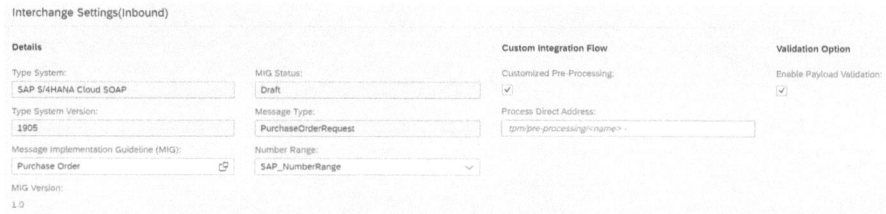

Figure 5-21. Interchange Settings

11. If you want to add any parameters to the template, select Parameters.

5.1.4.1 Basics of Identifiers in the Agreement

When exchanging messages and defining how to share documents, an identifier specifically identifies the trading partner. The Company/Trading Partner profile's Identifiers tab lets you access the identifier details. The identifier comprises a collection of fields that aid in identifying a certain business partner during B2B transactions. Those fields are as follows.

- **Identification:** A name or value that, according to the specified identification scheme, identifies the trading partner.

- **Type System**: The type system specifies how business papers are exchanged during a transaction. ASC X12, SOAP, UN/EDIFACT, and IDoc are among them. A B2B transaction's underlying standard is established by choosing the type.

- **Scheme Name** and **Scheme Code**: The Trading Partner can be uniquely identified in this reference framework. There are numerous ways to identify a trading partner, such as by name, identification number, or employee information. But, you need a frame of reference that implies that the number displayed is the identification number to uniquely identify the trade partner. Here are some common codes that can specify your identifier's identification.

- **Agency Name** and **Agency Code**: Establishes the name for the trading partners to be identified and the identification system. The scheme name includes both the name and the code.

Depending on whether a business transaction activity is inbound or outbound, an identifier may be referred to as the sender or receiver identifier.

These identifiers have specific meanings and standards and aid the system in understanding and identifying the partners engaged in a B2B transaction.

Next, let's look at an example of a transaction between the SOAP identifiers and SOAP types, as shown in Figure 5-22.

Figure 5-22. *Agreement using SOAP identifiers*

Assume there are two transactions included in the agreement.

- From the company to the trading partner

- The company receives one from the trading partner

As there are two separate message encodings for the two parties to the agreement, you must create a new identifier for the company and trading partner to make it easier for them to communicate. Here, the Identification as Company and Identifier as Trade Partner fields for the company and trading partner, respectively, come into play.

5.1.5 Activate Trading Partner Agreement

To push the details of your trading partner agreement into the SAP Cloud Integration partner directory, activate the agreement.

In the context of a wider business network, you can store data about business partners connected to the tenant using the partner directory, a tenant-specific storage option.

1. Open the Agreements tab.

2. Find the agreement you wish to activate and open it.

3. Choose **Activate**, as shown in Figure 5-23.

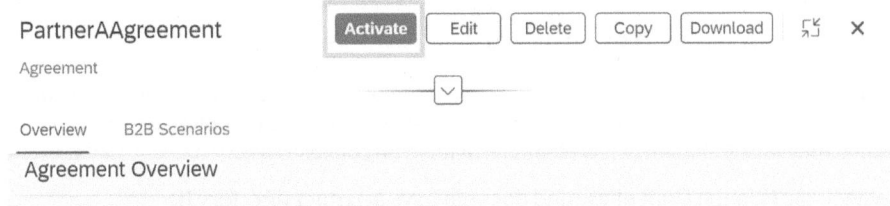

Figure 5-23. *Activate the agreement*

4. Using the Partner Directory Data field next to the transactions, you may also access the partner directory information in your B2B transactions.

5. The Partner Directory Data field shows (see Figure 5-24), depending on the transaction type—Inbound or Outbound.

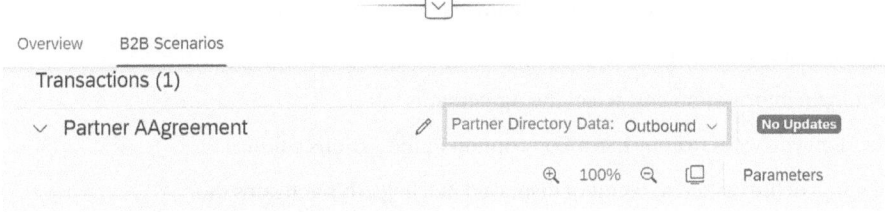

Figure 5-24. *Partner Directory Transaction Type*

6. Select either Copy PID or View Data from the drop-down menu adjacent to these buttons. Copy PID copies the transaction's partner directory ID. You may see the Partner Directory Data tab by clicking View Data.

Let's move on to the Copy Trading Partner Agreement. This agreement is used when you copy the trading partner assignments from one document to another. For example, if you have created a purchase order and assigned a trading partner, you can use the Copy Trading Partner Agreement to copy the same trading partner to the corresponding goods receipt and invoice documents.

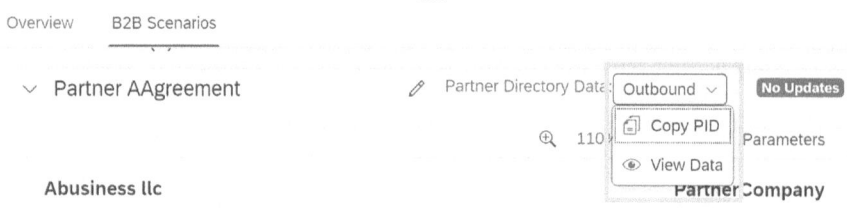

Figure 5-25. *Functions with partner directory transaction type*

5.1.6 Copy Trading Partner Agreement

There may be circumstances in which you desire to utilize an existing agreement. Now, rather than creating an agreement from scratch, you can reuse an existing agreement using the copy feature offered by the application. Before replicating an agreement, a few elements need to be considered.

- An agreement that is in the Draft phase and does not have a partner directory ID (PID) can be copied directly.

- You can only copy an agreement by selecting a different trading partner if the agreement you're seeking to copy is currently active. This is so because each PID is associated with a certain active trading partner agreement's trading partner. Thus, you must copy that agreement with a separate trading partner so that, when activated, the copied agreement won't replace the information already there in the partner directory.

- The function also offers the Open Draft option, enabling you to see the contract in draft form to edit or add information before copying.

The following explains the procedure.

1. Enter your login information and select the Agreements tab.

2. The agreement you want to copy has a Copy option in the Actions column.

3. With the name of the copied agreement displayed in a field, this activates the Copy Agreement dialog box. Give the agreement a name, then click Save.

4. You can also click **Save** after making changes and editing the information using the Open Draft option, as shown in Figure 5-26.

Figure 5-26. *Copy partner agreement*

Now that you know about the Copy Trading Partner Agreement in TPM, it is important to highlight the significance of maintaining accurate and up-to-date partner information in your partner directory.

The partner directory is a central repository that contains detailed information about your business partners, such as their contact details, addresses, bank information, and other relevant data. Businesses must maintain accurate and complete information in their partner directory as it is the foundation for various business processes, including procurement, sales, and payments. The next section discusses Partner Directory data.

5.1.7 Partner Directory Data

Partner Directory is a database that contains details on the trading partner agreement, including identifiers, names, and other crucial features. Generic integration flows use this information after that. These integration processes have parts that can be adjusted to pull data from the partner directory as needed. Partner Directory, in essence, enables you to build a single integration flow with dynamic processing based on the called parameters associated with a particular partner ID.

By looking at the Partner Directory Data page, you can solve consumer problems.

The whole text of an agreement is pushed into the partner directory when a trade partner agreement is activated. A new entry is made in the partner directory for every agreement-related business transaction activity and Interchange Envelope extraction (depending on the adapter type). The Interchange Extraction creates a camel exchange header object, which is an XSLT script that extracts the necessary parameters from the type system headers.

1. Go to the Partner Directory Data tab in your application. All partner directory IDs created in the system are shown on the tab. The convention TPM is used to begin the partner directory ID. Moreover, you can see the following information.

   - Activity Plan ID is the unique activity ID assigned to the contract's business transaction activity. The concatenated value of several head parameters set in the agreement makes up the ID.

   - Agreement Link is shown in the column. The related transaction activity in the agreement is opened by selecting this link.

   - Created Date shows the date that the agreement was put into effect.

   - Last Modified Date shows the time and date the agreement changed.

2. With the Search bar, you may look up specific partner directory information, as shown in Figure 5-27.

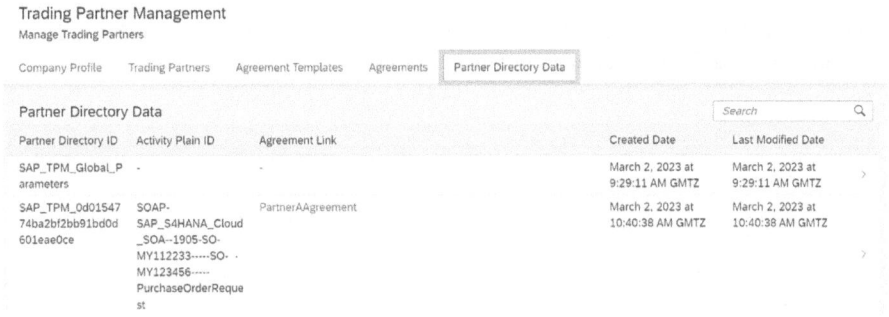

Figure 5-27. *Partner Directory Data*

3. Select the particular Partner Agreement after clicking on the Agreement Link. You can see the Partner Agreement thoroughly.

The next section discusses how to export the data.

5.1.8 Export Data

You can save the following files to your local PC using TPM.

- Company profile

- Trading partner

- Trading partner agreement template

- Trading partner agreement

You can download this information as a JSON file.

1. Go to the Business Profile tab by selecting the Design tab.

2. Click **Download** next to the company profile, as shown in Figure 5-28.

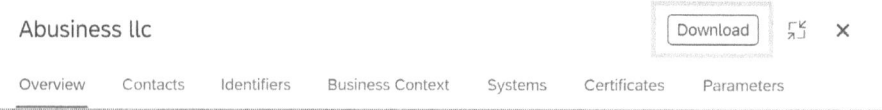

Figure 5-28. *Download company profile*

3. With the Download option, you can utilize the other application content similarly.

The next section covers integration flow configuration in SAP Cloud Integration.

5.1.9 Integration Flow Configuration

To test the end-to-end scenario in TPM, configure your integration flows.

When you need to run an end-to-end business scenario, you must configure the integration flows only once.

The required artifacts are included in the Cloud Integration: TPM package and must be configured. Figure 5-29 is an example of a general TPM integration flow process.

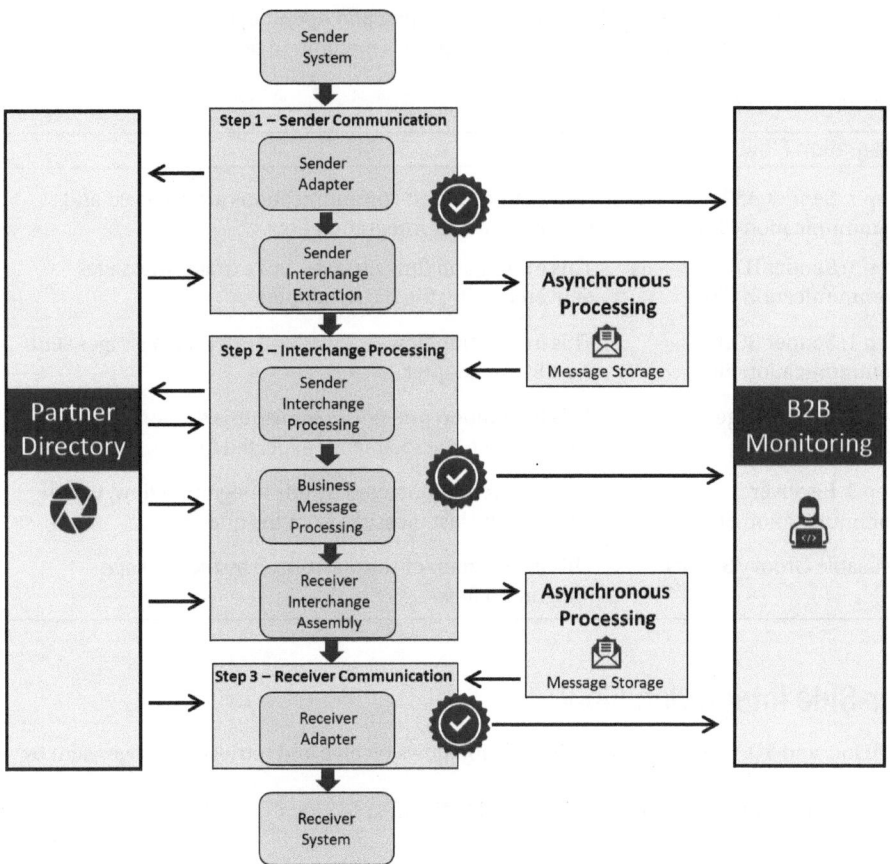

Figure 5-29. *Generic integration flow process for TPM*

The following are the procedures to copy the integration package to your Design area.

1. Enter your Cloud Integration tenant. Open the Cloud Integration: TPM package by searching for it in the Discover tab, as Figure 5-30 shows.

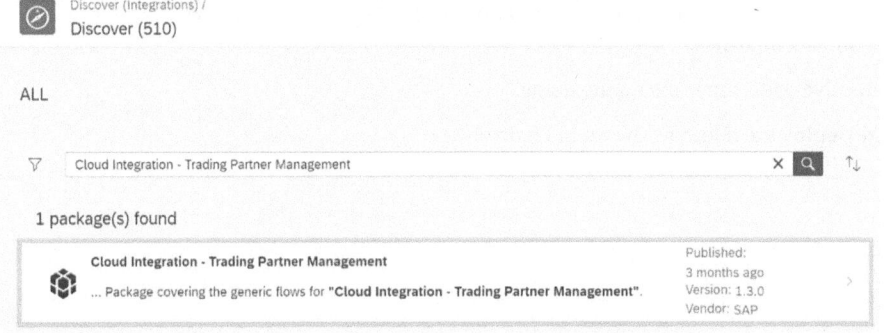

Figure 5-30. *Cloud Integration Trading Partner Management package*

2. Select the Design tab. Go to the Artifacts tab after selecting and opening the integration package. Table 5-14 summarizes the package's integration flows.

Table 5-14. *Package Integration Flow*

Type	Integration Flow	Description
Sender	Step 1: Sender AS2 Communication Flow	AS2 adapter-based communications are received and extracted by this integration flow.
	Step 1: Sender IDOC Communication Flow	This integration flow receives and extracts messages delivered using the IDOC adapter.
	Step 1: Sender SOAP Communication Flow	This integration flow receives and extracts messages sent via a SOAP adapter.
Interchange	Step 2: Interchange Processing Flow	This integration procedure transforms the sender's message into the recipient's expected format.
Receiver	Step 3: Receiver Communication Flow	The receiver is contacted by this integration flow, which retrieves the last message from the queue.
	Reusable Groovy Scripts	This is the group of scripts utilized by the package's integration flows.

5.1.9.1 Sender-Side Integration Flow

The AS2, AS2 MDN, IDoc, and SOAP adapter-based integration flows receive and retrieve messages sent by these adapters.

The configuration of these integration flows is explained in the next sections.

5.1.9.1.1 Sender AS2 Communication Flow

Sender AS2 communication flow in TPM refers to the process by which a sending partner exchanges Electronic Data Interchange (EDI) messages with a receiving partner using the AS2 (Applicability Statement 2) protocol. The following steps configure the communication flow.

1. Click the integration flow's Action button in the Artifacts tab. Select Configure in the Sender AS2 Communication Flow menu.

2. Update the Address, User Role, Message ID, Partner AS2 ID, Own AS2 ID, and Message Subject fields.

3. On the Receiver side, enter the queue name.

4. **Save** and **Deploy** the iFlow, as shown in Figure 5-31.

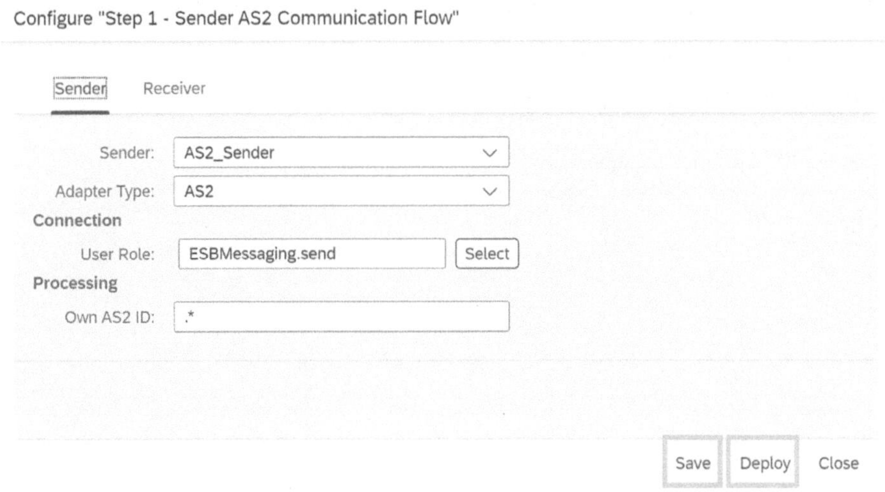

Configure "Step 1 - Sender AS2 Communication Flow"

<u>Sender</u>   Receiver

Sender:	AS2_Sender ∨
Adapter Type:	AS2 ∨
Connection	
User Role:	ESBMessaging.send Select
Processing	
Own AS2 ID:	.*

Save   Deploy   Close

Figure 5-31. Sender AS2 Communication Flow

5.1.9.1.2 Sender AS2 MDN Flow

Using the AS2 adapter, technical acknowledgment is captured using this integration flow. The B2B Monitoring tab also has a Technical Acknowledgment section. Figure 5-32 shows the Sender AS2 MDN integration flow. You can deploy the integration flow and configure it according to the business requirement.

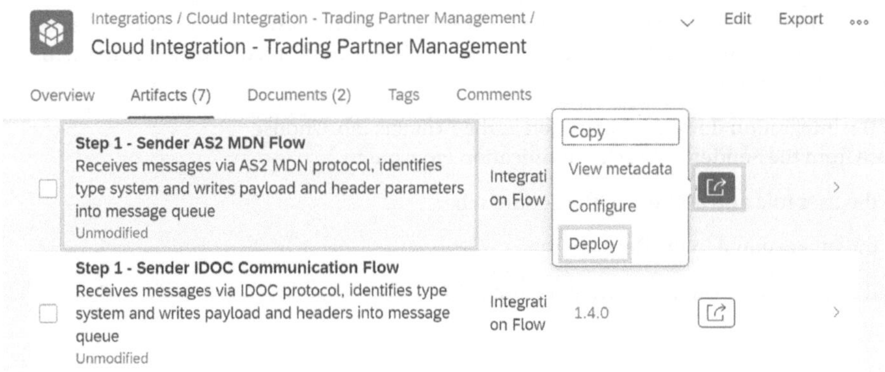

Integrations / Cloud Integration - Trading Partner Management /   ∨   Edit   Export   ○○○
Cloud Integration - Trading Partner Management

Overview   Artifacts (7)   Documents (2)   Tags   Comments

Step 1 - Sender AS2 MDN Flow Receives messages via AS2 MDN protocol, identifies type system and writes payload and header parameters into message queue Unmodified	Integrati on Flow	Copy View metadata Configure Deploy	>
Step 1 - Sender IDOC Communication Flow Receives messages via IDOC protocol, identifies type system and writes payload and headers into message queue Unmodified	Integrati on Flow	1.4.0	>

Figure 5-32. Deploy Sender AS2 MDN Flow

5.1.9.1.3 Sender IDOC Communication Flow

In TPM, the sender IDOC communication flow is the process by which an IDOC (intermediate document) is sent from the sending system to the receiving system via an IDOC interface. This communication flow involves multiple steps and components. The following steps configure the flow.

1.   Choose the integration flow's Action button in the Artifacts tab. Choose **Configure** from the Sender IDOC Communication Flow menu.

2. Update the address and user role in the Sender tab.

3. Update the queue name in the Receiver tab.

4. **Save** and **Deploy** the iFlow, as shown in Figure 5-33.

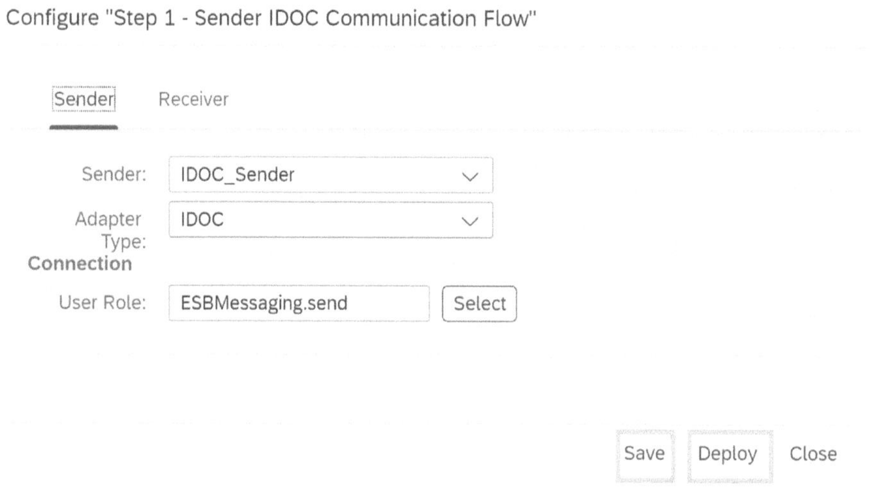

Configure "Step 1 - Sender IDOC Communication Flow"

Sender	Receiver

| Sender: | IDOC_Sender ⌄ |
| Adapter Type: | IDOC ⌄ |

Connection

| User Role: | ESBMessaging.send | Select |

Save   Deploy   Close

Figure 5-33. *Sender IDOC Communication Flow*

5.1.9.1.4 Sender SOAP Communication Flow

Sender SOAP Communication Flow in TPM refers to the process by which the SAP system sends messages to external trading partners using the Simple Object Access Protocol (SOAP) protocol. The following steps configure the SOAP communication flow.

1. Choose the integration flow's Action button in the Artifacts tab. Choose Configure from the Sender IDOC Communication Flow menu.

2. Update the user role and address in the Sender tab.

3. Update the queue name in the Receiver tab.

4. **Save** and **Deploy** the iFlow, as shown in Figure 5-34.

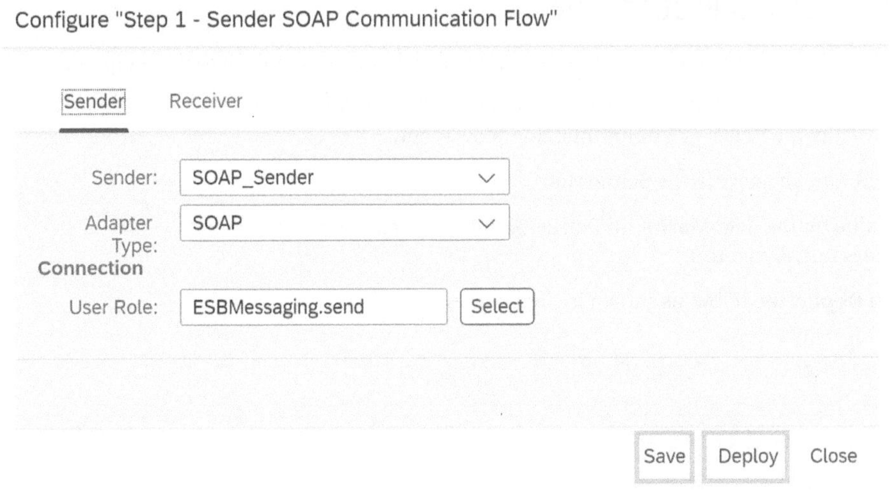

Configure "Step 1 - Sender SOAP Communication Flow"

Sender	Receiver

| Sender: | SOAP_Sender ∨ |
| Adapter Type: | SOAP ∨ |

Connection

| User Role: | ESBMessaging.send | Select |

Save Deploy Close

Figure 5-34. *Sender SOAP Communication Flow*

5.1.9.2 Interchange Processing Flow

Let's look at how to receive and retrieve messages via the integration process. In this integration flow, the message the sending partner delivers is changed to the structure that the receiving partner anticipates.

1. Choose the integration flow's Action button in the Artifacts tab. Choose Configure from the Interchange Processing Flow menu.

2. Update the queue name in the Sender tab and the Receiver tab.

3. In the More tab, update the Maximum Retries Number field.

4. **Save** and **Deploy** the iFlow, as shown in Figure 5-35.

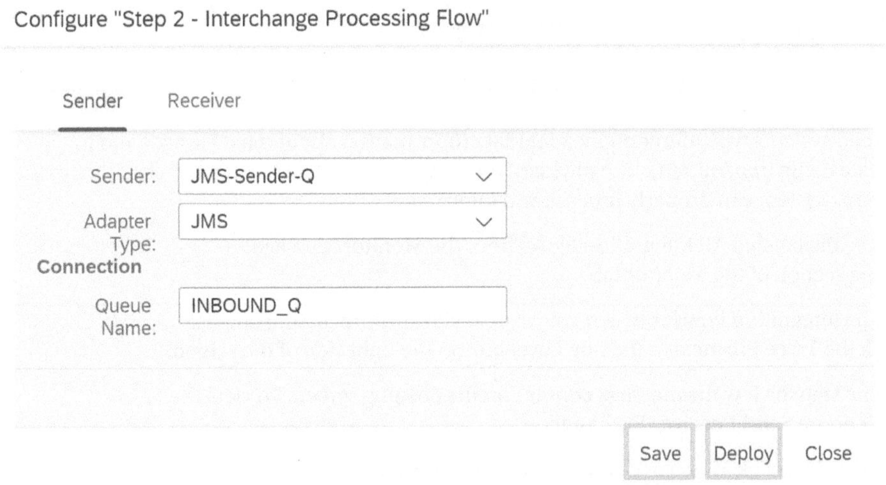

Configure "Step 2 - Interchange Processing Flow"

Sender	Receiver

| Sender: | JMS-Sender-Q ∨ |
| Adapter Type: | JMS ∨ |

Connection

| Queue Name: | INBOUND_Q |

Save Deploy Close

Figure 5-35. *Interchange Processing Flow*

5.1.9.3 Receiver Communication Flow

This integration flow obtains the final message from the queue, which then transmits it to the recipient.

1. Choose the integration flow's Action button in the Artifacts tab. Choose Configure from the Receiver Communication Flow menu.

2. Update the Queue name in the Sender tab.

3. Set the value for the field Maximum Retries Number by selecting the More tab. The value is initially set to 1.

4. **Save** and **Deploy** the iFlow, as shown in Figure 5-36.

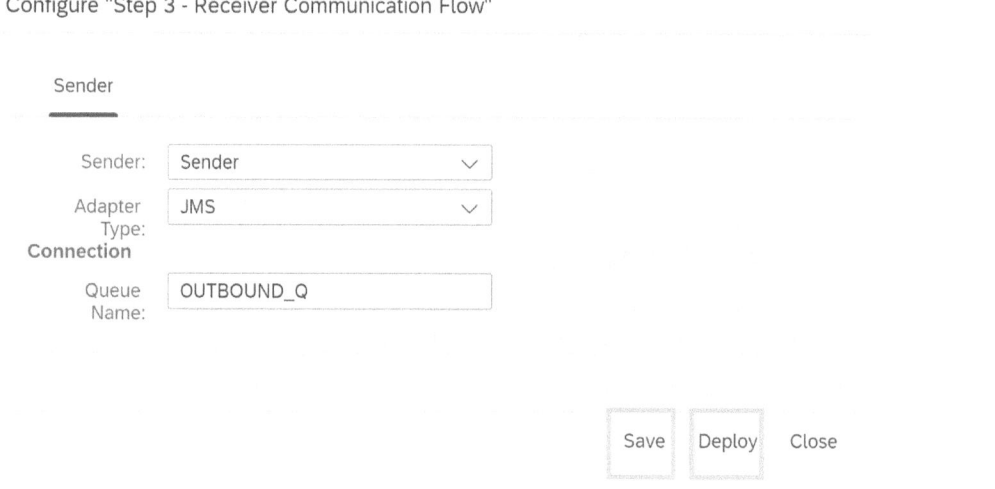

Configure "Step 3 - Receiver Communication Flow"

Figure 5-36. *Receiver Communication Flow*

5.1.9.4 Payload Indicator in Integration Flow

In an integration flow, a payload indicator is a component that identifies the data payload contained in a message.

A message between systems typically contains metadata (information about the message) and a payload (the actual data being transmitted). The payload indicator identifies the format and structure of the data so that the receiving system can properly process and interpret it.

1. Choose the tile labeled All Integration Flows from the Monitor Message Processing section of the Monitor tab.

2. Choose the integration flow for which you wish to view the processing details, then click the Trace link next to the Log Level box on the right side of the screen.

3. Choose the step to view the message content on the ensuing screen. To view the payload, choose Send Receiver Exchange.

4. On the right side, click Message Content. The Payload tab appears.

5. This shows the agreement transaction's payload. Depending on the purpose type provided in the company/trading partner profile, a payload indicator indicates whether the transaction is type test, dev, or prod.

The receiver type system and accompanying purpose indicator are listed in Table 5-15.

Table 5-15. Payload Indicator

Receive Type System	Purpose	Indicator
ASC X12	Production	P
	Development	I
	Test	T
UN/EDIFACT	Production	
	Development	
	Test	1
IDOC	Production	
	Development	X
	Test	X

5.1.10 Monitor B2B Scenarios

You may monitor the processing progress of your business-to-business (B2B) exchanges using the B2B Monitoring view.

For B2B transactions, the inbound payload is an interchange. A functional group, a special case of a bulk message, is what it is: a bulk message. The following steps get you to the Monitoring tab.

1. Go to Monitor in your Cloud Integration application's B2B Scenarios section.

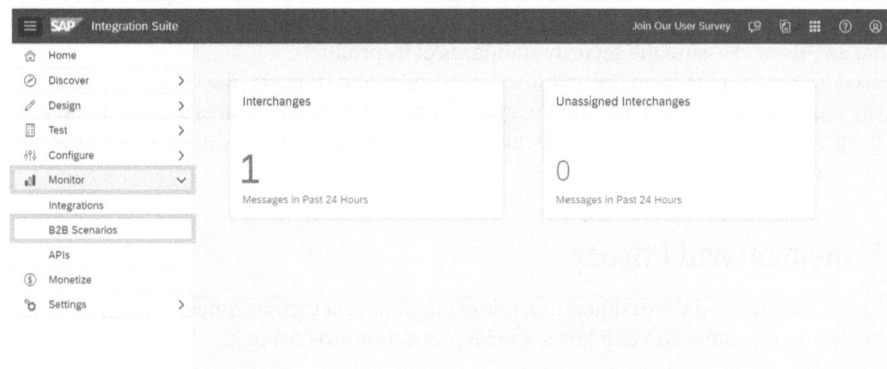

Figure 5-37. B2B Scenario Montor screen

2. To access the Standard monitoring view, select Interchanges. The upper portion of the screen shows the filter options. These filter settings allow you to search for a specific interchange.

3. Choose **Adapt Filters** if you wish to customize your filter settings.

4. Choose **OK** after selecting or deselecting the desired fields in the dialog box.

5. You can save customized filter settings by selecting the Save As option under the down arrow next to Standard (*).

6. From the list of interchanges, select and open one.

7. The list of business transaction events is shown on the Events page and includes the following information.

   - ID: Gives the event's special identifying number

   - Type: Displays the event's type

   - Date: Date of creation of the event

   - Monitoring Reference: A link to the message's details for reference

8. Choose the link for an occurrence in Monitoring Reference. In a separate window, it presents the integration flow's message processing specifics related to the transaction.

9. Go to the Monitor tab and choose the Unassigned Interchanges tile.

10. The interchanges that led to an error early in the processing are shown in this tile.

11. Click **Go** after setting the filter according to your needs.

12. From the list, select and open an exchange.

13. The resulting screen lets you see the error's specifics. The details of the integration flow involved in the transaction are shown in the error log via the link in the message processing log ID.

5.1.11 Security

You can get a comprehensive overview of all security-related aspects of TPM in the security guide, including information on process-related security, identity and access management, data storage, protection, and privacy, as well as what SAP does to assure the security standards of its products.

Using a cloud-based integration platform, the platform's host, SAP, and its users (the customers) are subject to strict security requirements. The integration platform's security-related features are discussed in this part, along with the precautions you may take to safeguard client data transmitted through the platform when an integration scenario is being carried out.

5.1.11.1 Data Protection and Privacy

The integration platform processes and stores different kinds of client data at various times. The highest level of security is provided for this data, and SAP takes special precautions to ensure it.

5.1.11.1.1 Types of Stored Data

While a B2B scenario is running, many types of data, including message content, may be saved.

In general, TPM deals with SAP Cloud Integration settings.

Due to the possibility of including personal information, this material must be considered sensitive data. The following are examples.

- **Profiles of trading partners with contact information**: Maintaining contact information for various trading partners is possible using trading partner profiles. Local sensitive data laws like GDPR may apply to this material. Make sure you manage such data per the relevant legislation. The company that owns the tenant is the one who owns, maintains, and uses the data. SAP does not directly access the data, just providing the framework for keeping such data.

- **Runtime Settings for Cloud Integration**: The profiles, agreement templates, and the actual agreement are all maintained using configurations kept in separate tenant databases.

- **Secure Content**: The suitable secure store of Cloud Integration houses the secure content required for the execution of the integrations. Usernames, passwords, secret keys, certificates, and so on are included.

- **Content for B2B Monitoring**: The runtime monitoring for Cloud Integration as well as the B2B monitoring storages, both house the processed messages, logs, and related content. Certain authorizations control sensitive data access.

5.1.11.1.2 Specific Data Assets

The main items that relate to TPM are as follows.

- Company profile
- Trading partner profile
- Trading partner agreement templates
- Trading partner agreements

5.1.12 Summary

SAP Trading Partner Management is a suite component that enables businesses to manage their B2B partner relationships. You were introduced to the features and components of TPM and learned the basic concepts involved in developing B2B scenarios.

The next chapter explores a new Integration Suite capability: Integration Assessment.

CHAPTER 6

■ ■ ■

SAP Integration Assessment

Integration has become crucial to enterprise software implementation in today's rapidly changing business landscape. As companies adopt new technologies to gain a competitive edge, ensuring that these technologies integrate seamlessly with existing enterprise systems is imperative.

SAP Integration Assessment provides a structured approach to assess integration readiness, identify requirements, and define the integration approach. This chapter walks through the key features of SAP Integration Assessment, including the solution advisory methodology, integration domains, essential characteristics, and technology mapping.

The chapter also provides a step-by-step guide to the integration assessment process, starting with the initial setup and configuration of the Integration Solution Advisory Methodology (ISA-M) and culminating in identifying and selecting an integration technology.

The chapter touches upon the critical aspect of security and discusses the different types of data storage available to ensure data security in the integration process. Whether you are an SAP consultant or a business analyst responsible for enterprise software implementation, this chapter equips you with the knowledge and tools needed to perform a successful SAP Integration Assessment.

6.1 Overview of SAP Integration Assessment

The Integration Solution Advisory Methodology used by SAP Integration Assessment defines the integration landscape. It helps in defining integration patterns and ensuring the consistent adoption of integration technology.

You can evaluate your integration strategy with the help of SAP Integration Assessment, which offers a guided method. You can benefit from recommendations on integration technologies that meet your company's needs.

You can evaluate your organization's integration plan in a systematic manner using SAP Integration Assessment. The technologies for integration that best match your needs can then be identified with its assistance. The SAP Integration Assessment dashboard is shown in Figure 6-1.

J. Bagga, *Introduction to Integration Suite Capabilities*, https://doi.org/10.1007/978-1-4842-9630-1_6

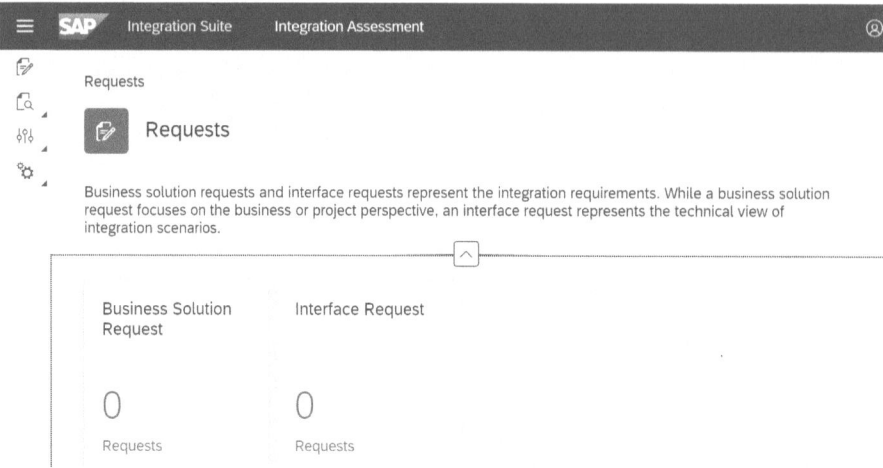

Figure 6-1. *SAP Integration Assessment dashboard*

6.1.1 Features of SAP Integration Assessment

The following describes how you can best work with SAP Integration Assessment features.

- Record the mapping and regulations for ISA-M technology. Uses the ISA-M master data included with the framework and adds further context.

- Make guided interface requests. Implement ISA-M rules and regulations while making an interface request.

- Choose the integration technology that best fits the integration strategy of your company with the aid of the interface request procedure. It offers direction at the time of technology selection.

- Use surveys. You can use questionnaires to guide your technological decisions. Also, you can design your own inquiries to choose the best integration technology.

- Take advantage of wise technical suggestions. Learn the technologies that are used to integrate applications. Each technology supports the integration use case patterns and the integration patterns employed.

- Keep tabs on the progress of interface requests.

- Know your integration environment. You can learn more about your integration landscape with the help of SAP Integration Assessment. A summary of the relevant integration domains, integration methods, use case patterns, defining traits, and deployment models are provided.

- Find and use SAP integration standard content.

6.1.2 SAP Integration Solution Advisory Methodology

The fundamental concept for ISA-M is that you establish a few technology-neutral integration requirements relevant to your organization's integration plan in the first phase (integration domains, integration methods, and integration use case patterns). You connect these criteria to the integration technologies in the second phase that best suit the context of your customers.

Changes could be made to the integration assessment's standard content throughout the life cycle of the standard content.

The next section discusses the primary steps of the assessment process.

6.1.2.1 Integration Domains

A hybrid landscape's *integration domain* refers to a location where integration is required, such as on-premises to cloud or cloud-to-cloud integration.

The starting point for the SAP Integration Solution Advicory Process is the identification of integration domains (ISA-M).

The definition of integration domains is shown in Figure 6-2.

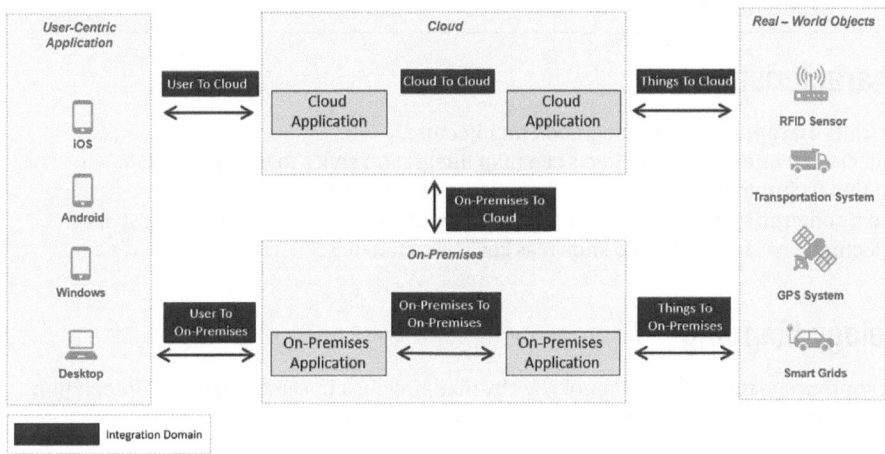

Figure 6-2. *Integration Domain architecture*

6.1.2.2 Integration Methodologies and Use Case Patterns

An integration style identifies a fundamental category or kind of integration, such as process integration or integration with things. You can improve the integration styles relevant to your organization's integration strategy by using use case patterns.

Some examples of integration style use case patterns are shown in Table 6-1.

Table 6-1. *Integration Style Use Case Patterns*

Integration Style	Use Case Patterns
Process integration	A2A integration B2B integration Master data integration
Data integration	Extract, transform, and load (ETL) scenarios need for data replication Data virtualization Data orchestration
Analytics integration	Embedded analytics Cross-application analytics
User integration	UI integration Mobile integration Chatbot integration
Thing integration	Analytical thing Processable item The data lake thing

6.1.2.3 Key Characteristics

Verify the important traits that apply to your integration architecture.

Enterprise architects and integration architects can map integration styles to the appropriate capabilities of integration technologies by using key characteristics as criteria.

The criteria used by enterprise architects and integration architects to connect integration styles to relevant integration technology capabilities are known as key characteristics.

6.1.2.4 Technology Mapping

Determine the most appropriate integration technologies by mapping the chosen collection of integration domains, integration methods, and use case patterns, as shown in Figure 6-3.

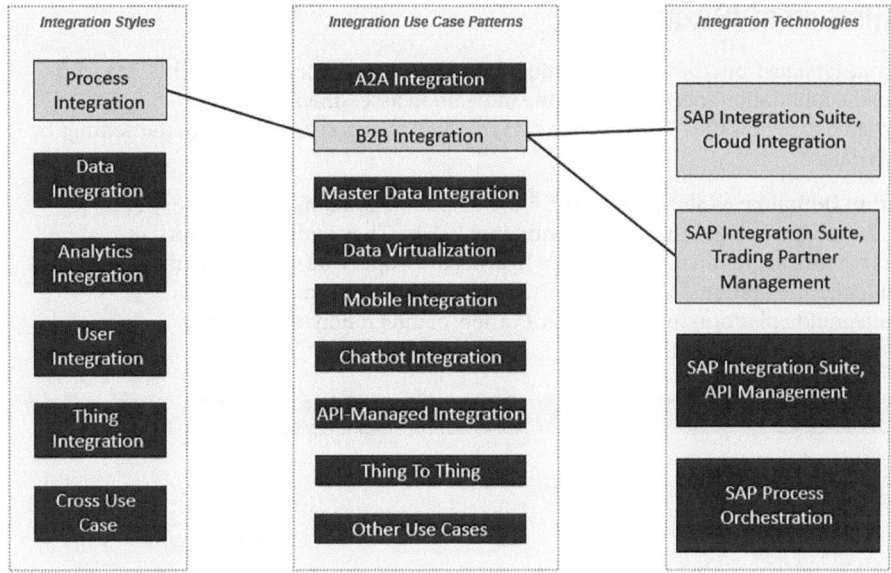

Figure 6-3. *Technology mapping architecture*

6.1.3 Integration Assessment Process

Create a strategy for your integration technology and then govern it.

The SAP Integration Assessment gives you the resources to choose the best integration technology plan.

The general procedure for evaluating your integration strategy with SAP Integration Assessment is shown in Figure 6-4.

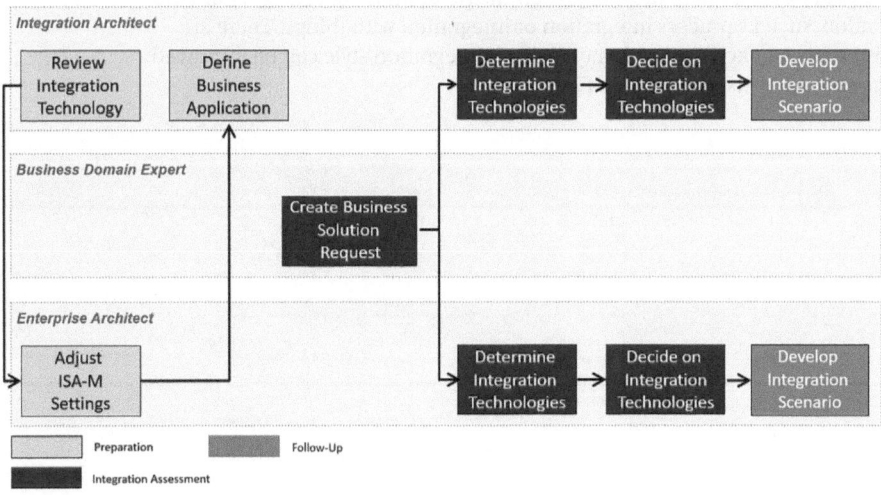

Figure 6-4. *The architecture of the Integration Assessment process*

6.1.3.1 Configuration of ISA-M

Settings contain the master data from the SAP Integration Solution Advisory Methodology (ISA-M). This information serves as the foundation for carrying out the integration assessment.

Visit the Settings area to review the ISA-M master data that SAP has preset or to change the settings by introducing your own data.

- **Integration Domains**: As shown in Figure 6-5, integration domains serve as the "big picture" for integration and the point of entry into ISA-M. They outline common locations where integration is required in a hybrid landscape. Due to their technical independence, integration domains can also assist in developing a design for a hybrid integration platform that combines a variety of integration services and technologies (SAP or non-SAP).

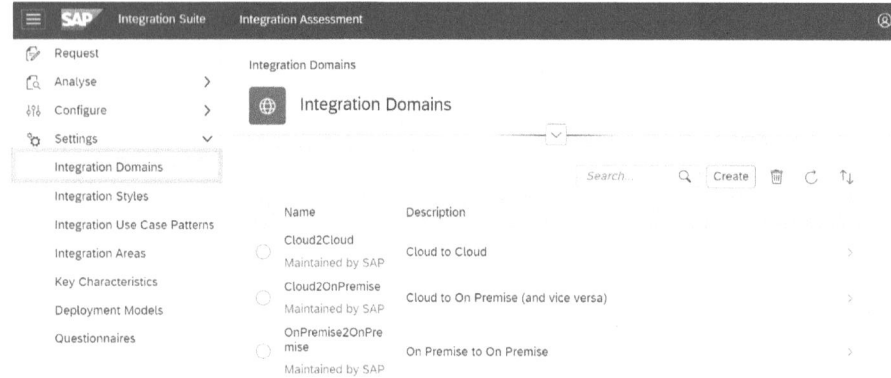

Figure 6-5. *Integration Domains*

- **Integration Styles**: An integration style identifies a fundamental category or kind of integration, such as process integration or integration with things. There are distinctive traits for each integration type. Each integration style can be improved using use case patterns, as shown in Figure 6-6.

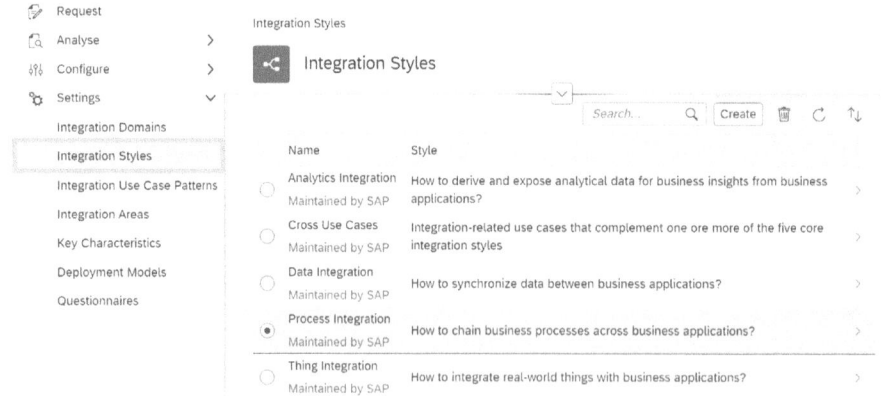

Figure 6-6. *Integration Styles*

- **Integration Use Case Patterns**: Using use Integration case patterns, as shown in Figure 6-7, which outline commonly occurring integration use cases in business landscapes, you may refine each integration style. The integration use case patterns relevant to your company or those you want to analyze further can be added to your integration architecture to assess it.

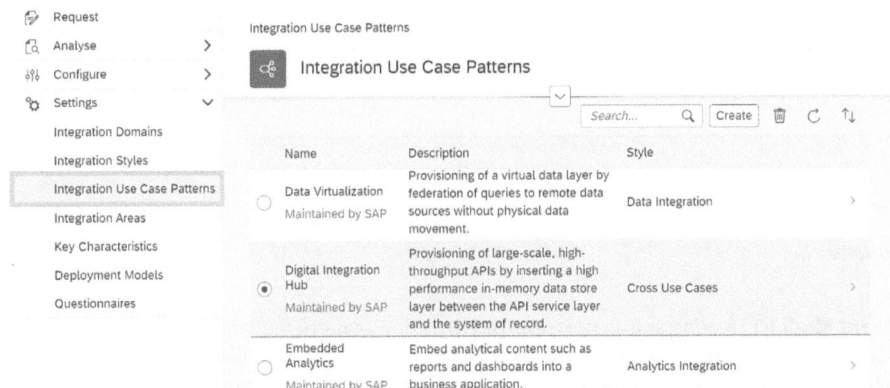

Figure 6-7. *Integration Use Case Patterns*

- **Integration Areas**: The combination of integration domains and styles is an integration area. For instance, the IP-3 integration pattern specifies the pairing of the process integration style and the cloud-to-cloud domain, as shown in Figure 6-8.

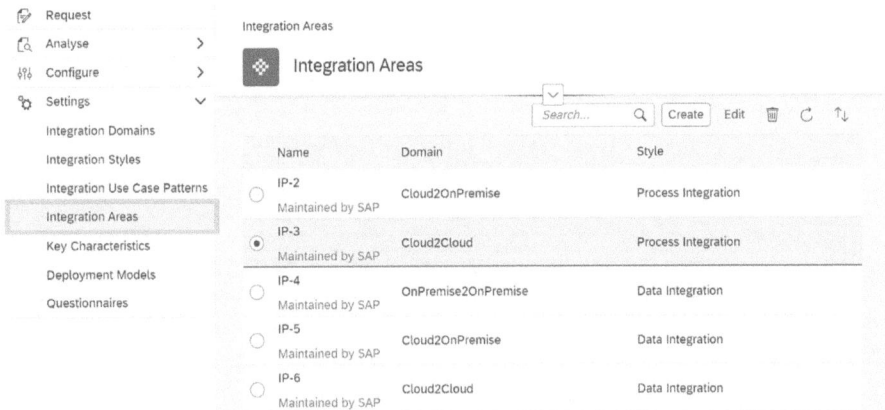

Figure 6-8. *Integration Areas*

- **Key Characteristics**: The criteria used by enterprise architects and integration architects to connect integration styles to relevant integration technology capabilities are key characteristics. The level of support for each technology's essential characteristic is specified by SAP Integration Assessment when you check the relevant integration technologies, as shown in Figure 6-9.

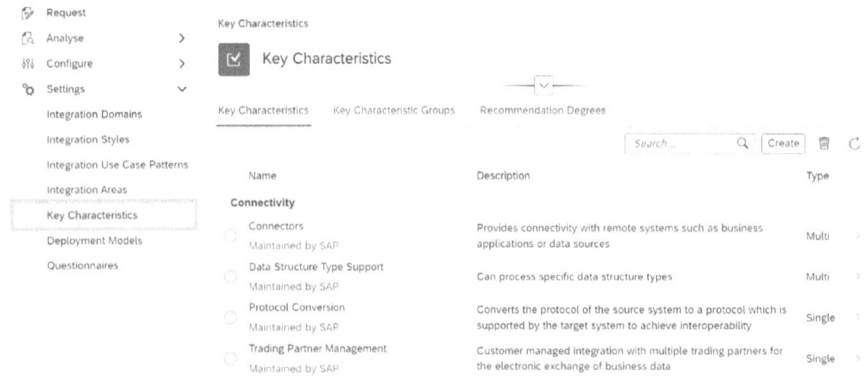

Figure 6-9. *Key Characteristics*

- **Deployment Models**: Hosting software is referred to as deployment. Several methods can be used to deliver cloud services. The integration and application technologies influence the deployment model. Select the Domain Determination tab to determine the integration domain with which the combination of source and target deployment models is associated, as shown in Figure 6-10.

Figure 6-10. *Deployment Models*

6.1.3.1.1 Questionnaires

A collection of questions is used as a questionnaire throughout the request procedure. The SAP Integration Assessment distinguishes between standard SAP surveys and unique questionnaires. There are two categories of surveys: business solution request questionnaires and interface request questionnaires (from a technical standpoint) (business standpoint).

The questionnaire covers your integration requirements. The questions are organized and focused on the important qualities that have been identified.

The interface request procedure is founded on questionnaires. They are either based on the functional scope or the ISA-M context.

You can record the requirements for the interface systematically by using customizable questionnaires. The system makes intelligent technology recommendations using the findings.

6.1.3.1.2 Create Questionnaires

A questionnaire in SAP Integration Advisor is a list of questions that aids the user in setting up an integration scenario. It is an effective tool for learning more about the configuration-required integration scenario.

The SAP Integration Advisor's Questionnaire Designer tool can be used to design a questionnaire. Users have the option to construct unique surveys that are suited to their integration scenarios. Users can add questions, input fields, and other components to the questionnaire using the drag-and-drop interface provided by the questionnaire designer.

The following steps create a questionnaire.

1. Open your Integration Assessment from the Integration Suite.

2. Navigate to the Settings tab from the left Menu panel. Select **Questionnaires**.

3. Choose **Create**, as shown in Figure 6-11.

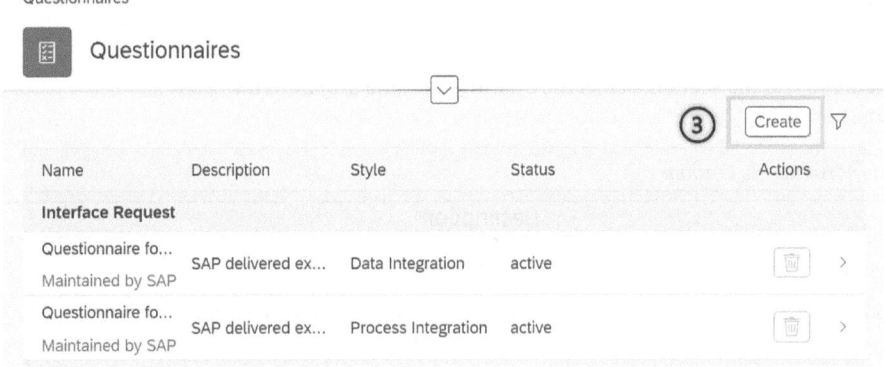

Figure 6-11. *Questionnaires*

4. Complete the Name and Description fields in the Create Questionnaire dialog box, and select Business Solution Request or Style Request in the Request Type menu. Select **Create**, as shown in Figure 6-12.

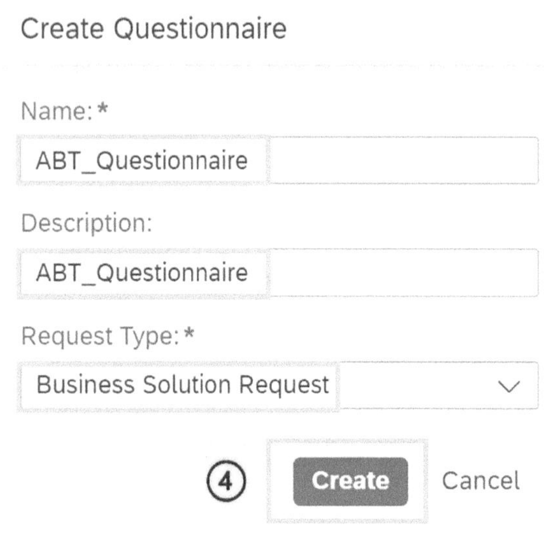

Create Questionnaire

Name: *

ABT_Questionnaire

Description:

ABT_Questionnaire

Request Type: *

Business Solution Request ⌄

④ **Create** Cancel

Figure 6-12. *Create Questionnaire*

5. Click **Create** to add the section. Select the option (see Table 6-2) from the drop-down box.

Table 6-2. *Add Sections Available Options*

Option	Description
Create Question Section	A fresh section
Copy Question Section	An option to replicate a piece from a ready-made questionnaire
Add Attachment Section (if you choose Style Request as the request type)	Includes a section for attachments
Add Integration Content Section (if you choose Style Request as the request type)	Creates a section for content for integration

The Add section, described in Table 6-2, is shown in Figure 6-13.

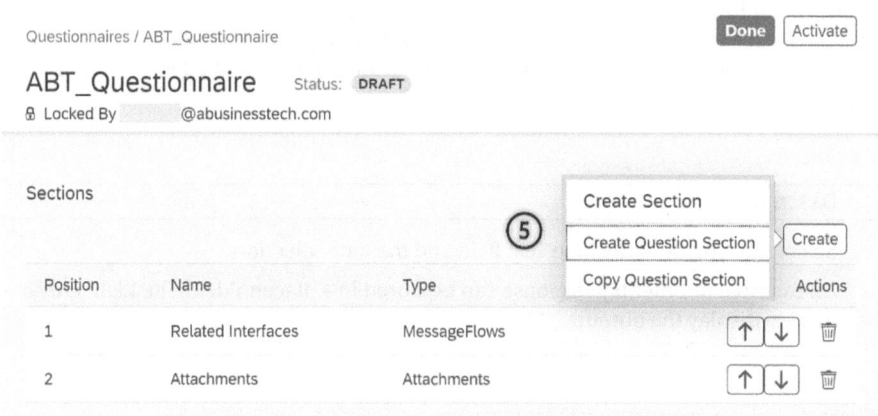

Figure 6-13. *Create Section*

6. Select the **Create Question Section** option to add a question to the section (Create). You can write new questions or copy current ones from a prepared questionnaire. Click on Create New Question as shown in Figure 6-14, Provide the new question name and Answer source.

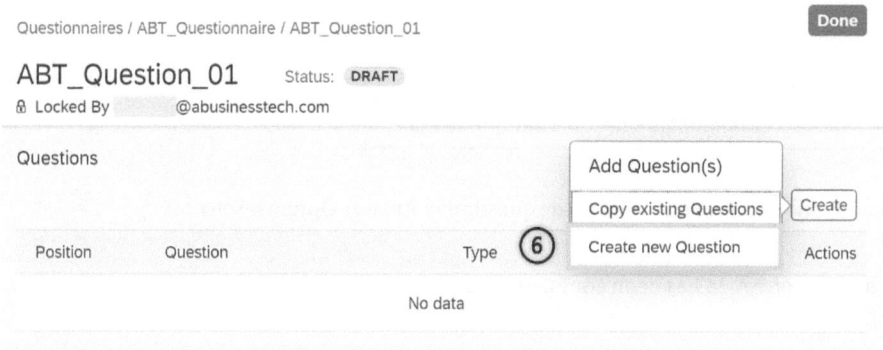

Figure 6-14. *Create New Question*

7. Table 6-3 describes the options for answer sources.

Table 6-3. *Answer Source Descriptions*

Answer Source	Description
Free Text	Free text can be entered in the box provided.
Key Characteristic	An important attribute provides the question's potential solutions.
Custom Options	The type is either single select or multiple select.

8. The question's definition is completed by revising the question and specifying whether it is required or not. You must specify several settings depending on the response source you've chosen. For free text, specify the parameters described in Table 6-4.

Table 6-4. *Free Text Parameters*

Parameter	Description
Type	Select the type from the Free Text field and the date selection.
Placeholder	An example or potential response can be stored in a placeholder field. Light gray is used to display the output.

- For Key Characteristics, you must specify parameters, as described in Table 6-5.

Table 6-5. *Key Characteristics Parameters*

Parameter	Description
Type	Select either single select or multiple select.
Key Characteristic	Decide the essential quality.

- For Custom Options, you must specify parameters, as described in Table 6-6.

Table 6-6. *Custom Options Parameters*

Parameter	Description
Type	Select either single select or multiple select.

9. You can add unique answer choices to the question in Answer Options. Select Activate.

Once the Configuration of ISA-M has been completed, you can evaluate the integration technology, as discussed in the next section.

6.1.3.2 Evaluate Integration Technology

Evaluate, modify, and add instances of associated integration technologies.

You can modify the setup to use your own integration technologies or use SAP's standard integration technologies (such as SAP Integration Suite and Cloud Integration).

1. Choose **Configure ➤ Integration Technologies** from the menu in the Integration Assessment screen. The list of available technology profiles can be found in the Technology Profile tab. You can create and customize your own technology profile or select from the existing SAP standard content.

2. Select Create. Enter the details shown in Table 6-7 in Create Technological Profile.

Table 6-7. *Create Technological Profile Attributes*

Attribute	Description
Name	Name the technology.
Vendor	Name a vendor.
Domains	Choose one or more domains.
Styles	Choose a single integration style or a few.

3. From the list, choose the integration technology. Choose Edit to add extra technology-specific details.

4. The tabs described in Table 6-8 allow you to specify various attributes.

Table 6-8. *Attributes*

Tab	Settings
General	The following attributes can be specified in this tab. • Technology instances: Provide the name of the instance and the deployment strategy. • Provide an attachment and its name and URL.
Domains and Styles	Choose domains and styles using this tab. When developing the technology, certain settings were already established.
Key Characteristics	Use this page to track how well the technology addresses the relevant key characteristics and their values.

The next section explains how to create the business application in SAP Integration Assessment.

6.1.3.3 Create a Business Application

Evaluate, modify, and add instances of connected applications. You can use SAP's default programs (such as SAP ERP) or customize the setup to use other applications.

1. Navigate to **Configure ➤ Applications**.

2. Choose the application profile you want to use from the Application Profile tab.

3. To create the application, select **Create**, as shown in Figure 6-15.

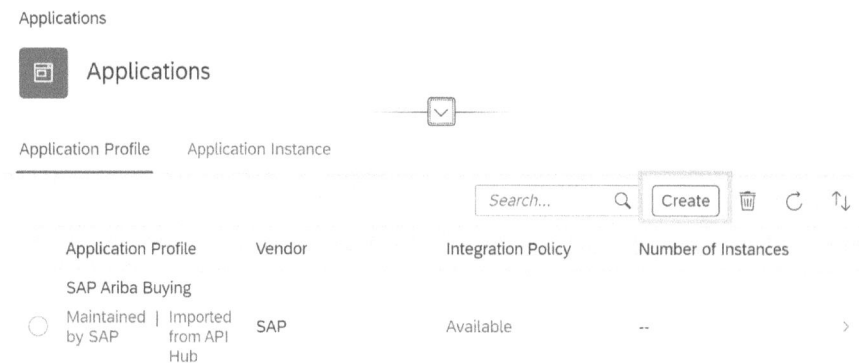

Figure 6-15. *Applications*

4. From the list of preconfigured application profiles, choose the desired profile. Click Create Application Profile and provide information to build a user-defined application profile.

5. Click Edit on the recently generated application profile to provide information about the unique technology.

6. Provide any other specifics about the custom technology.

You can create the integration policies now that you've created the business application.

6.1.3.4 Create the Integration Policies

There are policies that rule outlines how to utilize integration technology in a specific environment.

You can modify the configuration to utilize the integration technologies in place of SAP's standard integration tools (such as Cloud Integration and SAP Integration Suite).

A hierarchy of technologies is created by specifying each attribute by developing policies. The laws are ranked based on how strongly they are advocated for in accordance with a specific style and domain.

The top-ranked technology is chosen based on the guidelines offered by the regulations for use in particular areas, fashions, and with applications.

1. Click Analyze ➤ Integration Policies to begin. You'll see a list of the available integration technologies in the Integration Technology tab. You can create and customize your own integration policy or select from the existing SAP standard content.

2. To add a fresh integration policy, select Create.

3. In Establish Integration Policy, provide the details shown in Table 6-9.

Table 6-9. *Establish Integration Policy Attributes*

Attribute	Description
Style	Indicates the fundamental integration category or type
Domain	In a hybrid landscape, identifies the location where integration is required
Recommendation Degree	Indicates the degree of the recommendation
Rule	Provides a detailed description of the technology's use case
Applications	Lists the applications that are supported

All the attributes listed in Table 6-9 should be entered into the corresponding tabs, as shown in Figure 6-16.

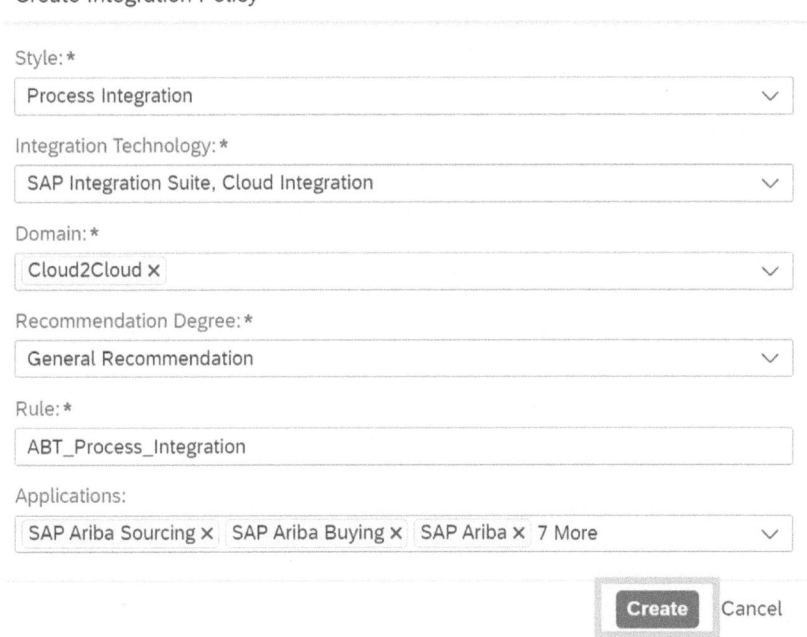

Figure 6-16. *Create Integration Policy*

Once you have created the integration policies, the next step is creating the business solution and interface requests, explained in the next section.

6.1.3.5 Create a Business Solution Request and Interface Request

SAP Integration Assessment requires coordination across many personas.

The business domain expert's responsibility is to develop a request for a business solution.

To create a request for a business solution, follow these steps.

1. Access SAP Integration Assessment by logging in, then go to the Requests area. Select the Business Solution Request tile, as shown in Figure 6-17.

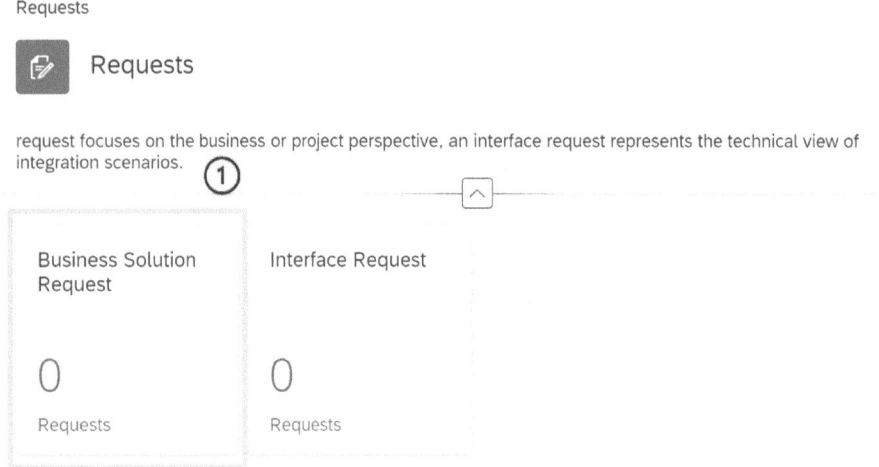

Figure 6-17. *Business Solution Request*

2. To add a new request for a business solution, select **Create**, as shown in Figure 6-18.

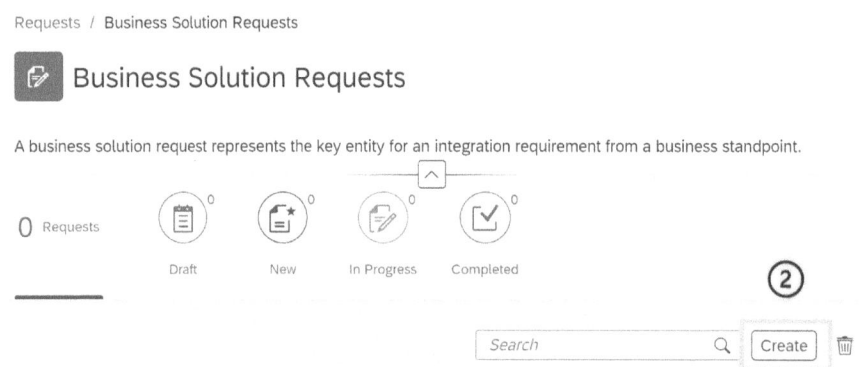

Figure 6-18. *Business Solution Requests*

3. Click **Create** after giving the request a name, as shown in Figure 6-19. The request's status is listed as Draft. Select Edit.

Create Business Solution Request

Name: *

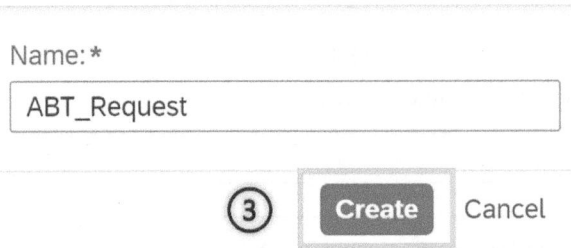

Figure 6-19. *Create Business Solution Request*

4. Fill out the questionnaire in the General tab for your professional background, as shown in Figure 6-20.

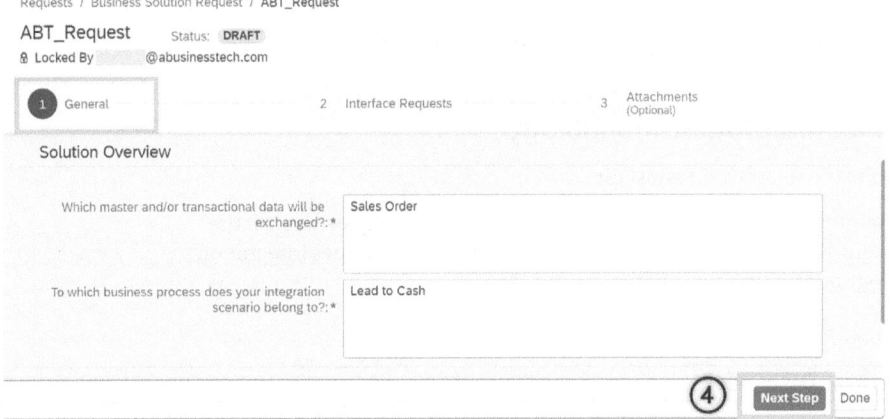

Figure 6-20. *Questionnaire for General tab*

5. Choose + (Create) in the Interface Requests tab to add the interface request information, as shown in Table 6-10.

Table 6-10. *Create Interface Request Options*

Field	Description
Name	Give the interface request a name.
Source Application Instance	Select after choosing a source application instance.
Target Application Instance	Choose after choosing a target application instance.
Style	Choose the integration style.
Use Case Pattern	Choose a use case pattern.

6. For your settings to be saved, select Add.

7. Hit the **Submit** Button.

8. The status of the interface request is now Requirement Analysis.

9. The status values shown in Table 6-11 can apply to interface requests.

Table 6-11. *Status Values for Interface Requests*

Status	Description
Requirement Analysis	When you complete the interface request questionnaire
Technology Selection	The moment you choose the technology
Completed	After the technology is established

10. When the technology has been chosen, the Business Solution status changes to Completed.

The next section explains how to define the integration technology.

6.1.3.6 Define an Integration Technology

1. Navigate to the Interface Request tile.

2. Select **Requirement Analysis**. Enter your responses to the questions on the subsequent screens, then select your action. Table 6-12 describes integration technology questions.

Table 6-12. *Integration Technology Questions*

Field	Description
Integration content	A scenario for integration
Connectivity	Describes the technology, adapters, business applications, and protocols
API design and management	Offers sender and receiver interface structure and connectivity with any additional business applications
Data format and validation	Provides the data format for the sender and recipient exchange in your integration scenario
Data Orchestration	Describes the aggregate messages, sequence, and routing rules
Security	Provides specifics about digital encryption
Monitoring and Operations	Describes the message and error processing process in full

3. Choose **Review** once you've completed providing your answers to the questions.

4. Analyze your responses and select Choose Technology from the menu.

5. Technology Selection is now the requested interface's status, as shown in Figure 6-21.

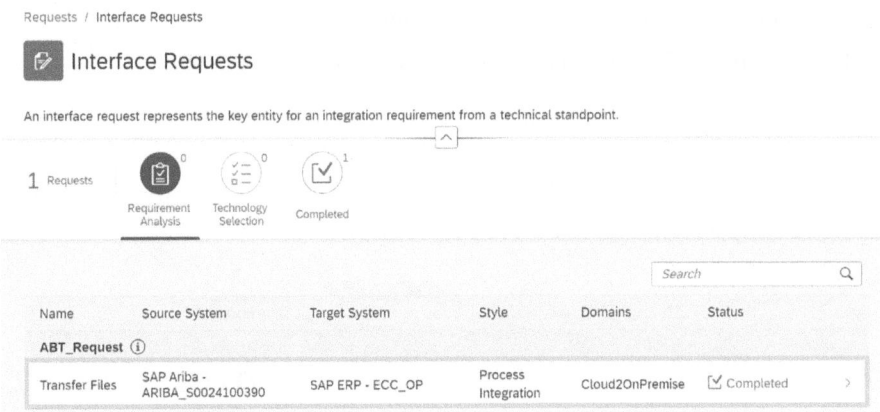

Requests / Interface Requests

Interface Requests

An interface request represents the key entity for an integration requirement from a technical standpoint.

Name	Source System	Target System	Style	Domains	Status	
ABT_Request ⓘ						
Transfer Files	SAP Ariba - ARIBA_S0024100390	SAP ERP - ECC_OP	Process Integration	Cloud2OnPremise	☑ Completed	›

Figure 6-21. *Interface Requests*

The next section covers deciding on the integration technology.

6.1.3.7 Decide on an Integration Technology

Evaluate the interface request results based on the parameters and the survey. Choose the technologies you want.

6.1.3.7.1 Complete the Interface Request's Technology Selection

You submitted an interface request, and it now has the Technology Selection status. The following explains how to complete the interface request's technology section.

1. Select **Technology** on the summary page of your interface request by going to that page.

2. The information in Table 6-13 is for each technology, as seen on the Edit Results page.

Table 6-13. *Edit Result Page Information*

Status	Description
Coverage	Percentage of key characteristics covered when combined with the chosen alternative technologies
Policy Rule	Displays the rule applicable to the chosen coverage (This field explains how to apply this technology to your situation.)
Policy-Based Recommendation	Shows the degree to which this technology is advised (type of advising)
Application Policy	Calculation selects highest-ranked associated integration policy

3. Evaluate the information, select the instance and technology, and click OK.

4. The suggested technology list is customized for your settings.

5. Choose **Submit**. The request's status is now Finished, as shown in Figure 6-22.

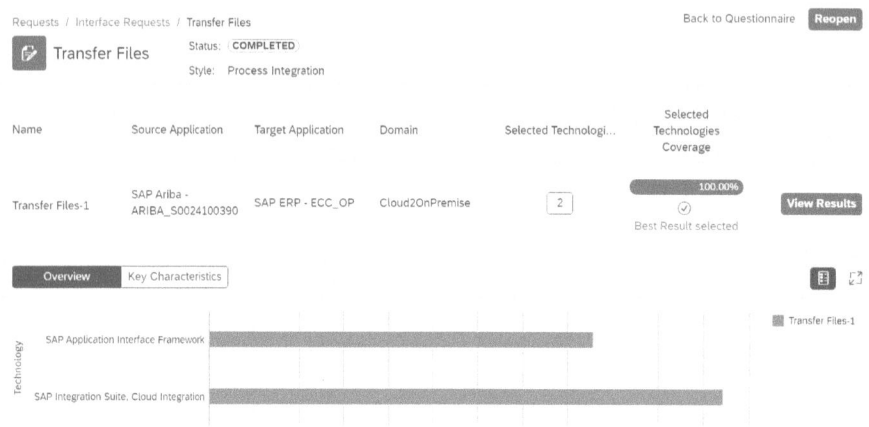

Figure 6-22. *Completed integration request*

6.1.3.7.2 Analyze Integration Patterns

The following steps analyze the integration pattern in SAP Integration Assessment.

1. Go to Analyze Integration Patterns to view the supported integration patterns.

2. The integration domain and the integration combinations that apply to the company are shown.

3. A color-coded system is used to identify integration styles.

4. Each combination has a status that is shown in Table 6-14. The integration areas are shown in Figure 6-23.

Table 6-14. *Combination Status*

Status	Meaning
Not Applicable	Combination cannot be used
Not Assigned	Applied for but not assigned due to some business considerations
Applied	Combination has been assigned

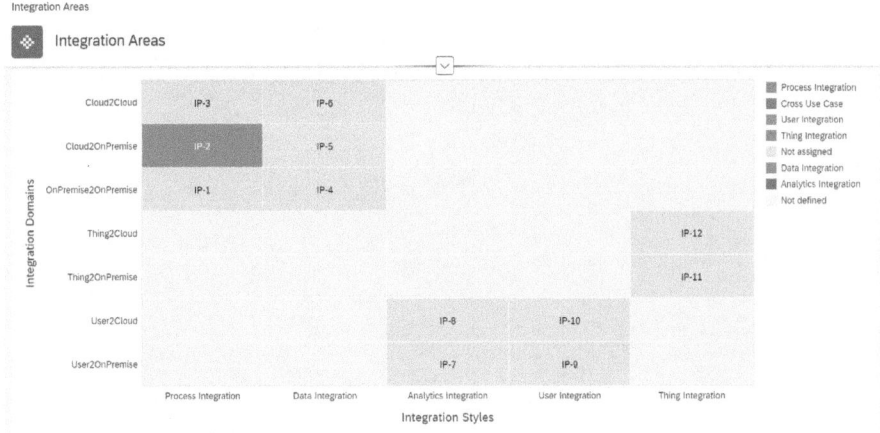

Figure 6-23. *Integration Areas*

6.1.3.7.3 Analyze Integration Policies

The following explains how to analyze and inspect the integration policies.

1. Go to Analyze Integration Policies to view and modify the supported integration policies.

2. You can provide integration rules for your company that cover tactics you'd like to steer clear of, standard recommendations, and acceptable alternatives.

3. These characteristics are for each integration policy, as shown in Table 6-15.

Table 6-15. *Analyze Integration Policies*

Guideline	Usage
Integration Technology	The technology's integration name
Recommendation Degree	Levels of recommendation
Rule	Where technology is employed
Supported Domains	Backed-up domains
Applications	Supported software

The analyze integration policies tab in SAP Integration Assessment is shown in Figure 6-24.

Figure 6-24. *Integration Policies*

6.1.3.7.4 Analyze Application Overview

Go to Analyze ➤ Applications Overview to view and modify the application.

The overview demonstrates if each collection of sources and targets is appropriate for the IT environment. The status is shown in Table 6-16.

Table 6-16. *Analyze Applications Overview Status*

Status	Meaning
Connection is there	The target and source applications are connected.
Connection is not there	The target application and the source application are disconnected.
Not applicable	Source and target applications cannot define a connection.

The Applications Overview tab is shown in Figure 6-25.

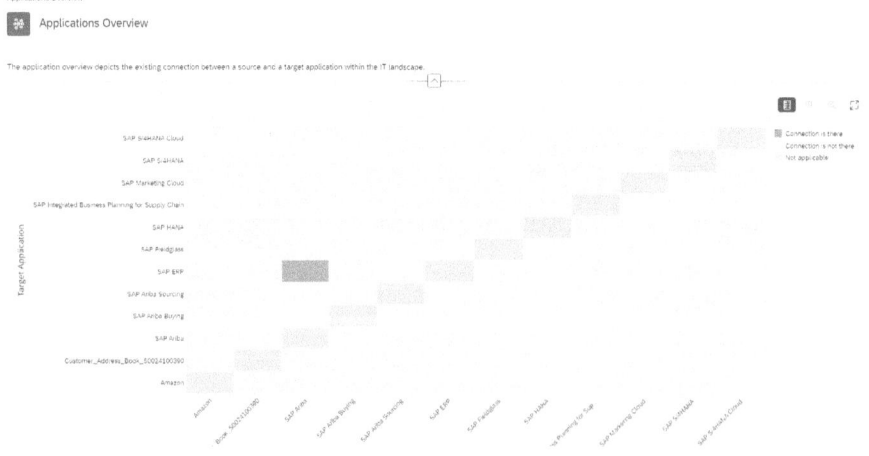

Figure 6-25. *Applications Overview*

6.1.3.8 Information Life Cycle

Add new versions of standard content to your dataset.

The most recent version of SAP standard content is accessible at onboarding. You can update to the most recent version as soon as a new version of SAP standard content is released, depending on your company's needs.

The integrated dataset is updated each time a fresh SAP version is available. The subsequent version of the material is produced each time the current version is selected or deselected. The records are not shown for editing after an update is completed and the most recent version is produced.

- **Versioning**: You can't just upgrade to the most recent version when there are multiple versions of the standard content. You can only move up to the next version.

- **Hierarchical Data**: The dataset has a hierarchical parent-child connection (also called *change execution sets*). The child object should always be considered when attempting the parent data object. There may be several child nodes under one parent node. Two categories of child nodes exist.

- **Error Scenario**: If an error occurs when updating, the update fails. Everything takes place in one transaction. Either it was updated correctly, or it wasn't. A displayed red icon shows the failure's specifics and cause. As a result, adequate action can be taken.

6.1.4 Security

The security-related features of SAP Integration Assessment are discussed in this part, along with the precautions you may take to safeguard the client information that SAP Integration Assessment uses during the integration assessment procedure. You may get an overview of all the application's security-related features in this part and details on data protection and privacy, access and identity control, and identity management.

6.1.4.1 Types of Data Storage

Integration Assessment maintains master data for the SAP Integrated Solution Advisory Methodology (ISA-M), including SAP-predefined master data, customer-adapted master data, and data related to business solutions and interface requests.

The following is a full list of all the data types that Integrated Assessment stores.

- Integration domains

- Integration styles

- Integration areas

- Applications

- Integration technologies

- Business solution requests

- Interface request

SAP Integration Assessment security was covered in Chapter 2.

6.1.5 Summary

This chapter was an overview of the features and methodologies used in assessing and selecting integration solutions for SAP systems. You learned about integration domains, integration methodologies and use case patterns, essential characteristics, technology mapping, initial setup, integration assessment processes, security, and types of data storage.

The chapter also explored security considerations and types of data storage. It explains the importance of data security and the different types of data storage available for SAP systems.

In the next chapter, you learn about another important SAP Integration Suite capability: SAP Open Connectors.

CHAPTER 7

■ ■ ■

SAP Open Connectors

In today's interconnected world, it is essential for businesses to integrate and connect with different systems to ensure seamless data exchange and automation. SAP Open Connectors is a cloud-based platform that enables easy and secure integration with hundreds of applications, systems, and data sources using prebuilt connectors, custom connectors, and APIs.

This chapter covers all the essential aspects of SAP Open Connectors, including its terms and resources, system references, connectors, formulas, common resources, and non-SAP connectivity options. It provides a comprehensive guide to working with connectors, creating custom connectors, authenticating connector instances, validating connector instances, and setting up OAuth proxies.

This chapter also explains how to create formula templates, execute formula instances, and access formula resources using the FaaR feature. It also delves into non-SAP connectivity, such as connecting with cloud integrations and SAP SFTP servers using cloud connectors.

7.1 Overview of SAP Open Connectors

With a data-centric approach, SAP Open Connectors is designed from the ground up to combine API administration and integration.

The developer experience is unified across all different applications and services thanks to the platform and special connectors. SAP Open Connectors create a uniform API layer and standards-based implementation across any environment, regardless of the application's back end (REST, SOAP, proprietary SDK, database, etc.). Doing this ensures that integration users—developers—are independent of the back-end services they rely on.

With SAP Open Connectors, you get quick access to a set of diverse APIs, formulas, common resources, and connectors. More than 150 out-of-the-box connectors are available. They can be consumed/exposed as APIs for your business processes.

7.1.1 Terms and Resources

There are several SAP Open Connectors terms that you should know.

- **Definition**: A single or many APIs can contact several API providers using an API hub, which provides consistent APIs to access a variety of services. For example, the customer relationship management (CRM) hub provides uniform access to many CRM application services, including those from Salesforce, Microsoft Dynamics, NetSuite CRM, Sugar CRM, Zoho CRM, and Autotask. By writing to *one* standard API, your application can write to *many* application services that fall within that category. Accounts, Contacts, Leads, Opportunities, and other CRM hub resources are only a

© Jaspreet Bagga 2023
J. Bagga, *Introduction to Integration Suite Capabilities*, https://doi.org/10.1007/978-1-4842-9630-1_7

few of the resources for which SAP has standardized the API calls. Hubs are gathering places for resources from different connections in that category of application services.

- **Connector instance**: An authenticated connector instance is a single authenticated connection made between an SAP account and an account with a provider of an API, such as Salesforce, Marketo, or NetSuite. A connector instance that has been approved can access any object, field, and piece of data connected to that account, including any custom data. An approved connector instance is created when a user successfully connects to the endpoint by providing an instance name, the required authentication data for that connector, and optional event configuration.

- **Bulk APIs**: With bulk APIs, you can move enormous amounts of data. Data can be routinely and in bulk downloaded and uploaded from an endpoint with the help of SAP Open Connectors. Employ the bulk endpoints of the provider when appropriate. However, provide a pseudo bulk service when the API provider does not support bulk endpoints for uploading and downloading data from the endpoint. For uploads, start with a file and create objects at the endpoint record by record. Use a query API against the endpoint to retrieve all the data for downloads and then loop through every result.

- **OCNQL**: Open Connectors Query Language, often known as OCNQL, is the search and filtering language SAP Open Connectors uses to simplify searching across all our various connectors. There is a standardized, common approach to search across all our connectors because many APIs provide some sort of search in their APIs, although they are almost all different. OCNQL occasionally offers more than the endpoint, but SAP transforms it to the endpoint's searching syntax.

- **Common resource**: A stand-alone resource defined and created by a user to enable a one-to-many transformation is called a common resource. One-to-many integrations can be facilitated using common resources that describe normalized fields rather than connectors. To construct transformations, you map the common resource fields to the fields from connector resources.

- **Discovery service**: One of SAP's several built-in data discovery services offers normalized metadata, like listing field names and categories. If provided by an endpoint, additional data may also be collected, including read-only, display name, and so on. Even if an endpoint doesn't offer discovery service APIs, SAP offers minimal resource-specific metadata (e.g., name and type). When a native discovery service is unavailable, SAP enables you to find custom fields by providing an object ID. The discovery and transformation services are combined to standardize responses from various endpoints.

- **Connector**: A connection to a particular API provider endpoint is made possible through connectors, which are prebuilt API integrations (e.g., Salesforce, Quickbooks, or Marketo). Authentication, resources, paging, errors, events, and search are among the standard set of capabilities that all connections have at the outset. Even if not all the connections in that category share a particular resource, you want to support the more extensive range of APIs that an application offers at the hub level. In contrast to many other CRM applications, Salesforce Sales Cloud, for instance, supports APIs. These APIs, which are only applicable to Salesforce, are listed in the connector's documentation.

- **Endpoint**: A unique URI that identifies a resource, the HTTP method, and the resource. GET, PUT, and PATCH methods, the resource's base URL, and more are all included.

- **Event**: An activity at an API provider connected to a connector instance that has been authenticated and tracked either by polling or webhooks.

- **Formula instance**: A specific example of a formula template that uses explicitly defined variables and is linked to a specific example of a connector.

- **Formula template**: The variables, triggers, and stages required for a formula instance to complete the workflow are contained in a formula template, which is a repeating workflow that is not reliant on a connection. A formula template's operations enable automating business workflows, migrating data, and maintaining system synchronization, among many other application cases across different services.

- **OAuth proxy**: With the help of the OAuth Proxy feature, you may use one endpoint application to access many environments, including development and QA. For instance, some suppliers limit callback URL usage to one per application. Several application endpoints can use the same callback proxy. Then, rather than using your own callback URL, you would use the proxy address. One callback URL enables many endpoint apps.

- **Payload**: Data obtained because of an API request. Alternatively, an API request's body.

- **Resource**: A object or entity that can be reached with a URI request.

- **Transformation**: Mapping a resource from an API provider to a common resource produces a transformation. This enables you to specify the fields that should be included in a particular resource, map those fields to those that the provider resource's fields include the same data, and then convert the data to fit the common resource.

7.1.2 Platform Reference

To better understand SAP Open Connector, let's dive into the platform.

7.1.2.1 User Profile

Your profile contains your fundamental details, including your name and email. It also contains vital details like your password and user secret to log in with SAP Open Connectors, as shown in Figure 7-1. You can update your user secret and password on your profile.

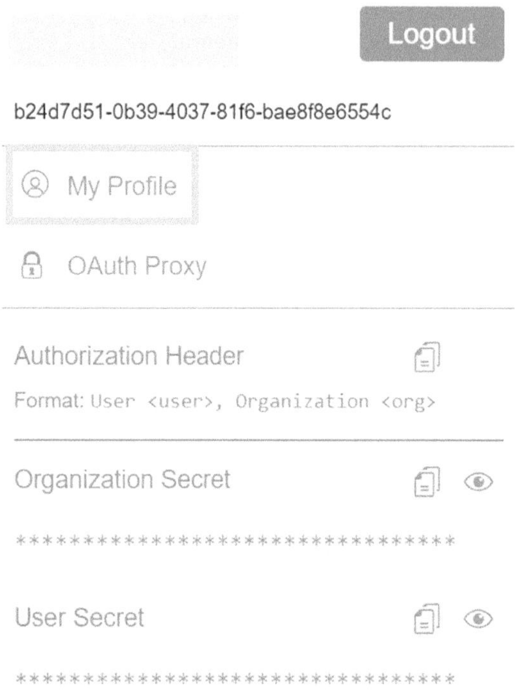

Logout

b24d7d51-0b39-4037-81f6-bae8f8e6554c

(8) My Profile

🔒 OAuth Proxy

Authorization Header   ⬠

Format: User <user>, Organization <org>

Organization Secret   ⬠ 👁

****

User Secret   ⬠ 👁

****

Figure 7-1. *User profile*

You can edit the following details on the My Profile page.

- First name
- Last name
- Email
- Login password
- Authorization method

7.1.2.2 Authentication

When you authorize SAP Open Connectors, each transaction must include your organization key and user secret as headers. Instead of an organization secret, you likely require a connector instance token when working with connector instances.

You are provided with a user secret and an organization secret when you open an account. A user (/user) is a member of an organization, whereas an organization is a customer account for SAP Open Connectors (/organizations). SAP Open Connectors accounts are represented by user and organization secrets.

Open the profile menu to access your organization and user secrets, as shown in Figure 7-2.

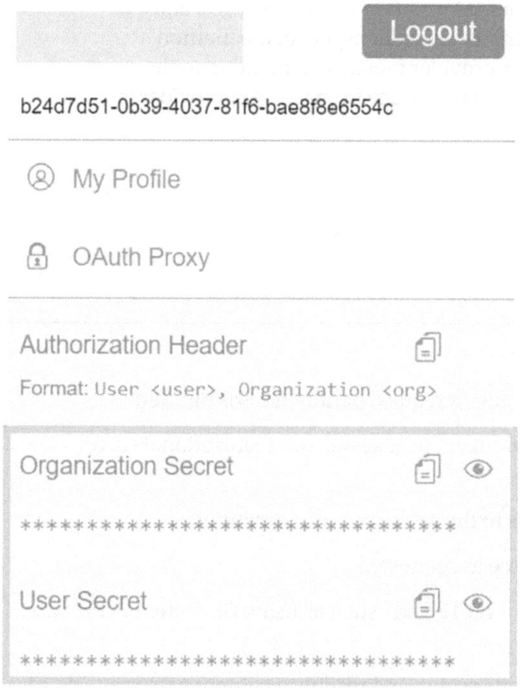

Logout

b24d7d51-0b39-4037-81f6-bae8f8e6554c

Ⓐ My Profile

🔒 OAuth Proxy

Authorization Header

Format: User , Organization <org>

Organization Secret

****

User Secret

****

Figure 7-2. User and organization secret

7.1.2.3 Base URL

A connector instance token, your organization and user secret, and HTTP Custom authentication are used to secure HTTP queries to the REST API. Most HTTP clients can understand basic HTTP features, such as HTTP verbs. All API replies, including errors, are delivered in JSON. The APIs employ HTTP response codes to signal API problems and have foreseeable, simple URLs.

7.1.2.4 Error Codes

There are three types of error codes.

- A **platform error** may be one of the following.

  - Response codes for HTTP requests generally fall into one of four categories: informative (100), success (200), redirection (300), and error (400).

  - requestId is a unique code that identifies the request. When you are troubleshooting, you can give this to support.

  - The message gives a narrative account of the request's outcome.

  You only receive a request ID and message if the problem comes straight from SAP Open Connectors. In this case, the HTTP Response code is transmitted by SAP Open Connectors.

- An **API provider error** is given with any further information in the message's body if a responder provides a message that contains a particular aspect of the method call. Error replies from API providers present a provider message in addition to the response code, request ID, and message. The error message from that API provider is what is in the provider message.

- **HTTP status codes** are described in Table 7-1.

Table 7-1. *HTTP Status Codes*

Response Code	Error	Description
200	OK	OK, no errors
400	Bad Request	Frequently because a request parameter was omitted
401	Unauthorized	False connector token, user secret, or organizational secret supplied
403	Forbidden	Provider access to the resource is not permitted
404	Not found	No such resource as requested
405	Method Not Allowed	Wrong use of an HTTP verb, such as using GET when POST was expected
406	Not Acceptable	The Accept header value and response content type are incompatible
409	Conflict	The existing resource being produced
415	Unsupported Media Type	Content type not supported by the server
500	Server Error	A problem occurred on the server

7.1.3 Connectors

Using a collection of prebuilt API connections, a connector in SAP Open Connectors enables users to connect to and interact with a particular system or application. It is a conduit for transferring information between SAP Open Connectors and the external system.

Connectors can be set up with the login information, security preferences, and other requirements needed to connect to the external system. After a connector is set up, it offers a set of API endpoints that may be used to carry out different tasks on the external system, including data retrieval, record updating, and entity creation.

Many prebuilt connectors for well-known platforms and programs, including Salesforce, Microsoft Dynamics, NetSuite, and others, are available through SAP Open Connectors. The SAP Open Connectors SDK enables users to interact with nearly any system offering an API by allowing them to build unique connectors.

7.1.3.1 Connectors Catalog

Lists of the API providers that are accessible using SAP Open Connectors are provided in the Connectors Catalog. You can create your own connector if the one you require is unavailable. Figure 7-3 shows the elements present in the SAP Open Connectors Catalog tab.

Figure 7-3. *Connectors Catalog*

1. Do a name search for connectors.

2. Create a unique connector.

3. Use connectors' connector cards to communicate with them. You can export the connector, add the resources to an existing connection, export an instance with the API provider, or view the API documentation.

4. Count the number of authenticated connector instances you used to connect to the API provider.

5. The connector's hub, beta status, and custom connector status are all shown via labels on the connector (private).

The Connectors Catalog displays a list of connection cards representing every connector offered to you. At the top of the inventory are connectors that already have authenticated connector instances. Each card displays the connection's name, ID, number of instances verified as valid, and the hub to which the connector is connected. Figure 7-4 shows the information on the tile in the Catalog tab.

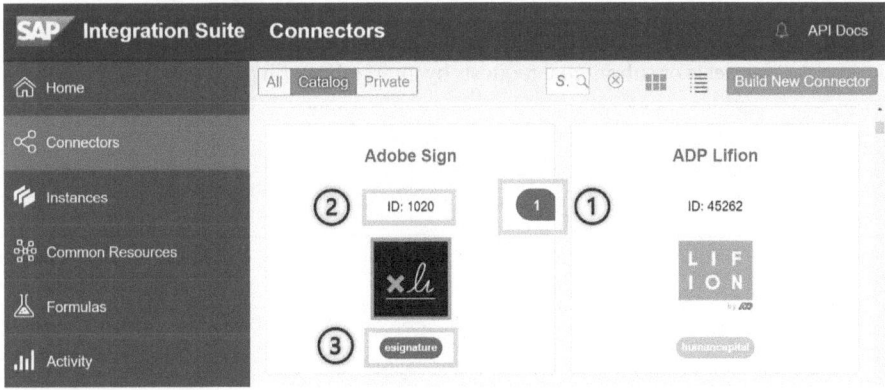

Figure 7-4. *Connectors Catalog description*

1. Number of authenticated instances

2. Connector ID

3. Hub

Hover your mouse over a connector card to start working with it. You can add your resources to a connector, authenticate a connector instance, view the connector's API documentation, and export the connector in JSON format.

Also, you can create unique connectors. They have a private label since they are only accessible to users inside your company.

7.1.3.2 Working with Connectors

Working with connectors in SAP Open Connectors enables businesses to streamline their workflows, reduce manual efforts, and improve efficiency by enabling data to flow seamlessly between different applications and systems. This allows businesses to leverage the strengths of different systems while still maintaining data integrity and security. SAP Open Connectors functions are covered in the upcoming sections.

7.1.3.2.1 Introduction to Connectors

Connectors are feature-rich, ready-to-integrate connectors for each cloud app you require.

- Connectors share common features including logging, interactive documentation, bulk download and uploading, bulk uploading and searching, and pagination.

- RESTful APIs provide access to normalized methods.

- You update every connector with changes made at the endpoint.

- Every connector is a multitenant connector that can accommodate an infinite number of authorized accounts without extra coding.

7.1.3.2.2 View Connector API Docs

You may learn about the available API requests for each connector by consulting the API documentation, as shown in Figure 7-5. Each request's description, a description of each field that can be used to fill out the request, and a list of necessary fields are visible. There are two views of API documentation: the standard API documentation and the API documentation linked to a particular instance. If you choose an authorized instance, you can test your API requests or submit real requests by interacting with the documentation.

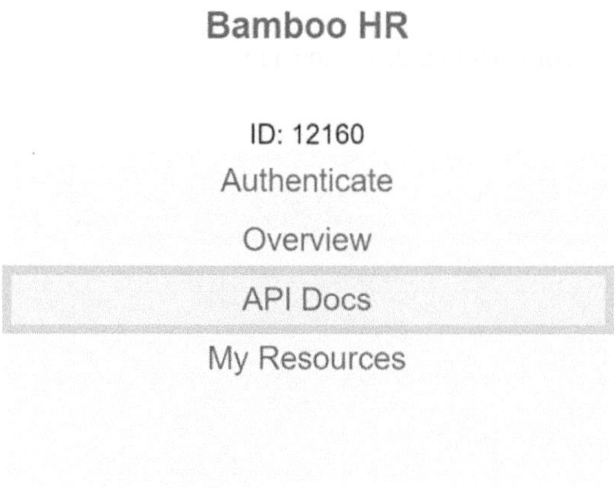

Bamboo HR

ID: 12160

Authenticate

Overview

API Docs

My Resources

Figure 7-5. API Docs

Using the API documentation, submit queries as follows.

1. On the left, pick an authenticated instance.

2. The endpoint to which you want to send a request should be expanded.

3. Click **Try It Out**.

4. Include any extra or necessary details.

5. Choose **Execute**.

7.1.3.2.3 Import/Export of Connectors

Connectors can be exported in JSON format from the Connectors Catalog. This enables you to copy open connectors or move your connectors between environments.

OData version 4, OData version 4, JSON, Swagger, SOAP, and Postman 2.1 files can all be used as imported connectors. A connector can be modified, have resources added to it, and have additional features like bulk and events configured when it is imported.

The following explains how to export the connector.

1. Click **Overview** after moving your cursor over a connector card, as shown in Figure 7-6.

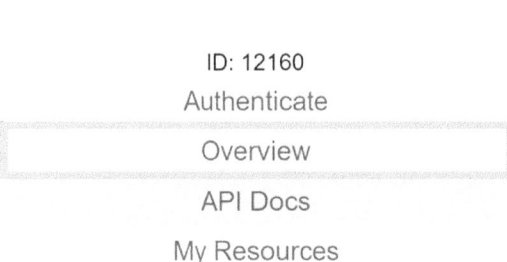

Figure 7-6. Overview

2. To download the connector as a JSON file to your computer, click **Export** on the Overview page, as shown in Figure 7-7.

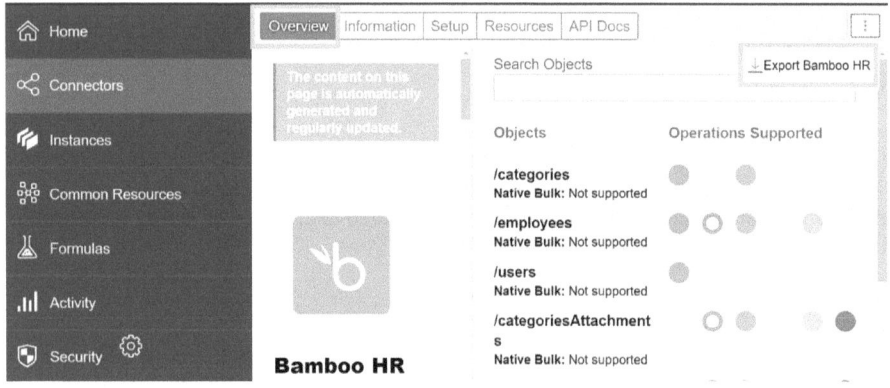

Figure 7-7. Export connector

The following explains how to import the connector.

1. Click **Create New Connection** on the Connectors page.

2. Click **Import** after carefully reading the Builder page's instructions, as shown in Figure 7-8.

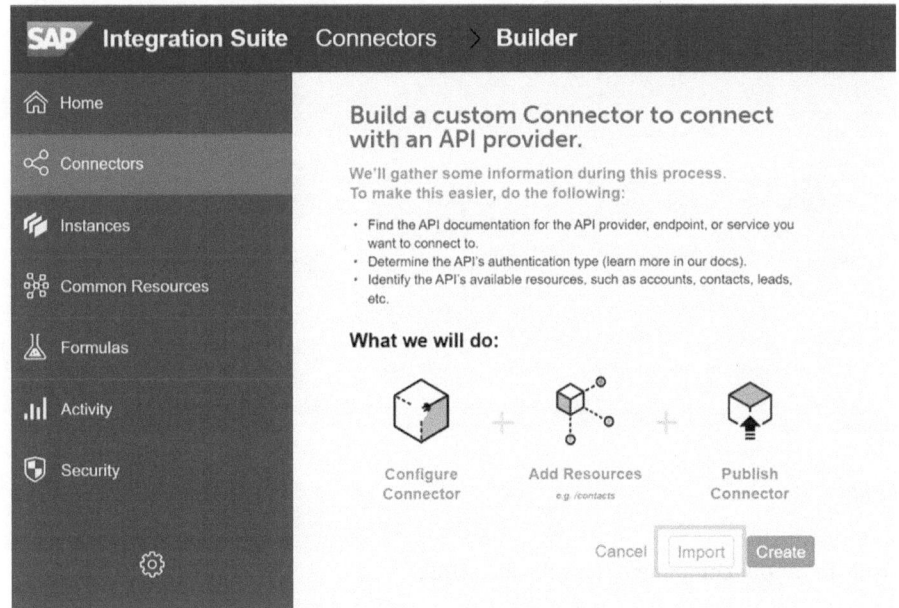

Figure 7-8. *Import connector*

3. Choose the import file type, as shown in Figure 7-9.

- Connector: A connector exports as a JSON file

- Swagger: An API's swagger specification file (OAS 2.0/openAPI 2.0)

- SOAP: A SOAP API description in an XML file

- OData v4: An Open Data Protocol (ODP)–compliant file

- Postman 2.1: A file that contains a Postman collection of requests

Import Connector

Select an import option

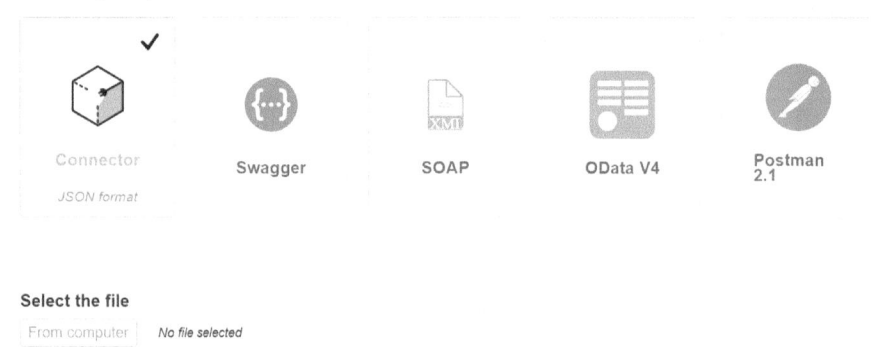

Connector
JSON format

Swagger

SOAP

OData V4

Postman
2.1

Select the file

| From computer | No file selected |

Figure 7-9. *Select the file*

4. Choose a file from your computer or type the file's URL.

5. Choose **Next**.

6. Choose the content you wish to import.

7. Click the **Import** button.

7.1.3.2.4 Authenticate the Connector Instance

In order to test API requests and get a sense of the capabilities offered by the connector, authenticate a connector instance using the user interface (UI).

1. To find the connector, first log in to SAP Open Connectors and then use the Connectors Catalog.

2. Authenticate by clicking after hovering on the connector card, as shown in Figure 7-10.

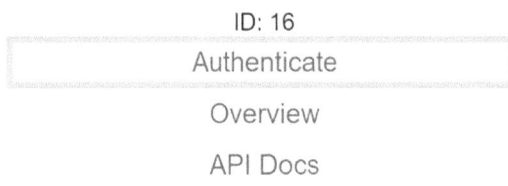

Amazon S3

ID: 16

Authenticate

Overview

API Docs

Figure 7-10. *Authenticate the connector instance*

3. Give the connector instance a name.

4. Any connector-specific information should be filled out.

5. You can choose or type one or more tags to add to the authenticated connection instance.

6. Click **Create Instance**.

7.1.3.2.5 Connector Instance Authentication Using Events (UI)

An Overview of the procedure for event-based connector instance authentication is given in this section. APIs or SAP Open Connectors are two ways to authenticate.

7.1.3.2.5.1 Connector Instance Authentication Using Polling

Events must be enabled and a few more parameters must be specified in order to authenticate a connector instance using events.

1. In accordance with the generic instructions in Authenticate a Connector Instance or the connector-specific document, enter the fundamental data necessary to authenticate a connector instance.

2. Click **Show Optional Fields**, and add a value to Callback Notification Signature Key to enable hash validation in the headers of event callbacks.

3. Activate events: Switch Events Enabled on.

4. Choose the event kind in Event Type if the connection supports both polls and webhooks.

5. If you want to select how frequently SAP Open Connectors should check for changes, use the Event poller refresh interval (mins) slider or enter a value in minutes.

6. Choose the resources you want to poll.

7. Optionally add one or more connector instance tags by typing them in or choosing them from a list.

8. Click **Create Instance**.

7.1.3.2.5.2 Connector Instance Authentication Using Webhooks

Events must be enabled and a few more parameters must be specified in order to authenticate a connector instance using events.

1. Enter the connector instance's name and any configuration information required for the connector instance to be authenticated.

2. Click **Show Optional Fields**, enter a key in Callback Notification Signature Key, and then click Enable Hash Verification to allow hashing verification in the header of event callbacks.

3. Activate events: Switch Events Enabled on.

4. Choose webhooks as your event type if the connector allows both polling and webhooks.

5. To receive information regarding the events, add an event notification callback URL.

7.1.3.2.6 Connector Instance Authentication Using API

Depending on the API provider's authentication criteria (OAuth 2.0, basic, OAuth 1.0, or custom) and any unique information they require, how the authentication of a connector instance with the APIs varies greatly from connector to connector. The payload you send, however, always contain the fundamental data, such as the name of the connector instance, a configuration array to pass the provider the data, tags to help you classify the instance, and an optional request to retrieve information about the objects retrieveObjectsAfterInstantiation.

Any of the APIs that can be used to authenticate a connector instance.

- POST /instances
- POST /elements/{id}/instances (do not include "key": "Element Key" in the body of the request).

7.1.3.2.7 Connector Instance Authentication Using Events

Authenticating a connector instance with events securely authenticates and manages access to API endpoints in the SAP systems. This method uses event-driven architecture to provide enhanced security and control over API access, ensuring that only authorized users and systems can access sensitive data and functions.

7.1.3.2.7.1 Using Polling to Authenticate a Connector Instance

Events must be enabled and a few more parameters must be specified in order to authenticate a connector instance using events.

The polling configuration must be added to the JSON body of your POST /instances request in order to authenticate a connector instance with polling events.

The following event and polling configuration settings should be added to the configuration JSON object when a connector instance is authenticated with polling events.

- event.notification.enabled: true
- event.vendor.type: polling
- event.notification.callback.url: <YOUR_CALLBACK_URL>
- event.notification.signature.key: <OPTIONAL_SIGNATURE_KEY>
- event.poller.refresh_interval: <NUMBER IN MINUTES>
- event.poller.configuration: <POLLING_CONFIGURATION>

7.1.3.2.7.2 Using Webhooks, Authenticate a Connector Instance

Events must be enabled and a few more parameters must be specified in order to authenticate a connector instance using events.

The webhook configuration must be added to the JSON body of your POST /instances request in in order to authenticate a connector instance with webhook events.

The following event and webhook configuration settings should be added to the configuration JSON object when connector instance is authenticated with webhook events.

- event.notification.enabled: true
- event.vendor.type: webhooks

- event.notification.callback.url: <YOUR_CALLBACK_URL>

- event.notification.signature.key: <OPTIONAL_SIGNATURE_KEY>

7.1.3.2.8 Information on Authenticating Connector Instances

A connection between one user and an API provider when they have an account is represented by an authenticated connector instance. In SAP Open Connectors, a connection instance is created when an API Provider is authenticated. Apply formulas to the connector instance to apply logic to that though, map the connector instance to a common resource, and test the instance in the API docs by using the authorized connector instance.

You grant SAP Open Connectors access to your data at the API provider when authentication is done through SAP Open Connectors. Use the /instances API if you need to grant access to your own application for an API provider.

7.1.3.2.8.1 Authenticate a Connector Instance

Every API provider is unique. A connector instance can be verified using the instances API or SAP Open Connectors. Because each connector differs from the others, SAP Open Connectors offers thorough documentation for each one. The connection documentation provides additional details regarding bulk, querying, and events in addition to instructions on how to prepare for authentication, authenticate using SAP Open Connectors or the instances API.

7.1.3.2.8.2 View Connector Instance Information

Each authenticated connection instance can be recognized by its distinct connector instance ID and connector instance token. To refer to an instance in formulas and scripts, use the connector instance ID. In order to establish trust with an API provider, SAP Open Connectors use the connector instance token. In the header of any API queries you make to the connector, you must include the connector instance token.

Your connector instance token and ID can be found as follows.

- A connector card's instances banner should be clicked.

- At the top of the card, take note of the connector instance ID.

- See the token by clicking after hovering on the connector instance card.

7.1.3.2.8.3 Test Connector Instance

You can test a connector instance using the API documentation after authenticating it. You may test every single one of the requests in the API documentation as soon you've got an authenticated connector instance.

While evaluating your connector instance, do the following.

1. On a connector card, click the instances banner.

2. After clicking API Documentation, hover over the connector instance card.

3. You should enlarge the endpoint to which you wish to send a request.

4. Choose Try it Now.

5. Provide any additional or necessary information.

6. Choose **Execute**.

7.1.3.2.8.4 A Connector Instance Update

You can modify your connection instance or request a new API provider authentication. You can update the connector without re-authenticating if you are only making a few minor changes. But, you must re-authenticate if you want to add events or change a configuration parameter.

Update a connector instance that has been authenticated.

1. A connector card's instances banner should be clicked.

2. Click **Edit** after moving your cursor over the connector instance card.

3. After making your changes, click **Update**.

The following describes how to re-authenticate an authorized connector instance.

1. Click a connector card's instances banner.

2. Click **Edit** after moving your cursor over the connector instance card.

3. After making your modifications, click **Re-Authenticate**.

7.1.3.2.8.5 Delete

You can permanently remove a connection instance from SAP Open Connectors by deleting the instance and its connector instance token.

Getting rid of a connection instance

1. A connector card's instances banner should be clicked.

2. Click Delete after moving your cursor over the connector instance card.

3. Verify the delete instance.

In the next section, you create custom connectors.

7.1.3.3 Create Custom Connectors

You can create your own connector for your APIs using the Open Connectors from the SAP Integration Suite. You may have already grouped the individual API queries using Postman collections to test or try out an API. You may now create a custom connector using Connector Builder's most recent innovation by importing your current Postman collections.

SAP API Business Hub offers a bunch of APIs that allows users in many different countries to access data on regional companies and customer reviews.

1. Open your SAP Cloud Connector. Navigate to the Connectors section. On the top-right side, click **Build New Connector**, as shown in Figure 7-11.

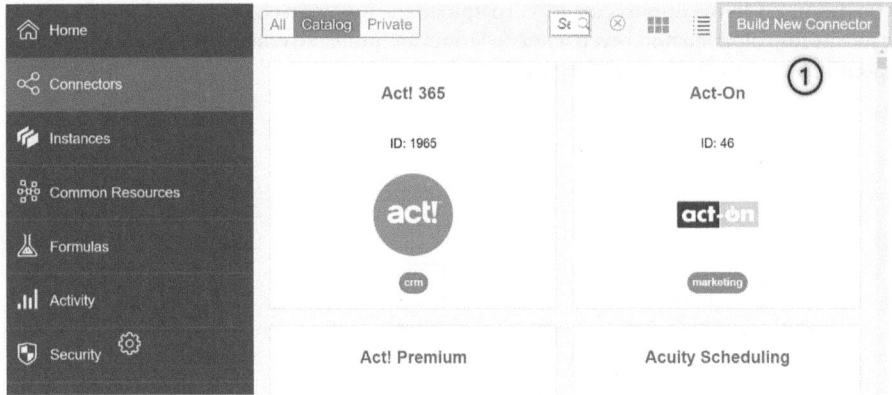

Figure 7-11. Build New Connector

2. To construct a custom connector using the Postman collection, select **Import**, as
   shown in Figure 7-12.

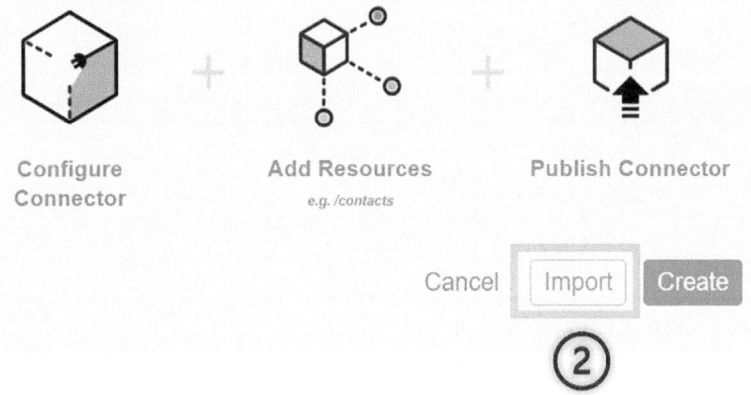

Figure 7-12. Import Custom Connector

3. Choose **Postman 2.1** in the import category. To upload the postman collection that was locally downloaded and saved, choose From Computer. To continue, choose Continue Import, as shown in Figure 7-13.

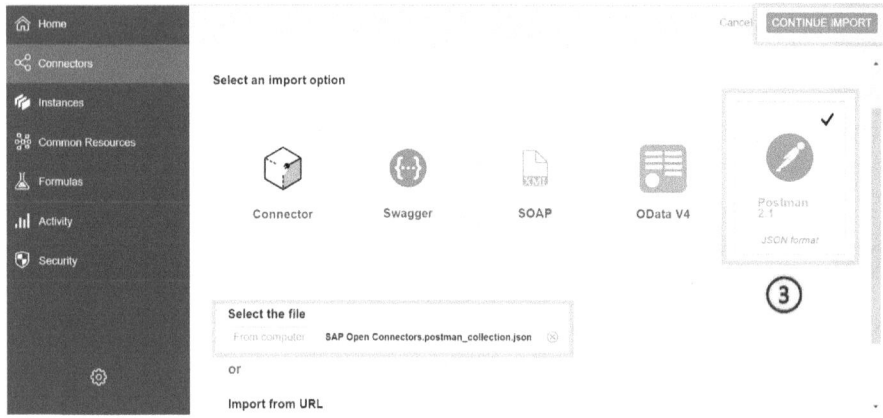

Figure 7-13. *Import Postman collection*

4. Resources would automatically be produced according on the imported Postman collection. The resources you want to use to create the connector can be chosen. Choose **Import** to create the connector after selecting Import All Resources to select every resource, as shown in Figure 7-14.

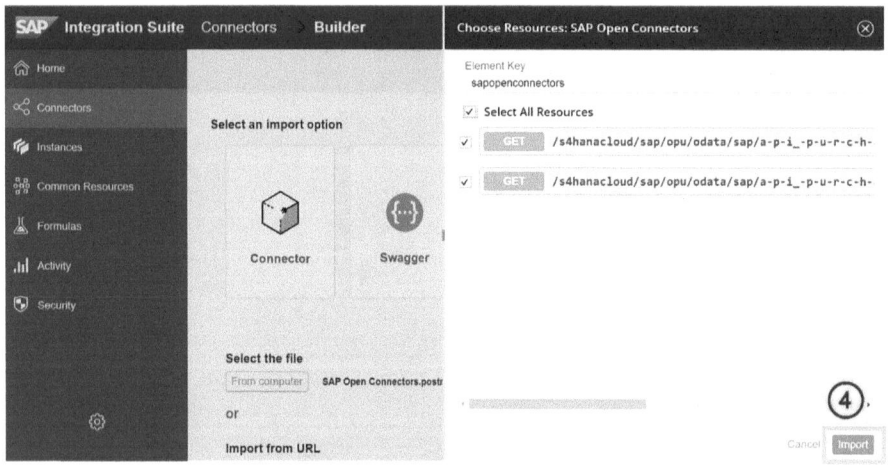

Figure 7-14. *Import available APIs in Postman collection*

5. Based on the imported Postman collection, the connector is generated, as shown in Figure 7-15.

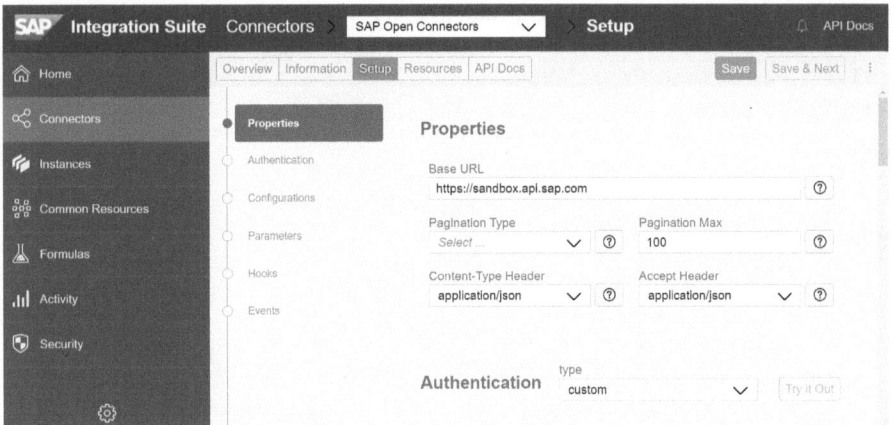

Figure 7-15. Information page

7.1.3.3.1 Properties

On the Setup page, after you've finished filling out the connector details, customize the properties. The base URL, information about pagination, and header formats are all properties.

1. Enter the URL for the base URL, which is used for all API requests. Resources that you add are positioned in relation to the Basic URL. The base URL is typically mentioned in the Introduction, Getting Started, or Overview sections of API providers' websites.

2. Choose the pagination method that the API provider supports in Pagination Type.

3. Enter the maximum number of returned records that the API provider support in Pagination Max. Pagination Max is referred to as a limit by some API providers. Pagination or Paging information can be found in the API documentation.

4. Choose the media type that the API provider anticipates is requested in the Content-Type header.

5. Configure response body media type that you anticipate from the API provider in the Accept header.

7.1.3.3.2 Authentication

Depending on the kind of authentication, different information must be provided when setting up an account with the API provider. Configurations, options, and hooks can be used to override the default data if the API provider demands complicated authentication.

SAP offers the attributes required to support a typical OAuth 2.0 flow. You might need to add more configuration to the parameters because each API provider supports OAuth 2.0 differently. You must first build an SAP app at the API provider before configuring the OAuth 2.0 details. Use the app's default information. Users log in to that app once they have authenticated through SAP.

There are the following options available for authentication, as shown in Figure 7-16.

- OAuth2
- OAuth2Password
- JWTOAuth
- OAuth1
- OAuth2ClientCredentials
- AWSv2
- AWSv4
- Basic
- Custom
- WSSecurity
- Tis

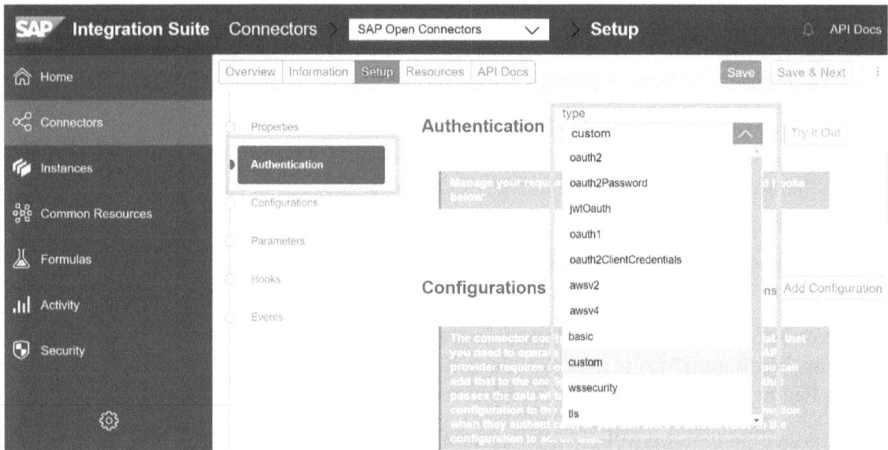

Figure 7-16. *Authentication*

7.1.3.3.3 Configuration and Parameters

In order to address the many problems that API providers bring, connector setup and parameters must operate together. The data that you wish to save with the connector is specified when you set up the connector setup. When a user authenticates a connection instance, you can collect data from them as part of the information. Moreover, variables that you can control using parameters and hooks can be stored in the configuration. Also, you have the option of storing data that you must send to the API provider with each request.

With the help of parameters, you may specify the data you must include in each request to an API provider and how they should receive it. Variables that are added to the settings, user-provided data, particular values, and more can all be sent.

The following explains how to configure a system.

1. Click **Add Configuration** after navigating to Configuration.

2. Type in the configuration's name. SAP Open Connectors display the configuration name exactly as you define it here if you choose to display this in the user interface.

3. You have the option of updating or leaving the value that key automatically generates from the configuration name. The connector configuration's configuration attribute is identified by the configuration key. The configuration key is also used in parameters and hooks to refer to the configuration.

4. Choose the configuration type in Type. Passwords, booleans, or text strings can be used as configurations.

5. Enter any default value in Default for the setting. A user may change the default value if it is displayed in the user interface.

6. Provide a brief description of the setting in the Description field. The description is accessible as hover help if the configuration is displayed in SAP Open Connectors.

7. Switch on Required to make configuration a prerequisite for authentication.

8. To stop the settings from showing up on the UI after user authentication, turn on Hidden UI. The setup shows up on the interface by default.

9. Click **Save**, as shown in Figure 7-17.

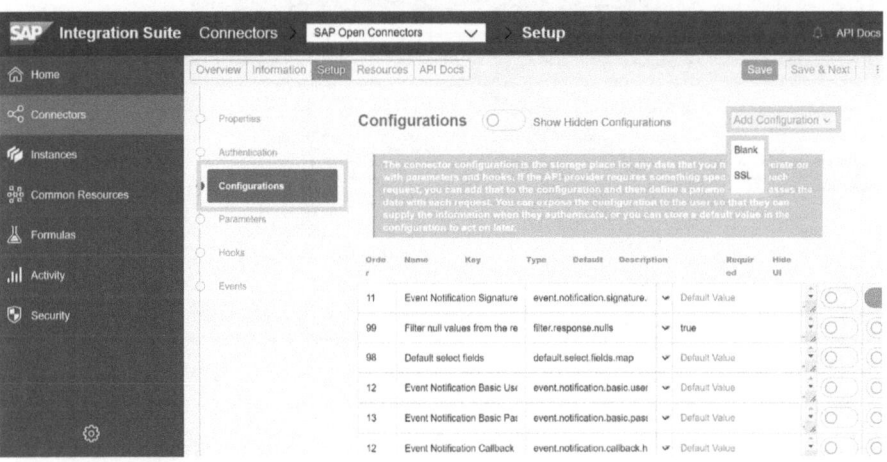

Figure 7-17. *Configuration*

The following explains how to specify the parameters.

1. To add a parameter, click Configuration and then Add Parameter.

2. Fill out the parameter information in the top row of SAP Open Connectors.

   - Type the parameter's name in the Name field. The name should correspond to any pre-existing values as it is listed in the API documentation. For example, the name of the configuration key that it refers to must match the type of configuration you selected.

   - Choose the parameter's source in Type.

   - If you wish to change the default workflow so that the top row of parameters represents the API provider's answer rather than SAP Open Connectors' request, choose Source and then choose Response.

   - Give the parameter a brief description in the Description field.

3. Fill out the API provider parameter field in the bottom row as follows.

   - The parameter's name should be mapped to the name of the API provider in the Name field.

   - Choose the parameter delivery method for the API provider in Type. Refer to the Connector Parameter field's API Provider Parameter Type table.

4. Click **Save**, as shown in Figure 7-18.

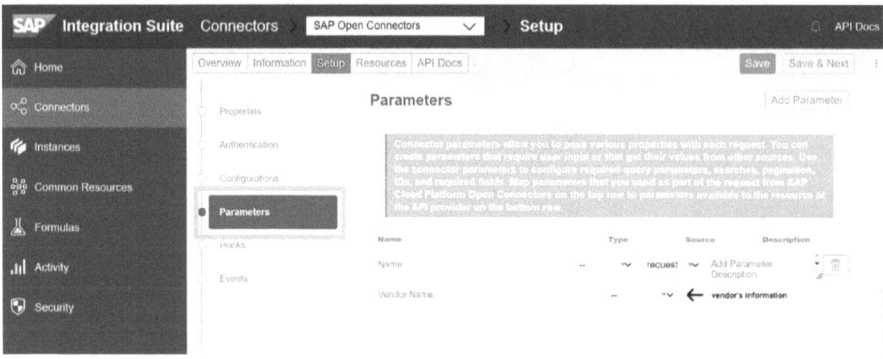

Figure 7-18. *Parameters*

7.1.3.3.4 Hooks

Prerequest hooks allow you to run custom JavaScript code before an API request and after the API provider returns a response (post-response hook). While building a connector, you have the option of using global hooks or resource hooks. Whereas resource hooks only activate on requests made to and responses received from certain endpoints, global hooks activate on each request and response.

To operate on a setting or to modify any portion of a request or response, use hooks. Because the endpoint expects authentication, you might need a hook. You might need to deliver a value to an endpoint, but the data type that is needed is not one that SAP Open Connectors can handle.

Click the PreRequest Hook or the PostRequest Hook to add the snippet of the code, as shown in Figure 7-19.

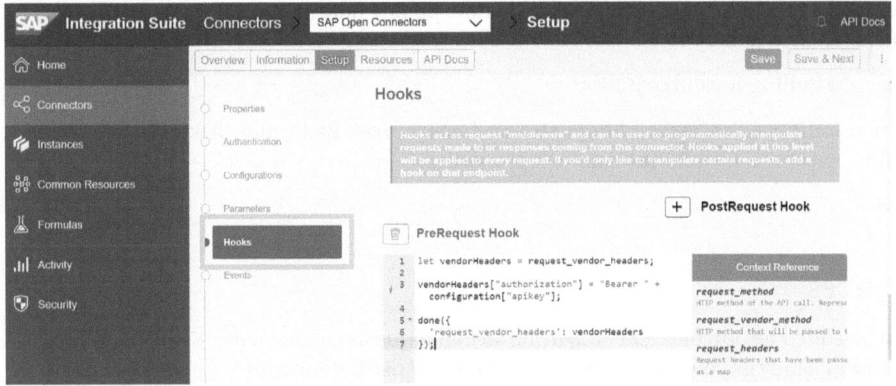

Figure 7-19. *Hooks*

7.1.3.3.5 Events

You frequently want your connector to receive an update whenever an event happens at an API source. For instance, when a user adds a new contact to a CRM connector or when they upload a file to a Dropbox account. Webhooks and polling are the two types of events that SAP Open Connectors offer. By polling, SAP Open Connectors ping the API provider to see whether any changes have been made at predetermined intervals. The API provider alerts SAP Open Connectors via webhooks whenever updates are made. A connector can be configured to support webhooks or polling events.

Configure Webhook Events

Evaluate the API documentation to check if the API provider offers webhooks. Webhooks frequently call for particular setup procedures at the API provider. An API provider sends us webhooks, which you must translate into the expected normalized format. In order to do this, you must use JavaScript in event hook to alter an API provider's answer.

The following explains how to configure webhooks.

1. Open the Events tab on the Setup page.

2. Switch Enable Events on.

3. Go to Event Types and choose Webhooks.

4. To ensure that the data received from the webhook is in the format you anticipate, click Add an Event Hook and use JavaScript to create an event hook as demonstrated in the function reference.

5. Click **Save**.

The following explains how to configure poll events.

1. Open the Events tab on the Setup page.

2. Switch Enable Events on.

3. Go to Event Type and choose Polling.

4. To choose how frequently SAP Open Connectors should check for changes, use the Default Interval Polling Time (in minutes) slider or input a number in minutes.

5. Provide a resource's name that includes appropriate polling data in the Resource Name field (created and updated data).

6. Click the **Add Polling Resource** button.

7. Add extra resources as necessary: After entering the resource's name and clicking Add Polling Resource, fill out the resource's properties.

8. Click **Save**.

7.1.3.3.6 Resources

Your connection can be added resources after configuring so that requests can be made. Keep the API provider's documentation handy as you create new resources. You'll use it frequently.

Depending on the hub, some resources are automatically available. For instance, SAP Open Connectors creates endpoint resources like accounts, contacts, leads, and opportunities if you construct a connector in the CRM hub.

7.1.3.3.6.1 Define New Resource

Multiple steps are involved in defining a resource, but you are not required to finish them all. For instance, only difficult or advanced use cases require the addition of hooks. The resources you intend to download or upload in bulk require configuration for bulk.

When defining resources, start by introducing the resource and giving some general details about it. Provide a description and include any endpoint-level configuration for each endpoint. Add additional arguments to each request sent to each endpoint if necessary. Write the hooks necessary to carry out sophisticated activities for use cases that are more complex. Define the models for each request and response if the method (POST, PUT, or PATCH) requires them.

7.1.3.3.6.2 Add Resource

You can add a resource to make it possible to simultaneously establish several endpoints for that resource. Moreover, you can add each endpoint separately. When resource names match, SAP Open Connectors automatically groups individual endpoints.

The following explains how to add a resource.

1. Select **Add a New Resource**.

2. Determine whether the resource is a child resource by looking for things like / users/{id}/tasks or /contacts/{id}/tasks.

3. Provide the resource's name as you want it to appear in SAP Open Connectors in the field for resource names. For instance, to add a /deals resource to a CRM connector, enter deals. The name you select generates an endpoint in the hub and appears in the API documentation. For instance, the new hubs/crm/deals endpoint allows access to the new GET /deals endpoint, which is documented as GET /deals.

4. Add the path to the resource's location at the API provider to the vendor resource name. Your entered value is immediately added to the base URL. Do not add a slash before the resource name if your base URL ends with a /. Add a forward slash (/) before the resource name if your base URL doesn't already conclude with one.

5. Enter the resource's unique identifier in the Primary Key field. Primary keys are frequently resource-specific ID fields. A main key in this instance could be id.

6. Enter the properties that indicate the created and modified dates in the Created Date Key and Updated Date Key fields. The terms for the generated and modified dates can be created, createdate, timecreated, lastModified, or dateModified.

7. Choose the techniques to add. While configuring the endpoints, you provide the methods you choose. Make certain that the API provider supports the methods you choose.

8. Click **Go**.

7.1.3.3.6.3 Configure Endpoints

Through an editable API documentation format, SAP Open Connectors displays the endpoints connected with each method you selected to configure for the resource once you enter the information to build it. The resource and methods you chose in the previous step were combined to form each endpoint. To customize each endpoint, use the Endpoints tab. You can add or remove endpoints on the Endpoints page if you didn't choose the appropriate methods in the previous step.

The following explains how to set up endpoints.

1. Click **Edit**.

2. Expand the Configuration section.

3. If you wish to change how the API documentation treats the endpoint, change the resource type from API in the resource drop-down menu.

4. Add a Root Key to the resource to identify the top-level field and to control what you submit (Request Root Key) and receive (Response Root Key).

5. To override the connector's default pagination settings, choose Pagination Type. Supported is the default if you make no selections.

6. If your use case demands sending requests to various endpoints one after another, choose a different endpoint in Next Resource.

7. Type a description here. This is stated in the API description and should make it clearer to the user what is being requested and the expected outcome. The GET /deals endpoint's description can be "Brings up a list of offers. Filter using relevant fields like company and contact using OCNQL."

7.1.3.3.6.4 Add Parameters

You can pass multiple characteristics to the endpoint using endpoint parameters. Configure searches, pagination, necessary fields, and IDs using the endpoint options. For most endpoints, you can set both mandatory and optional parameters. For each method, with the exception of delete, SAP Open Connectors offers a set of common default settings.

1. Click **Add New Parameter** if you need to add a parameter.

2. Enter the parameter's name in the Parameter Name field. In some circumstances, the name might be a value supplied to the API provider, or it can be found in the API documentation.

3. Put the name of a parameter you want to map to in Vendor Name. As an illustration, if you are introducing an id parameter, Vendor Name should be the resource's exclusive id field, similar to deal Id.

4. Choose the parameter's source in Parameter Type.

5. Choose the method of parameter receipt for the API provider in Vendor Type.

6. You can choose the parameter's data type in Vendor Datatype and Parameter Datatype.

7. Enter a model's name that already exists if the parameter type is body. To build an endpoint model, refer to Add Models later in this section.

8. Choose Request in Parameter Source to change the default workflow so that the parameters on the left are a part of the request. ID, GET,

9. By turning on Required, you can make a parameter a necessary component of the request.

10. Describe the parameter in detail in Parameter Description. This explanation also displays if the argument is mentioned in the API documentation.

11. Click **Save**.

7.1.3.3.6.5 Add Hooks

Creating a resource hook involves the following steps.

1. Depending on what you need to change, select Add a Post-Response Hook or Add a Prerequest Hook in the endpoint edit pane.

2. For your use case, create the script that is required. For further details on the tools and libraries available, see Custom Hooks.

3. Click **Save**.

7.1.3.3.6.6 Add Models

The following explains how to add a model.

1. Look up the specifications for the request body in the API provider's documentation or get a sample payload.

2. A request model or response model appears.

3. Give the model a name.

4. Provide the details.

5. Click **Save**.

7.1.3.3.6.7 Configure Bulk

Each resource can be configured to support bulk data uploads and downloads in either CSV or JSON formats.

The following explains how to do bulk configuration.

1. Click on the Bulk tab.

2. Choose whether to allow bulk download, upload, or both.

3. Choose the endpoint's preferred format for each field.

4. Click **Save Bulk Configuration**.

7.1.3.3.6.8 Add Endpoints

A resource's new endpoint can be added by following these steps.

1. Click **Add a new endpoint** next to the resource's most recent connected endpoint.

2. Choose a method for the newly created endpoint. The endpoint's SAP Open Connectors side is shown on the left.

3. Enter the URL of the endpoint after choosing the vendor method associated with it on the right side.

4. Click **Go**, then proceed to specify the resource by doing so.

7.1.3.4 Validate Connector Instances

Now, you may simultaneously validate many Connector instances to make sure they all function properly. Formerly, manually validating each instance would have required the user to create an instance first, then use an API to see if it was functioning properly. This process can be completed in seconds by simply clicking a button.

1. Navigate to the Instance page in the SAP Open Connectors.

2. Choose the instances you want to verify.

3. To validate instances, click the button. After choosing at least one instance, this button becomes visible.

4. A progress signal for the instance validation appears.

5. Instances that have passed validation are marked with a green check, while those that have failed are marked with a red exclamation point. Until you re-authenticate the instance, the one that failed validation remains highlighted in red. The green checkmarks only show up while you are still on the page. Only a 30-minute window is used to save instances' validity. After 30 minutes, you must validate those instances again to see if they are still operational.

6. The API Docs page for the connector also displays the failed validation. When you hover over the instance that failed validation, an error message displays, and it is highlighted in red.

7.1.3.5 Set up OAuth Proxies

It is required to obtain the API key and API secret for your application before employing an OAuth proxy with SAP Open Connectors. The website of the application provider has access to that data.

1. Click the Account Profile icon after logging in to your SAP Open Connectors account.

2. Go to See My User Secret and View My Organization Secret, respectively. You need your organization and user secrets later in this setup procedure, so note them.

3. Choose **OAuth Proxy, as shown in** Figure 7-20.

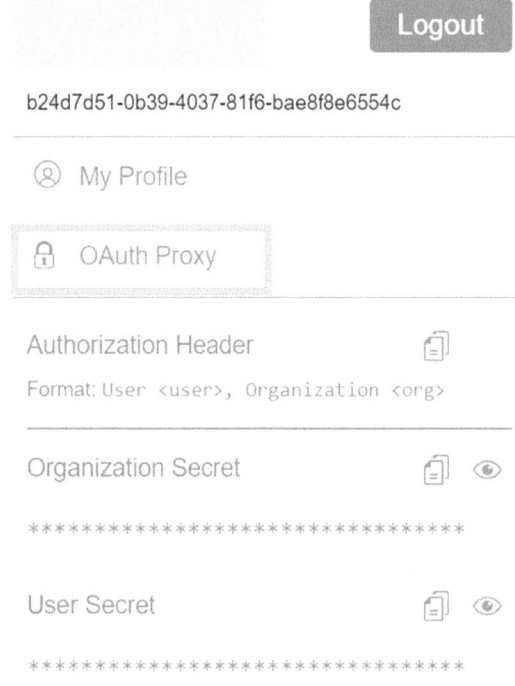

Figure 7-20. *OAuth Proxy*

4. Click the Add button located above the OAuth Proxy list.

5. Fill out the Create page with the following details.

   - OAuth Proxy Name

   - API Key

   - API Secret

6. In the Redirect URL field on the application provider's website, paste the distinct URL produced by SAP. The box is used in this illustration, as shown in Figure 7-21.

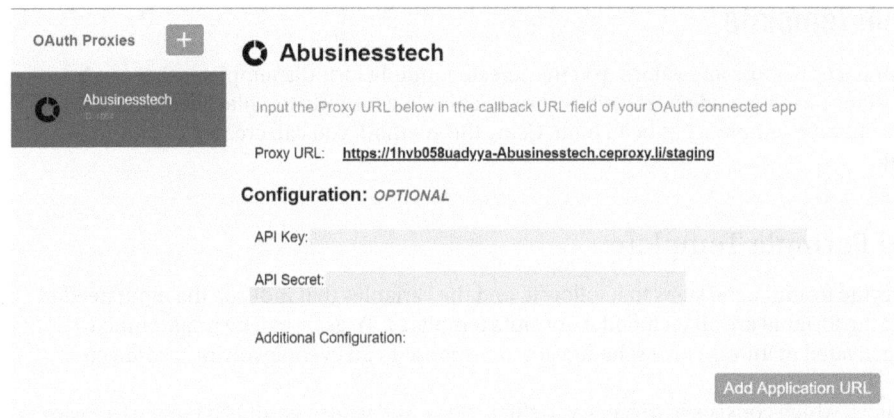

Figure 7-21. *OAuth Proxies*

7.1.3.6 Failed Token: Disable Connector Instance

OAuth tokens are automatically refreshed by SAP Open Connectors on demand and regularly every night to reduce the need for re-authentication. Vendor authentication endpoint failures are uncommon, although they do happen occasionally.

- The vendor's service isn't working or isn't available.

- The vendor's end has canceled or invalidated the user account.

- There is a network/service issue that is unrelated yet has an impact.

When there is a network or service failure, SAP Open Connectors waits ten seconds before attempting the authentication up to five more times, with generally positive results. You might, however, keep trying with each request sent to the connector instance if there are access or account problems with vendor-side services.

After five unsuccessful attempts, the authentication token refresh is frozen to stop further resource use and increase delay (with ten seconds between each attempt). If this happens, the connection instance immediately be disabled and require manual re-enablement and endpoint reauthentication. You are notified via email if a connector instance linked to your account is automatically disabled.

7.1.4 Formulas

You can create formula templates and reusable workflow templates in SAP Open Connectors without using API providers. Formula templates feature triggers that start a series of steps, such as events or timetables. Formulas support a wide range of usage scenarios across several services. They can, for instance, move data between systems, keep systems in sync, or automate business procedures.

Formula instances can be made using the templates you've created after building the template. You use actual connectors and values in formula instances to swap out the variables in the templates.

Formulas can easily transfer logic from your apps to SAP Open Connectors. This keeps your code simpler and easier to maintain so you can concentrate on reaching your goals.

The next sections continue the discussion on formulas.

7.1.4.1 Formula Template

With the use of different connectors and values, you may create multiple formula templates that can be reused. A formula instance can be made by replacing the variables in a formula template with actual instances and values after the template has been built. Using this method, you can create formulas that are effective and versatile.

7.1.4.1.1 Build Formula Template

The trigger that starts the formula, the steps that follow it, and the variables that indicate the input needed to run an instance of the formula are all included in formula templates. Triggers can be programmed into formulas to be activated manually, on a schedule, or in response to an event involving a connection instance.

A formula's context, which consists of triggers, variables, steps, and values produced by steps, is also built as you compose it. In the formula's later steps, you can refer to that context. Think of every component as a separate construction block with a name and details you can use to reference it. For instance, you construct the email body in a single step using a formula that sends an email notification for a new contact. You should refer back to the step where you built the body when you delivered the message in a subsequent step.

1. Go to the Formulas page, as shown in Figure 7-22.

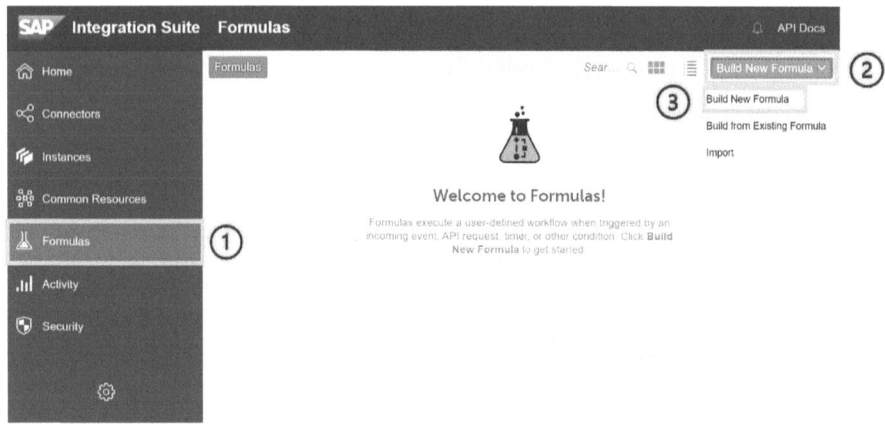

Figure 7-22. *Build New Formula*

2. Select **Build New Formula**.

3. Click **Create** after giving your formula a name.

4. Choose a trigger.

   - For a formula that is activated by an event set up on a connection instance, select Event.

   - For a formula that runs whenever a certain request is made to a connection instance, select Connector Request.

- If you want a formula to run at a set time or interval, select Scheduled.

- Choose Manual as the formula's trigger by using the POST /formulas/instances/:id/executions API request.

5. Depending on the trigger choice, complete the trigger attributes.

   a. A connector instance variable should be provided in Event.

   b. Provide an API method, API endpoint, and connector instance variable in Formula Request.

   c. Specify a CRON schedule in Scheduled.

   d. You don't need to supply any more properties in Manual.

6. Click **Save**.

7.1.4.1.1.1 Create Variables

You can create two variables: Value variables, which are replaced by a queueing system in the formula instance, and Connector Instance variables, which are substituted by connector instances in the formula instance. When creating a variable, note the Formula Step Variable Name, which is how the variable is referred to in the context of the formula.

1. Open the formula template. Click **Open** after hovering the cursor over the formula card on the Formulas page.

2. Click the Variables button.

3. Choose the kind of variable you want to create.

4. Give the variable a name. The Formula Step Variable Name (the variable in the formula context) can have spaces. But remove them to create the Formula Step Variable Name (referred to as key in JSON).

5. Press **Save**.

7.1.4.1.1.2 Edit and Delete Variables

A variable's name can be changed, and it can also be taken out of a formula template. Here is the process to follow in order to edit or delete a variable.

1. Open the formula template. Click Open after hovering the cursor over the formula card on the Formulas page.

2. Click the Variables button.

3. Select **Edit Variable**.

4. Choose the variable you want to change or remove. Click **Save** after changing the name, or click **Delete**.

7.1.4.1.1.3 Review and Replay Execution

Three columns of executions are displayed.

- **Formula Executions**: It is a list of all the times the chosen formula template has been used. Unsuccessful executions are highlighted in red.

- **Steps**: The formula's steps and an indicator indicating its success or failure.

- **Execution Values**: It provides information about the specified step's requests and responses.

The following explains how to evaluate the executions.

1. Click **Open** after hovering the cursor over the formula card on the Formulas page.

2. Click **Executions**.

3. Click Execute to review.

4. Click on the step to review it.

The following explains how to repeat executions.

1. Click **Open** after hovering the cursor over the formula card on the Formulas page.

2. By switching on the Debugging switch, enable debugging.

3. Click **Choose Instance**.

4. Click the + button next to an existing formula instance to choose it, or click New Formula Instance to start a new instance and define its variables.

5. Click **Run** after choosing an instance.

7.1.4.1.1.4 Copy Formula Template

You can create a copy of a formula template by building one from an existing template or using a sample formula from SAP Open Connectors.

1. Click **Formulas**, then click **Build New Formula** on the Formulas page.

2. To build from an existing formula you can select the following options.

   - Choose the template to use one of your current formula templates.

   - Choose the formula by clicking CE Sample Formulas to use an SAP Open Connectors formula template.

3. Click **Create** after giving your formula a name.

7.1.4.1.1.5 Import and Export Formula Template

The following explains how to import the formula template.

1. Click Formulas, then click Build New Formula on the Formulas page.

2. Choose Import by clicking.

3. Choose the JSON file you want to import.

4. Click **Create** after you could give the file another name.

A formula is exported by doing the following.

1. Open the Formulas page.

2. Click Export after moving your cursor over the connector card. Click Save.

7.1.4.1.1.6 Parallel Execution

They make use of multithreaded executions to let formula instances run as quickly as feasible. In other words, a single formula instance can be used by several executions to send requests simultaneously. Some API providers forbid concurrent requests from the same account. If this occurs, you can modify a formula template or a specific formula instance so that it runs in stages. As a result, the formula is less effective and takes longer to execute.

The following explains how to switch a formula's execution from the standard multithreaded to single threaded.

1. Get the formula template open. Click Open after hovering the cursor over the formula card on the Formulas page.

2. Select **Edit**.

3. Click **Show Advanced**.

4. Turn on/off the Single Threaded Execution switch.

5. Click **Save**.

7.1.4.1.1.7 Rename a Formula Template

The following explains how to rename a formula template.

1. Activate the formula template. Click **Open** after hovering the cursor over the formula card on the Formulas page.

2. Press **Edit**.

3. Rename the formula template's name.

4. Click **Save**.

7.1.4.1.1.8 Add Description to Formula Template

A formula template's description has further details. While constructing a formula instance, specify the formula template use case and anything else that another user could need using the information.

The following explains how to describe a formula template.

1. Click the area in Formula Description after opening the formula template.

2. Click **Save** after adding your formula description.

7.1.4.1.1.9 Delete a Formula Template

If there are no instances connected to a formula template, you can delete it. Delete the occurrences first if the formula template has any.

The following explains how to remove a formula template.

1. Open the Formulas page.

2. Click **Delete** after highlighting the Formula Template card.

3. Confirm the removal.

7.1.4.1.1.10 Deactivate the Formulate Template

A formula template can be made inactive to stop any executions of its formula instances. Here is how to do that.

1. Open the formula template card after hovering over it on the Formulas page.

2. Click **Edit**.

3. Turn off the Active slider.

4. Click **Save**.

The next section discusses formula triggers.

7.1.4.1.2 Formula Triggers

Formulas include the triggers that start formulas, the actions the trigger takes, and the variables used to specify the formula's inputs.

Triggers that wait for a connection instance event can be built up. You must use a Connector Instance variable to create this trigger, which, when provided in a connector instance, corresponds to a formula instance set up to listen for events using webhooks or polling.

7.1.4.1.2.1 Event

A Connector Instance variable must be specified to create an event trigger.

In the Modify event trigger pane, click the Add an Event Trigger button. To indicate the connector instance that launch a connector instance when an event occurs, you can either establish a variable or set up the trigger in the UI.

The following shows how to set up the trigger with JSON. Choose "event" as the type. Include the Connector Instance variable (in the $config.variableName format) that causes the formula for elementInstanceId.

```
{
  "triggers": [
    {
      "type": "event",
      "properties": {
      "elementInstanceId": "${config.crmElement}"
    },
  "onSuccess": ["step1"]
    }
  ]
}
```

When a connector instance for an event trigger is configured for polling rather than webhooks, each item discovered during polling results in a single formula execution. For example, five distinct formula executions begin if the event detects five changes.

7.1.4.1.2.2 Connector Request

Whenever a particular API call is performed to a specific Connector Instance, you must use a Connector Instance variable that, if provided in a connector instance, corresponds to a formula instance to configure this trigger.

The following explains how to set up the UI.

1. Click (+) on the Edit Formula Request trigger pane after clicking to add a connector Request trigger.

2. Locate or make a variable that serve as the representation of the connection instance that launch when an event happens.

3. Enter the API call's method, such as GET, POST, PUT, PATCH, or DELETE, in the Method field.

4. Enter the endpoint in API, for example, hubs/crm/contacts.

The following explains how to set up the trigger with JSON.

1. Choose elementRequest as the type.

2. Among properties, do the following.

   a. Add the Connector Instance variable (in the format $config.variableName) that causes the formula for elementInstanceId.

   b. Choose the appropriate API verb for the method (e.g., GET, POST, PUT, PATCH, or DELETE).

   c. Enter the endpoint for API, for example, hubs/crm/contacts.

7.1.4.1.2.3 Scheduled

At the times that a CRON job has defined. I advise you to browse the numerous internet CRON work reference pages, like Crontab Guru.

To specify the time the event should occur, enter the CRON values after clicking to add a Scheduled trigger.

1. From the formula template, open the trigger for your formula, then choose Add Trigger.

2. To add a schedule trigger, select it.

3. Choose the appropriate interval and click Save in the Create Scheduled Trigger settings.

The following shows how to set up the trigger with JSON. Specify the type as scheduled. For properties. cron, enter a CRON string.

```
{
    "triggers": [{
        "type": "scheduled",
        "properties": {
            "cron": "0 0 12 ? * MON *"
        },
        "onSuccess": ["step1"]
    }]
}
```

7.1.4.1.2.4 Manual

POST /formulas/instances/:id/executions is manually called to initiate the trigger. Manual triggers don't need to be configured in any particular way. As synchronous API calls, you can use manually triggered formulas.

Moreover, you can stop a trigger by clicking Stop Execution after selecting the trigger's execution on the Executions tab, or by using PATCH /formulas/instances/executions/executionId.

7.1.4.1.3 Formula Step Types

Your formula can be made up of a number of different kinds of phases. Any step can be referred to using the $steps.stepNamesyntax. As each step is referred to by name within each formula, they all need to be distinct. A step name can be used again in a different formula, though. Both the entire formula and each of its separate steps can have Readmes and descriptions added to them; for more information, see Formula Readmes and Descriptions.

- **ActiveMQ Request**: The ActiveMQ Request (amqpRequest) step type posts a message to a MQ server like RabbitMQ using the AMQP protocol.

- **Connector API Request**: A specific Connector Instance is called via an API request in the Connector API Request (elementRequest) stage.

- **HTTP Request**: Any URL or endpoint can be called using HTTP/S during the HTTP Request (httpRequest) stage.

- **JS Filter**: If your Java script has to return true or false, use the JS Filter (true/false) (filter) step. As with every step, a name must be included.

- **JS Script**: Write your own Javascript using the JS Script (script) step, and be sure to include a proper JSON object that is passed to the done function. As with every step, a name must be included.

- **Loop Over Variable**: You can loop through a list of items from a previous step or trigger by using the Loop Over Variable (loop) step. Set the first step of the loop's onSuccess function. Set the onSuccess field to the loop step once you have completed the loop's final step to repeat the loop for the following object. Set onFailure to the next step to be executed after the loop closes when continues after it. When a loop step has run for each object in the list, onFailure is called.

- **Platform API Request**: The Platform API Request (request) phase calls one of the platform APIs via an API call.

- **Retry Formula on Failure**: When a formula instance execution fails, Retry Formula on Failure (retryFormulaExecution) retries it with the same input data. There is a limit of four retry attempts that you can configure.

- **Stream**: A file is moved from one Connector Instance to another using the Stream File (elementRequestStream) steps. Instead of merely making one API call, Stream Files configures two. The data is downloaded from one connector instance by one request, and uploaded to another by another. Use the download request's response body as the upload request's request body.

- **Subformula**: Another formula instance is run via subformula (formula) steps.

7.1.4.1.4 Formula Variables

When you execute a formula instance, you must specify the following two categories of variables.

- A variable that, when a formula instance is run, is substituted by a particular connector is known as a Connector Instance variable.

- A variable that is changed when a formula instance is run to a customizable value.

Formula variables are constrained to the formula and are not allowed to share the same name. However, you can use the same names for variables in various formulas, such as "originInstance" or "destinationInstance."

7.1.4.1.5 Formula Template: CRM to Messages

This example uses a messaging connector to send an email containing the event data after listening for it on a CRM connector.

1. Create a formula template with the trigger set to Event.

2. Add a Connector Instance variable that references a CRM connection because the trigger is a change to a CRM connector.

    a. Click the (+) button.

    b. Click **Connection Instance** after clicking **Add New Variable**.

    c. Give your CRM variable a name. Use crmElement in this illustration.

   d. Click **Save**.

   e. On the Edit event: "trigger" page, as shown in Figure 7-23, choose the newly generated variable (crmElement), and then click Save.

Figure 7-23. *Trigger page*

3. For the message connector, add yet another Connector Instance variable.

   a. Click the Variables button.

   b. Click the connector instance.

   c. Provide your name. You'll refer to it as messagingElement in this tutorial.

   d. Click **Save**, as shown in Figure 7-24.

Figure 7-24. *Variables*

4. To add another step, click **Add a Step** in the formula visualization.

5. Add a step in your JS script that creates a message when the trigger occurs.

   a. Choose JS Script.

   b. For the script's name, type it in. It is known as constructBody.

   c. Enter a script that creates a message.

   d. Click **Save**.

6. Establish a Connector API Message (you produced in the previous stage). It should be sent in the request step. Click Add OnSuccess after selecting the constructBody step.

   a. Select Connector API Request.

   b. Name the step sendEmail.

   c. Choose the earlier-created messagingElement variable by clicking the + button in the Connector Instance variable field.

   d. Choose POST in Method because the formula sends a POST request to the messaging hub to send an email.

   e. Put the email-sending API's information into API. Enter /messages in this situation.

   f. Select Show Advanced.

   g. Enter the link to the email created in the Body section. Type $steps.constructBody in this instance.

   h. Click **Save**.

7.1.4.1.6 Introducing the V3 Engine Upgrade for a Formula

You have personally contacted each current user of the V1 formula engine with this information. The V1 formula engine's lifespan is about to expire, and at the conclusion of Q2, 2021, support is discontinued. Although SAP Open Connectors introduced the V3 formula engine in 2017, you nevertheless supported V1 formulas for a while after that. However, as of June 30, 2021, the V1 engine was officially retired. Note that the knowledge base's formula-related documentation is built using the V3 formula engine.

1.   From your production account, export current V1 formulas.

2.   Create a staging account and import the V1 formulas there.

3.   Examine your V1 formula's design.

4.   Change the formula steps in Staging.

5.   Improve your formulas by using the V3 Engine.

6.   Test your formulas in staging to make sure they function properly when the V3 engine is used to execute your use case.

7.   Upgrade to the V3 engine for your production formulas.

7.1.4.1.7 Readmes and Descriptions for Formulas

By including a customized or automatically created Readme to your formula and descriptions to your formula and each of its individual steps, you may provide context, background, and any other information about both your formula as well as its steps. Readme files, step descriptions, and formula Readmes are all contained in the formula itself and are presented in the user interface and in any exported formulas.

7.1.4.1.7.1 Formula Readme

You should include a Readme in your formula to explain its use, describe how it works, and provide any other information that could be useful to users. Open the formula template, then choose Generate Readme to open the Readme file window, as shown in Figure 7-25.

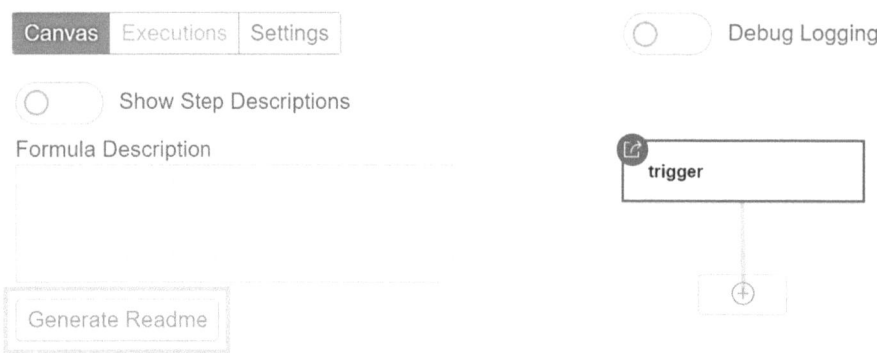

Figure 7-25. *Generate Readme*

The Readme is presented on the Preview tab of the Readme File window. The initial Readme is created automatically using the formula's description and any step descriptions; you may also use the Edit tab to edit the current Readme's text in its original markdown format. You should know that any custom text you place on the Edit tab is not added anywhere else in the formula.

The material from the Edit tab is overwritten with the Readme when you click Save. To return to a Readme that was automatically generated from any formula and step explanations present in the formula, click Regenerate. Make sure to save any crucial text somewhere else before regenerating a formula Readme because no custom content entered from the Edit tab is kept.

7.1.4.1.7.2 Formula and Step Descriptions

Formulas and formula stages should have descriptions to serve as documentation for your formula and, potentially, as materials for an automatically created Readme file.

Open your formula template, then click the area next to Formula Description to create a formula description. Click Save after adding the description.

Hover over a formula step in the template and choose Edit Description to add a description. Select Save after editing the description, as shown in Figure 7-26.

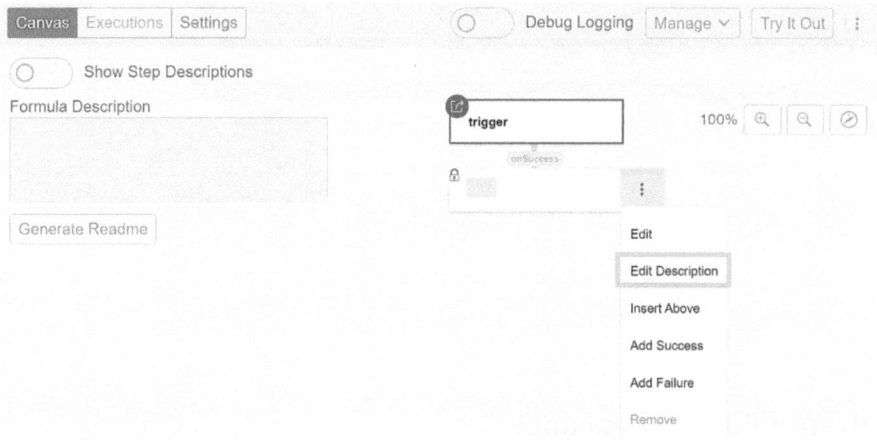

Figure 7-26. Edit Description

The next section discusses formula instances.

7.1.4.2 Formula Instances

In formula instances, particular connector instances and values are used instead of the template's Connector Instance and Value variables. Formula instances can be configured to send emails or a webhook URL to you in case of an error.

7.1.4.2.1 Create Formula Instance

A formula instance is a specific instance of a formula that has been created and configured for use in integration. When you create a formula instance, you define the specific parameters and settings used to transform data as it passes through the integration.

The following explains how to create a formula instance in SAP Open Connectors.

1. Open the Formulas page.

2. Click Create Instance after moving your cursor over the connector card.

3. Give the instance a name.

4. The Connection Instance and Value variables must be specified.

   • Choose or build a connector instance to replace a variable by selecting the variable in the Variables section in Instances.

   • Put a value there to replace the variable in the Values section in Variables.

5. Choose Show Advanced to add error notifications via email or webhook.

   • Enter an email address list that is separated by commas.

   • Enter a list of URLs separated by commas in the Webhook URL.

6. Click **Create Instance**, as shown in Figure 7-27.

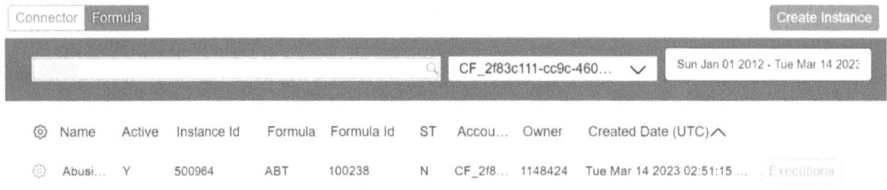

***Figure 7-27.** Instance*

7.1.4.2.2 Deactivate a Formula Instance

The following explains how to deactivate the formula instance.

1. Go to the instance of the Formula. Go to the Formulas page, and click the Instances banner after hovering over the Formula card.

2. Click **Edit** after hovering over the Formula Instance card.

3. Turn off the Active slider at the top right corner of the page.

7.1.4.2.3 Review Executions

The following explains how to review the executions in SAP Open Connectors.

1. Go to the instance of the Formula. Go to the Formulas page, and click the Instances banner after hovering over the Formulas card.

2. Click **Executions** after gliding your cursor over the Formula Instance card.

3. Review by clicking the execution.

4. To check it out, click the step.

7.1.4.2.4 Delete a Formula Instance

The following explains how to delete the formula instance.

1. Go to the instance of the Formula. Go to the Formulas page, and click the Instances banner after hovering over the Formula card.

2. Click Delete after hovering the mouse pointer on the Formula Instance card.

3. Confirm the removal.

7.1.4.2.5 Edit a Formula Instance

The following explains how to edit the formula instance.

1. Go to the instance of the Formula. Go to the Formulas page, and click the Instances banner after hovering over the Formula card.

2. Choose **Edit** after moving your cursor over the Formula Instance card.

3. Once you've made your changes, click Update to save the name, a Value variable, or any notifications.

4. The pencil icon must be clicked before choosing a new connector instance to edit a Connector Instance variable. Press Update.

7.1.4.2.6 Formula Instance ID

Each formula instance is identified by a different number called the Formula Instance ID. A formula instance card's title or the Instances page displays the ID, as shown in Figure 7-28.

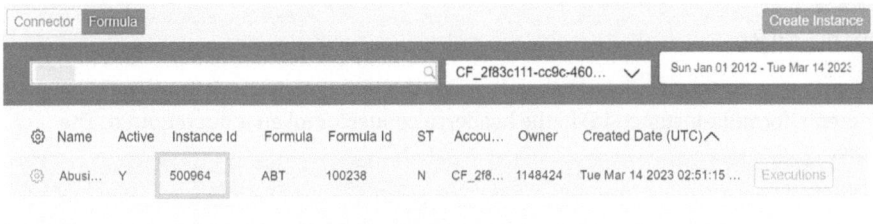

Figure 7-28. *Instance ID*

The next section discusses formula as a resource.

7.1.4.3 Formula as a Resource

Formulas with manual triggers can be made available as resources, called formula as a resource (FaaR). Thanks to this, you may now use the formula to send a synchronous API call. You can perform API queries to a formula after updating it to be used as a resource at `https://api.openconnectors.us2.ext.hana.ondemand.com/elements/api-v2/resourceName>`. You can further eliminate logic from your apps by using formulas as resources, and you can chain requests together more effectively.

Using FaaRs, you can combine numerous requests for APIs that only need one request to complete an operation that might be straightforward in other systems. Being a canonical resource, the FaaR executes multi-step procedures in this situation.

As the API queries are synchronous, a response is necessary before proceeding. The processing time restriction for SAP Open Connectors is thirty seconds to preserve efficiency. The answer alerts you if the request exceeds that threshold.

7.1.4.3.1 Set up FaaR

The following explains how to set up a FaaR.

1. Open the formula template. Hover your mouse over the Formula Card on the Formulas page, then click Open.

2. Click **Edit**.

3. Select **Show Advanced**.

4. The API Method and API URL should be updated in the Execute Formula via the API section.

5. Choose the API method used to call the formula, such as GET, POST, PUT, PATCH, or DELETE, in the API Method field.

6. Enter the formula's resource name, such as /account-enhanced, in the API URL field.

7. Press **Save**.

7.1.4.3.2 Execute FaaRs

In addition to the typical User and Organization information, an API request to the FaaR must contain the formula instance (elements-formula-instance-id) in the header; a connector token is not required. The following is an example.

```
curl -X GET \
  https://api.openconnectors.us2.ext.hana.ondemand.com/elements/api-v2/formula1 \
  -H 'authorization: User <USER_SECRET>, Organization <ORGANIZATION_SECRET>' \
  -H 'elements-formula-instance-id: 28683' \
```

7.1.4.3.3 Access FaaR API Docs

The following explains how to access FaaR.

1. Click API Documentation by first selecting the Formula Card on the Formulas page.

2. On the API Documentation page, select Try It Out.

3. Enter a Formula Instance ID in Elements-Formula-Instance-Id before making the API call.

7.1.4.3.4 Status Codes

You can enter status codes and explanations when you define FaaR, resulting in an answer to a FaaR request with a status code.

1. Include a JS Script step in the formula.

2. Include the following script in the step.

```
done({
  statusCode: xxx
  result: {
    label: 'message'
  }
})
```

Table 7-2 describes the formula status codes.

***Table 7-2.** Status Code*

Property	Description
statusCode	This is the status code—for example, 200, 401, or 502—that you want to include in the response. A valid status code must be the value.
result	The body of the answer, which may be text, an array of objects, a single object, or anything else connected to the status code. (The illustration includes an array with a key/value pair, label, and message.)

The next section is about the common resources in SAP Open Connectors.

7.1.5 Common Resources

Several vendor endpoints that provide the same data class are wrapped in a single API by common resources, such as /Contacts. Shared resources give developers a one-to-many coding experience. You write code against a single API and handle the routing and mapping of your requests to various REST, SOAP, SDK, and database-driven endpoints.

Common resources enable you to handle the data you care about in the optimal structure for your application or business and place your data models at the core of your application ecosystem.

A canonicalized representation of your data objects is provided by common resources, which also do away with the need for point-to-point data mapping to every new application. Our shared resources use enhanced API models to simplify mapping from the resources you describe to the endpoints you need.

After mapping the common resource to connectors, you convert the connector resources into a consistent request and response payload. Types and values of the payload can also be transformed. One system might use high, medium, and low severity levels, while another might use 1, 2, or 3.

Common resources come in two types: templated resources and resources you've created from scratch or altered based on a prebuilt resource. A template for typical mappings to connectors is provided via prebuilt resources. The prebuilt resources can be copied to add or remove connectors, fields, and mappings to suit your needs. Table 7-3 describes common resources.

Table 7-3. *Common Resources*

Resources	Description
normalize	A unique object that your business uses to examine data on that object, such as a customer, employee, or product. The object's fields are mapped to equivalent entities at API providers, allowing you to build integrations with the customized object.
field	Relates to the conversion of vendor information into a universal language within SAP. Normalization, as applied to common resources, is the process of converting fields and objects from several connector resources to a uniform individual of the field or object inside a common resource.
map	Data contained in a resource. A common resource's fields mapped to a connector instance's fields change the connector fields.
resource	The act of linking resources from different vendors together so that the vendor's objects can be modified.
transformation	A thing or thing that can be reached with a URI request. Several APIs contain the same resources, including Accounts, Contacts, and Customers. These various resources are normalized by SAP using common resources.
connector instance resource	The outcome of mapping a resource from an API provider to a generic resource.

7.1.5.1 Create Common Resource

Creating common resources involves setting up the configurations for the various components that are commonly used in integrations. The following explains how to create common resources in SAP Open Connectors.

1. Go to the Common Resources page.

2. Click **Build a New Common Resource**, as shown in Figure 7-29.

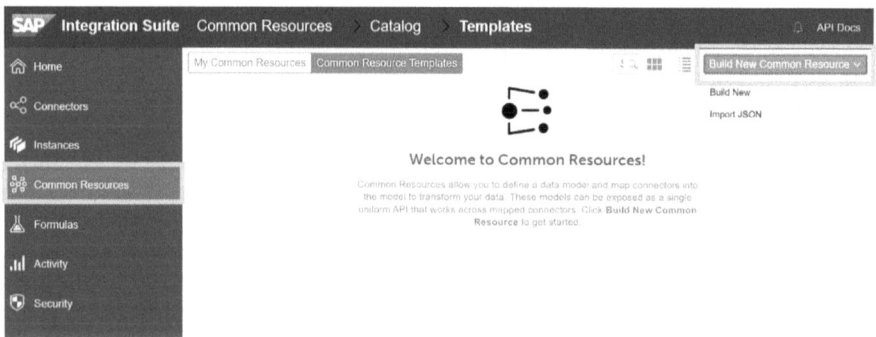

Figure 7-29. *Build New Common Resource*

3. Give your shared resource a name.

4. Include a display name if you choose. When creating user interfaces, the display name may be used instead of the field name.

5. Click **Add** in the Account Level Fields section.

6. Click **Save**.

7.1.5.2 Mapping and Transforming Fields

The term "transformations" describes the process of matching the fields in a resource from an API provider to the fields in another resource. After a common resource is created, fields in connection instance resources are mapped to the new common resource, which results in transformations.

You must first map each connector instance's fields to a shared resource before you can alter the fields. On the left are the common resource fields, while on the right are the resource fields for the connector instance.

You can map each field individually or add multiple fields at once to the shared resource and map them later. One field at a time is mapped according to these instructions.

This section outlines how to map to a shared resource using the UI at the account level.

1. Go to the Transformations page.

2. Click **Create New Transformation** on the Transformations page.

3. Prior to choosing the Connector Instance Resource, choose the Connector Instance.

4. Choose a field to map to id on the right, starting with the default field id.

5. To add a new field, click Add + Field adjacent to the Account Level Fields.

6. If the field is something other than a string, enter a name before selecting the data type.

7. To better match a field's appearance in the UI of an API provider, you can optionally include a display name.

8. Choose the appropriate field on the right for the new field to be mapped to.

9. Once you've finished adding all of your resources, click **Save**.

10. Click **Transformations** in the page's breadcrumbs to map the resource to a different instance.

7.1.5.3 Resources and Transformations Migration Across Environments

You might need to transfer your shared resources and transformations from one account or environment to another if you are developing in various environments or accounts.

7.1.5.3.1 Migrate Common Resources

Common resource migration involves getting the definition of the resource from one account or environment, then posting it to another.

The following explains how to migrate common resources.

1. Call GET /accounts/objects/objectName>/definitions in the source account or
   environment, substituting the name of the shared resource for objectName>.

```
{
  "fields":[
    {
      "type":"string",
      "path":"birthdate"
    },
    {
      "type":"string",
      "path":"FirstName"
    },
    {
      "type":"string",
      "path":"id"
    },
    {
      "type":"string",
      "path":"LastName"
    }
  ],
  "level":"account"
}
```

2. Your common resource might contain a subobject that is stored independently
   if you created it in an earlier software version. Restart the call with "objectName"
   set to the name of the subobject.

3. Make a POST /accounts/objects/definitions API call in your target account or
   environment, substituting objectName with the name of the shared resource and
   the fields item with the fields object from the preceding step.

7.1.5.3.2 Migrate Transformation

The process of migrating transformations involves getting the definition of the transformations through one
account or environment and posting it to another.

1. Use the GET command in the source account or environment, substituting
   keyOrId with the connector key and objectName with the name of the shared
   resource.

2. Do a POST /accounts/elements/keyOrId/transformations/objectName API
   request in your destination account or environment, substituting the connector
   key for keyOrId and the resource name for objectName. The JSON payload from
   the previous step should be included.

7.1.5.4 HTTP Request Made Using Custom JavaScript

Before discussing employing HTTP requests inside a shared resource, you must remember that common
resources operate dynamically; thus, each call that goes via one makes an additional HTTP call.

Performance is a crucial consideration. When you are willing to accept the longer call time for common resources, you should use HTTP requests. Although you likely won't notice much of a difference when using the shared resource with the additional HTTP request, it is nevertheless essential that you are mindful of the possibility before it is put into practice.

Let's discuss the circumstances under which you should send an HTTP request to a common resource. When a unique identifier is ready to be used in the next HTTP call, you should ideally make another one.

You now wish to retrieve the vendor's contact information and their information (e.g., phone numbers, fax, etc.). From the example, you can see that the vendor payload simply returns a contact ID. The good news is that by using that contact ID, you can make a subsequent request to get the contact details and add them to the ultimate payload that will be returned when accessing that shared resource object.

7.1.5.5 Use Common Resource Template

SAP offers a collection of common resource templates to help you more rapidly combine fields offered by some of the most popular connector types into a single, normalized resource. To learn more about common resources, their creation, and their use.

A common resource template must be copied from the common resource templates catalog into a new common resource before you can alter the fields and choose your connectors and services.

The following explains how to access the common resource template catalog.

1. Go to the Common Resources page after logging in to SAP.

2. Choose the Resource Templates tab on the Common Resources Catalog page.

3. Whenever you hover over a template for a common resource, choose from the following options.

   a. Click **Overview** to explore the common resource's description, fields, and data types.

   b. Select **Mappings** to copy the common resource or view its name, URL, system-level attributes, and mapped connections.

The following explains how to clone the common resource template.

1. Choose **Clone** by going to the Common Resource Template's Mappings tab.

2. Click **Next** after searching for and choosing the connectors you wish to be a part of the shared resource. Moreover, you can choose all the listed connectors by selecting the Connector Name checkbox.

3. Click **Next** after deciding at what level the system fields in the shared resource should be created.

4. Click **Save** after giving your shared resource a distinct name.

5. Your newly created common resource is now visible in the My Resources list and available for modifications.

The created common resource is similar, as shown in Figure 7-30.

Figure 7-30. *Common Resources*

7.1.6 Non-SAP Connectivity

Non-SAP connectivity is the capacity to link and integrate SAP systems with applications, platforms, and data sources not created or run by SAP. Non-SAP connectivity enables communication and information sharing between SAP systems and other external systems, including enterprise resource planning (ERP) systems, customer relationship management (CRM) tools, supply chain management (SCM) tools, and other databases and applications.

Application programming interfaces (APIs), web services, and custom connectors are a few ways to develop non-SAP communication. The integration of business processes, automation of workflows, and improved decision-making are all made possible by these techniques, which enable seamless data interchange and communication between SAP and non-SAP systems.

7.1.6.1 SAP Open Connectors with Cloud Integration

To use the capabilities of SAP Cloud Platform Open Connectors in integration scenarios, SAP Cloud Integration has a receiver adapter called OpenConnectors. In the integration scenarios, you can use prebuilt connectors to connect to non-SAP cloud applications and use their APIs.

You create the integration scenario using the adapter as OpenConnectors.

7.1.6.1.1 Create Instance

Creating an instance in SAP Open Connectors is the primary step. The following explains how to create the instance in SAP Open Connector.

1. Open the SAP Cloud Connector and navigate to the Instance section, as shown in Figure 7-31.

2. Click **Create Instance**, as shown in Figure 7-31.

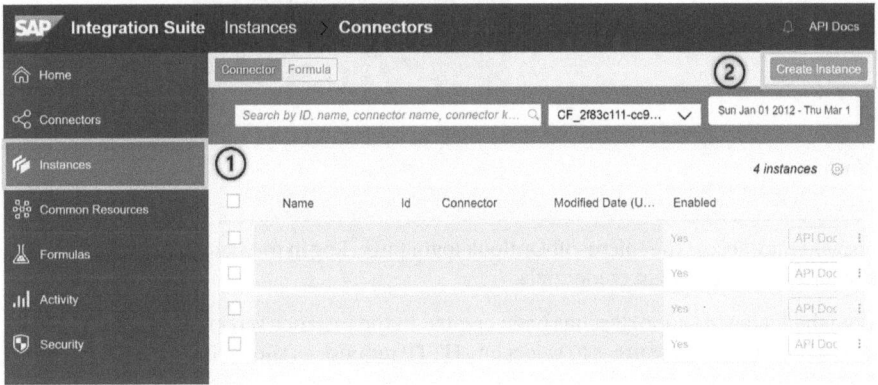

Figure 7-31. *Create Instance*

3. A list of various connectors comes up. Search for Outlook and click +, as shown in Figure 7-32.

Figure 7-32. *Outlook Connector*

4. Specify the name and click **Create Instance**, as shown in Figure 7-33.

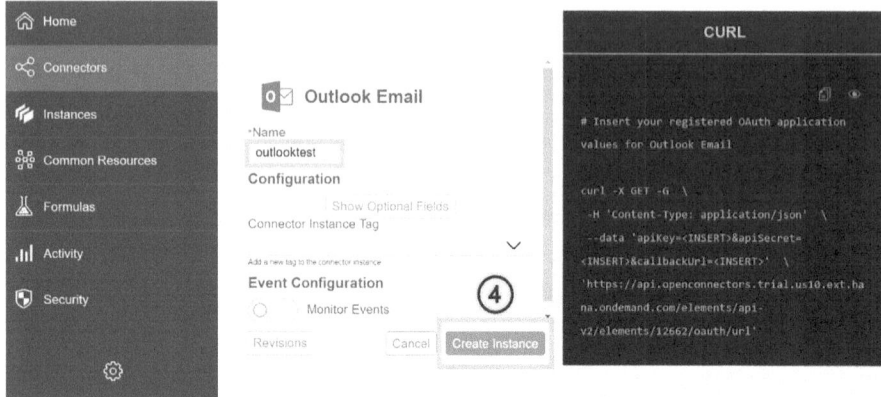

Figure 7-33. *Create Instance*

5. Using this, you may access the Microsoft Outlook login page. Log in to Microsoft Outlook by providing your login credentials.

6. You can see the Instance Oulooktest has been created in the Instance screen. Navigate to API Docs and Resources to select any HTTP request, as shown in Figure 7-34.

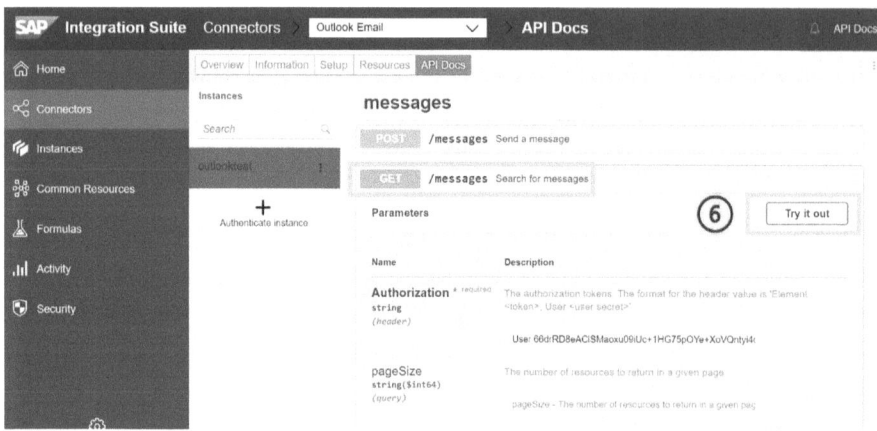

Figure 7-34. *Try it out*

7. Copy the Authorization request header. User, Organization, and Element are its three constituent parts. Without the comma and spaces, copy each of these three values separately. Use these values when constructing the OpenConnectors alias, as shown in Figure 7-35.

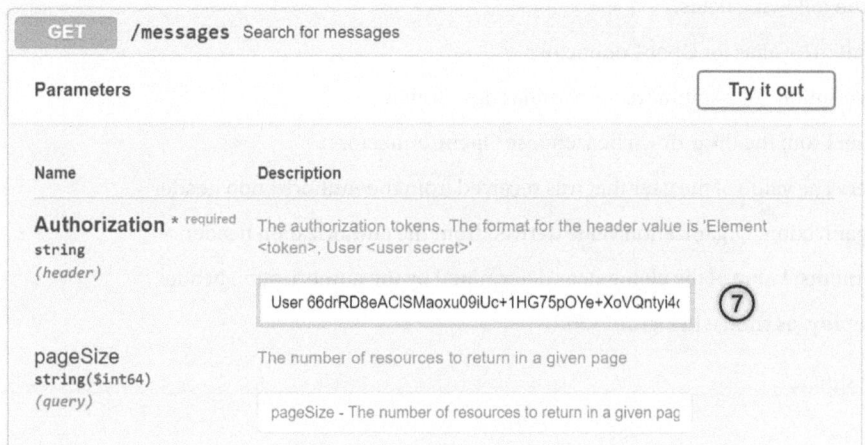

Figure 7-35. *Authorization header*

The authorization string is in the format depicted in Table 7-4.

Table 7-4. *Authorization String Values*

Authorization	Value
User	User 66drRD8eAClSMaoxu09iUc+1HG75pOYe+XoVQntyi4c=
Organization	Organization 9df4c885eeba78c8e136418147c4de93
Element	Element P4CQVEJoBbychxFdgqZp8TbHHrJSGM67V/OsSWH9DxA=

7.1.6.1.2 Create Integration Flow Using Cloud Connector

You can use the instance created in SAP Open Connectors in the SAP Cloud Integration. You can create the integration flow to replicate the data you see in SAP Open Connectors when testing the API.

The next section explains how to use the instance in SAP Cloud Integration.

7.1.6.1.2.1 Create OpenConnector Alias

In SAP Cloud Integration, creating an open connector alias allows you to reuse a set of connection parameters for different endpoints within the same integration flow. This means you can simplify the configuration of multiple endpoints by creating an alias containing the common connection parameters and referencing that alias in each endpoint configuration.

The following explains how to create an open connector alias in SAP Cloud Integration.

1. Open your Cloud Integration web browser.

2. Navigate to the Monitor tab and select Security Material from Manage Security Material.

3. Click **Create ➤ User Credentials**.

4. Define the following fields.

- Name: The alias for OpenConnectors

- Description: Text summarizing the alias description

- Type: From the drop-down box, choose OpenConnectors

- User: The value of the user that was received from the Authorization header

- Organization: Organization value derived from the authorization header

- Elements: Value of the element, as determined by the authorization header

5. Click **Deploy**, as shown in Figure 7-36.

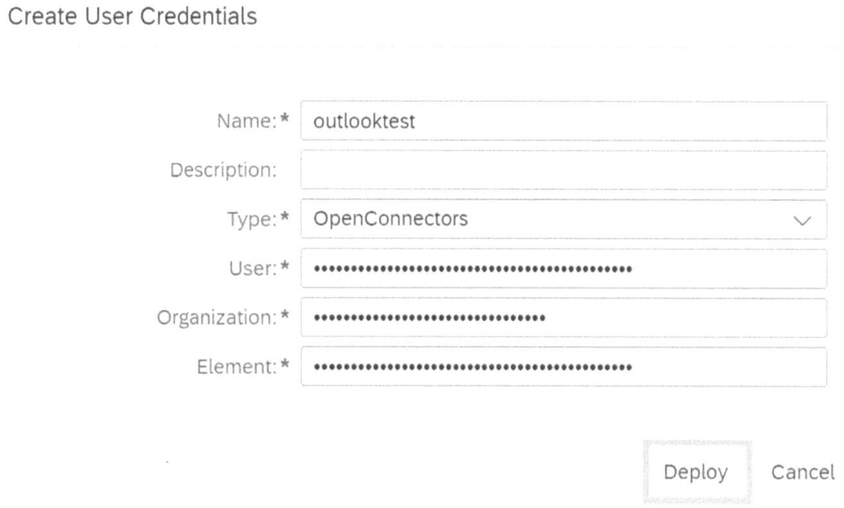

Create User Credentials

Figure 7-36. *Create User Credentials*

7.1.6.1.2.2 Create Integration Flow

1. Open the Design tab in Integration Suite. Create a new Integration package.

2. Add the Integration Flow artifact into the package and switch to Edit mode.

3. Connect the Sender and the Start message with HTTPs, by dragging the arrow from sender to the Start message.

4. Include a Request Reply step in the integration flow.

5. Use the OpenConnectors adapter type to connect the Request Reply to the Receiver, as shown in Figure 7-37.

322

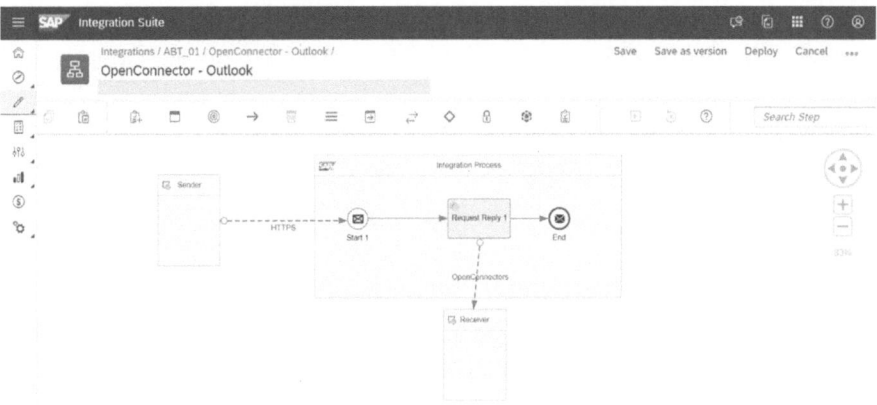

Figure 7-37. *Integration flow*

6. Open the Connection tab of the OpenConnector. Provide the following details in the Connection tab.

   • Base URI: The OpenConnectors service can be used to get this URI. Open the service URL for open connections. Go to API Documentation and select any HTTP method to try it out. You receive the base url when you execute it. The URL is in the following format: `https://open connectors service url>/ elements/api-v2`.

   • Alias: Provide the alias name used in the "outlooktest" security material.

   • Resources: This is a reference to the resources that are supported by your Open Connectors instance. When you select Resources from the Open Connectors service, you may get a list of all the available resources.

   • Method: Use the POST method to send an Outlook email.

   • Request Format: This is the format for the inbound request you should send to the OpenConnectors adapter, and it matches the data type.

   • Content-Type: This option displays only when you choose XML as the request format.

   • Response Format: You have the option of JSON or XML. JSON is the format in the example.

   • Timeout: Use the default setting of 60000 milliseconds.

7. Enter the credential name you configured in Manage Security in section 7.1.6.1.2.1. Figure 7-38 shows the details in the Connection tab.

Figure 7-38. *OpenConnectors in the Connection tab*

8. Save and deploy the integration flow.

9. Copy the endpoints from the Manage Integration Content on the Monitor screen in SAP Cloud Integration, as shown in Figure 7-39.

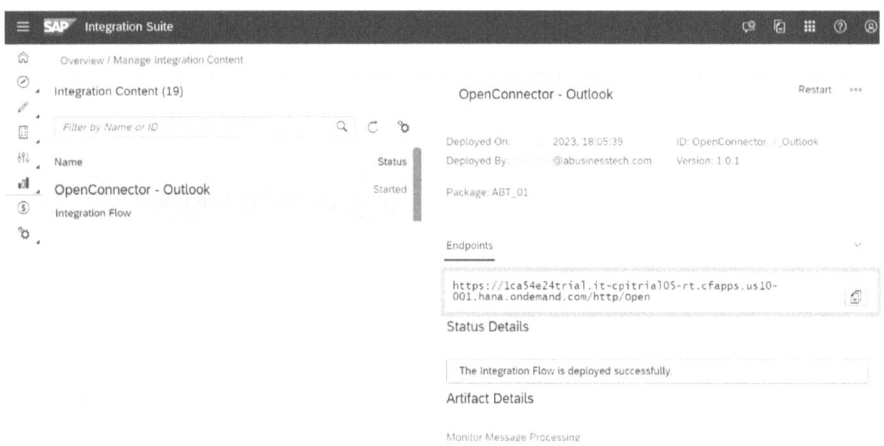

Figure 7-39. *Copy endpoint of integration flow*

10. Run the endpoints in Postman. The successful result is shown in Figure 7-40.

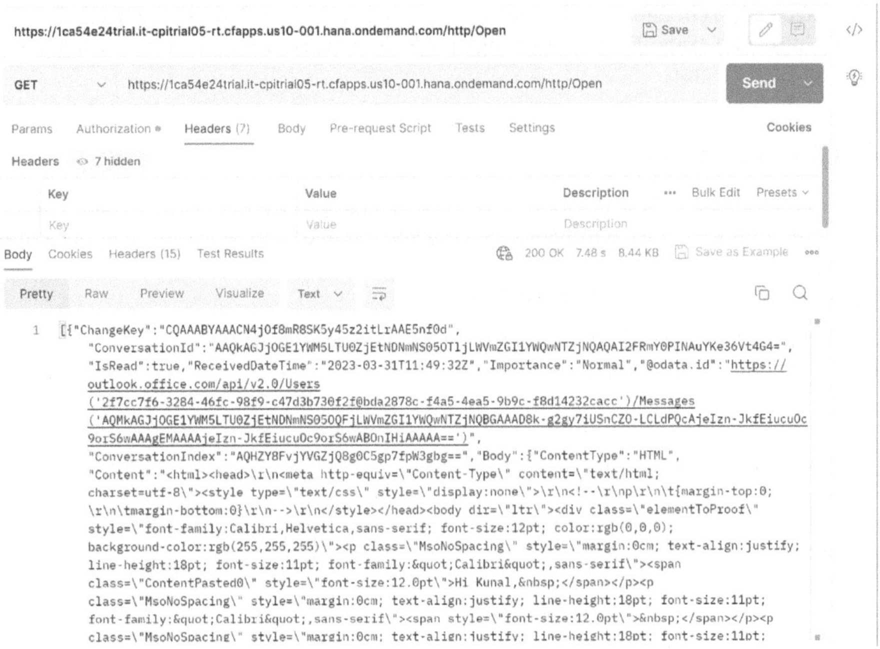

Figure 7-40. *Successful test result*

You have successfully used the SAP Open Connector in SAP Cloud Integration and created and tested the integration flow. Now you have done the non-SAP connectivity using the SAP Cloud Integration. The next section discusses SAP On-Premises Connectivity.

7.1.7 SAP On-Premises Connectivity

SAP on-premises connectivity describes the capacity to link and integrate SAP systems set up and run on-premises with other applications and data sources inside and outside a company's network.

On-premises connectivity, put simply, enables the communication between SAP systems deployed on an organization's servers and other systems, such as cloud apps, mobile devices, or other third-party applications, either through a direct link or via a middleware.

7.1.7.1 SAP SFTP Server with Cloud Connectors

You may link on-premises systems to the SAP Cloud Platform with the SAP Cloud Connector. The secure link it establishes between your on-premises systems and SAP Cloud Platform allows you to access SAP Cloud Platform services from your on-premises systems or access on-premise systems from SAP Cloud Platform.

You may integrate your on-premises systems using services from the SAP Cloud Platform, such as SAP SuccessFactors, SAP HANA Cloud, and SAP S/4HANA Cloud. This enables you to develop hybrid cloud scenarios and use the benefits of both on-premises and cloud systems.

The SAP Cloud Connector can be downloaded and installed on a computer with access to your on-premises systems. It is compatible with Linux, macOS, and Windows. After installation, you can log in using your SAP Cloud Platform account to configure the connection between your on-premises systems and the SAP Cloud Platform. Let's look at an example of transferring the file from the source to the target using the SFTP adapter and SAP Cloud Connector.

7.1.7.1.1 Set up Cloud Connector

SAP Cloud Connector is a component of the SAP cloud platform that allows secure communication between cloud applications and on-premises systems. It provides a secure and reliable channel for cloud applications to access on-premises data and services.

The following explains how to set up SAP Cloud Connector.

1. Download and Install SAP Cloud Connector from `https://tools.hana.ondemand.com/#cloud`, as shown in Figure 7-41.

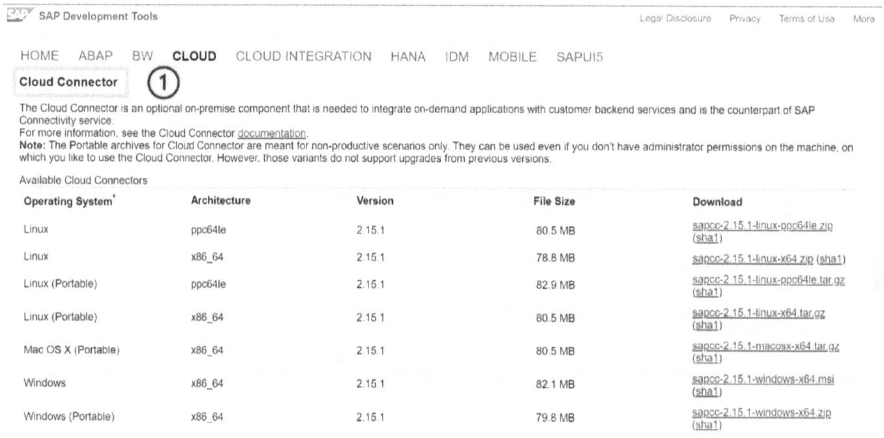

Figure 7-41. *SAP Cloud Connector*

2. After the successful Installation of SAP Cloud Connector. Open `https://localhost:8443`. This link is insecure, so it must be open with unsafe browsing. You are directed to the SAP Cloud Connector login page, as shown in Figure 7-42.

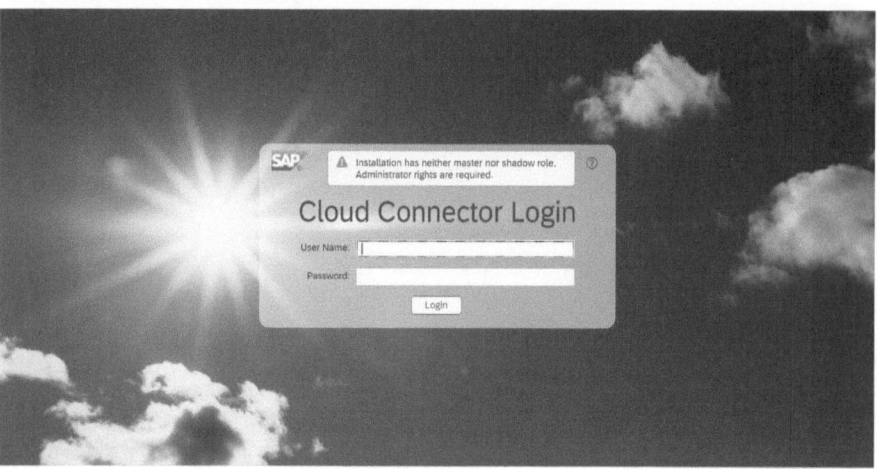

Figure 7-42. *Cloud Connector Login page*

3. *Administrator* is the default username, and *manage* is the default password in Cloud Connector.

4. Enter the credentials to log in to Cloud Connector. You are directed to the page where you must change the default password, as shown in Figure 7-43.

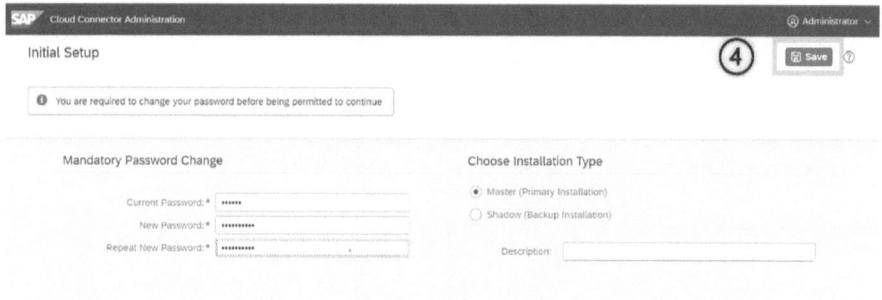

Figure 7-43. *Initial Setup*

7.1.7.1.2 Connect Cloud Connector with your Cloud Integration Tenant

The following explains how to connect the SAP Cloud Connector with your SAP Cloud Integration tenant.

1. A subaccount is also necessary for SAP Cloud Connector because it uses a different platform than Cloud Integration. You must first set up a subaccount, as shown in Figure 7-44.

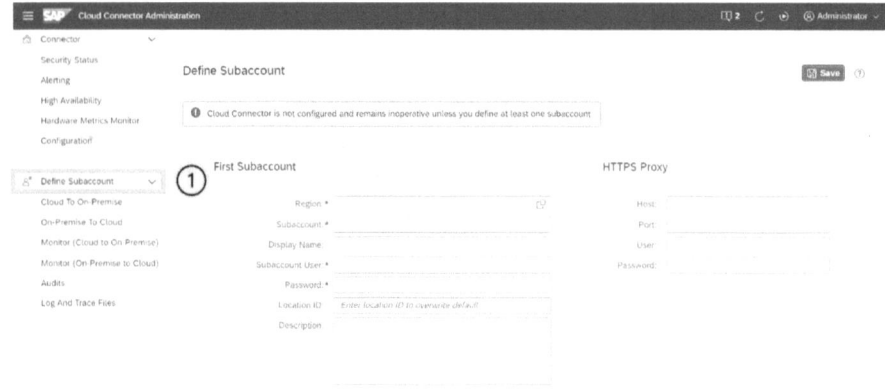

Figure 7-44. *Define Subaccount*

2. For the information to be entered during the SAP Cloud Connector subaccount setup, open your SAP BTP cockpit at `https://account.hanatrial.ondemand.com/trial/`. From the SAP BTP cockpit, copy the Subaccount ID, as shown in Figure 7-45.

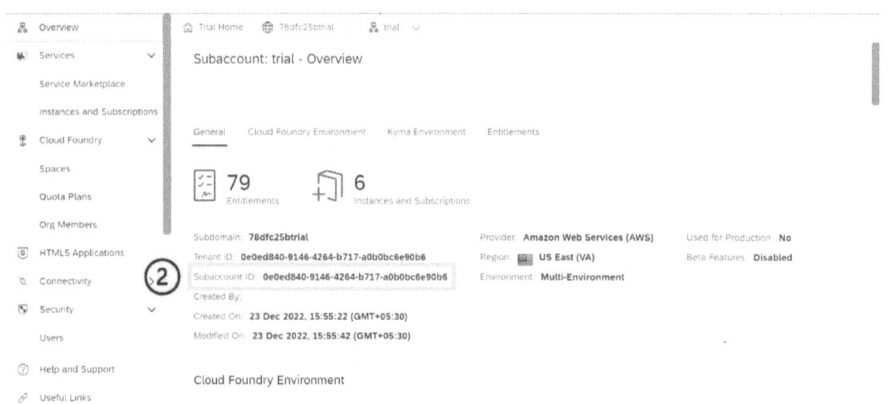

Figure 7-45. *Subaccount ID*

3. Fill out the SAP Cloud Connector subaccount with the subaccount information and additional information. The login email and password must match those used to access the SAP BTP cockpit, as shown in Figure 7-46.

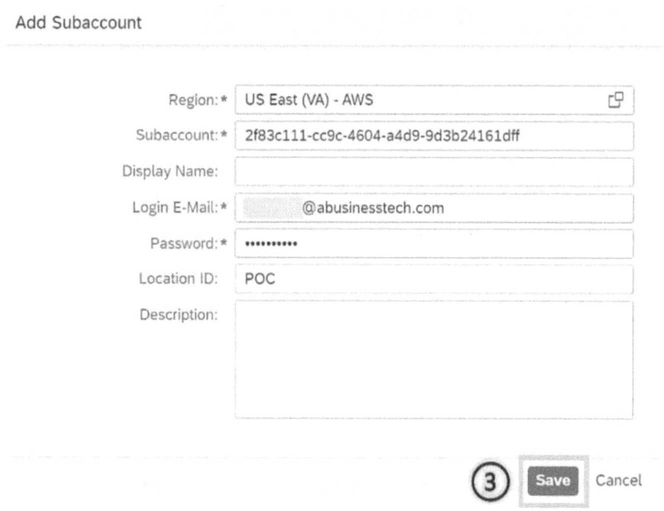
Add Subaccount
Region: *
US East (VA) - AWS
Subaccount: *
2f83c111-cc9c-4604-a4d9-9d3b24161dff
Display Name:
Login E-Mail: *
@abusinesstech.com
Password: *
Location ID:
POC
Description:
3
Save
Cancel

***Figure 7-46.** Add Subaccount*

4. As can be seen, the subaccount for SAP Cloud Connector has been set up successfully and is linked to SAP Cloud Integrator, as shown in Figure 7-47.

SAP Cloud Connector Administration
2
Administrator
Connector
Security Status
Alerting
High Availability
Hardware Metrics Monitor
Configuration
2f83c111-cc9c-4604-a4...
Cloud To On-Premise
On-Premise To Cloud
Monitor (Cloud to On-Premise)
Monitor (On-Premise to Cloud)
Audits
Log And Trace Files
Select Subaccount
2f83c111-cc9c-4604-a4d9-9d3b24161dff
2f83c111-cc9c-4604-a4d9-9d3b24161dff
4
Disconnect
Import
Connected since 2023-03-16 18:20:32 +0530 — no active resources available (check Cloud To On-Premise/Access Control)
Subaccount Overview
Region: US East (VA) - AWS
Region Host: cf.us10.hana.ondemand.com
HTTPS Proxy:
Subaccount Certificate: Certificate valid until 2024-03-16 18:20:28 +0530
System Certificate:
Subaccount: 2f83c111-cc9c-4604-a4d9-9d3b24161dff
Display Name: 2f83c111-cc9c-4604-a4d9-9d3b24161dff
Initiated By: @abusinesstech.com
Location ID: POC
Description:

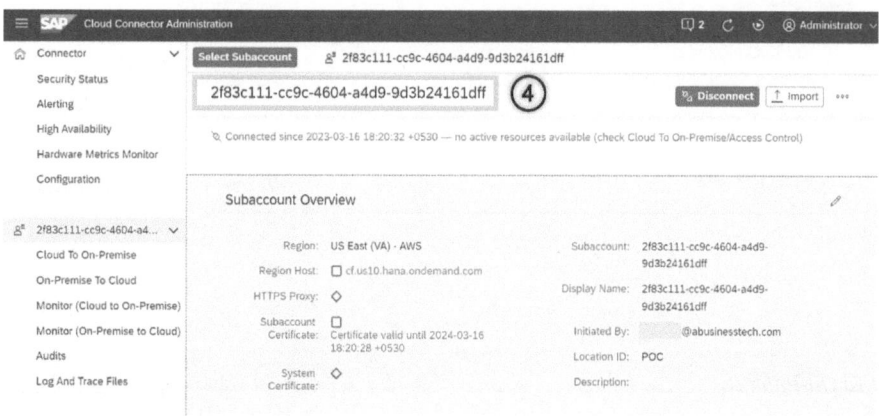

***Figure 7-47.** Subaccount created*

5. You may check whether the Cloud Connector is linked in the BTP cockpit. Select the Cloud Connection tab in the SAP BTP cockpit, as shown in Figure 7-48.

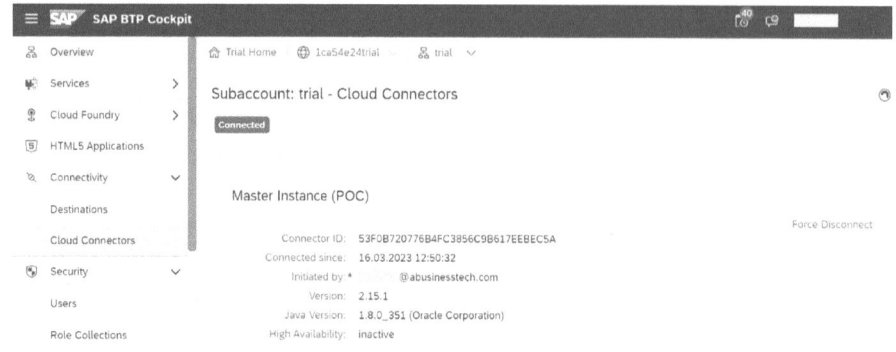

Figure 7-48. *Cloud Connectors subaccount trial*

7.1.7.1.3 Connect SAP Cloud Connector to On-Premises SFTP Server

The following explains how to connect the SAP Cloud Connector to an on-premises SFTP server.

1. Launch the SAP Cloud Connector for Cloud To On-Premise. Choose +, as shown in Figure 7-49.

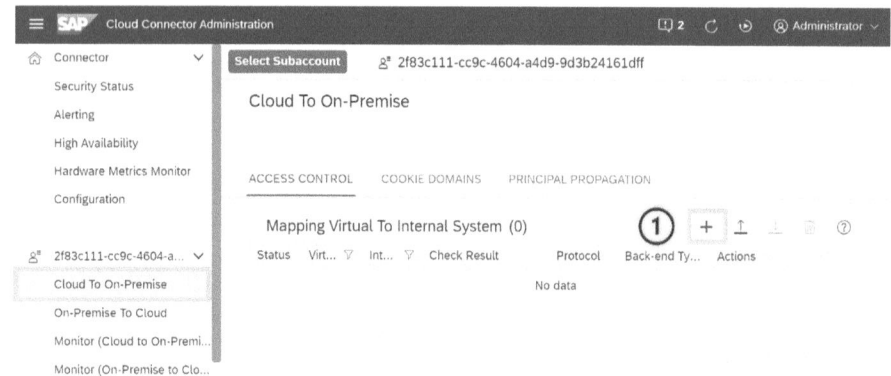

Figure 7-49. *Cloud To On-Premise*

2. Provide the following details in the Add Mapping System.

- Back-end Type: Non-SAP System

- Protocol: TCP

- Internal Host: 192.168.29.181 (Internal Host from SFTP Server)

- Internal Port: 2222 (Internal Port from the SFTP Server)

- Virtual Host: (Provide any name to your Virtual host) for example, my-sftp-virtual-host

- Virtual Port: 2222 (Port from the SFTP Server).

3. Click **Finish**. The SAP Cloud Connector and the On-Premise SFTP Server are now linked. Also, you can contact the server. If the SFTP Server and Client are connected, the Not Reachable Result is displayed, as shown in Figure 7-50.

Cloud To On-Premise

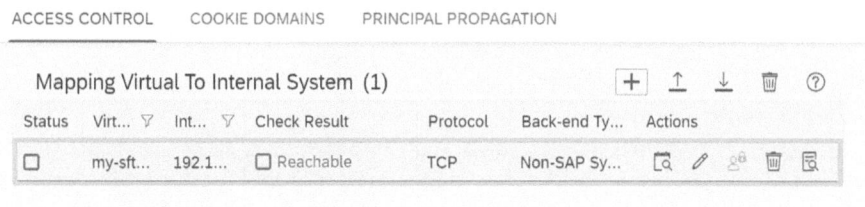

Figure 7-50. *Cloud To On-Premise system*

7.1.7.1.4 Connectivity Test from SAP CI to On-Premise SFTP Server

A connectivity test from SAP Cloud Integration to an on-premise SFTP server is a process of checking the connection and communication between the cloud-based integration platform and an SFTP server that is hosted on-premise. This test is important to ensure that the connection is established successfully and that data can be transmitted securely between the cloud and on-premise systems.

The following explains how to perform a connectivity test from SAP Cloud Integration to an on-premise SFTP server.

1. Launch **SAP Cloud Integration**, then select the Monitor tab. Open the Connection Test tile in Manage Security.

2. Choose the SSH tab. Provide the information from the Cloud Connector.

3. If you select None for Authentication and click Next, the Test is Successful message appears.

4. Testing connectivity with user credentials is our top priority. At Cloud Integration ➤ Monitor Section ➤ Manage Security ➤ Security Material, you can get that information. Establish user credentials.

5. Provide the following details for creating the user credentials, as shown in Figure 7-51.

    • Name: SFTP_UserCred (Name of the User Credentials)

    • Types: User Credentials (Select the type from the drop-down box as User Credentials)

    • User: tester (Username of the SFTP Server)

    • Password: password (Password of the SFTP Server)

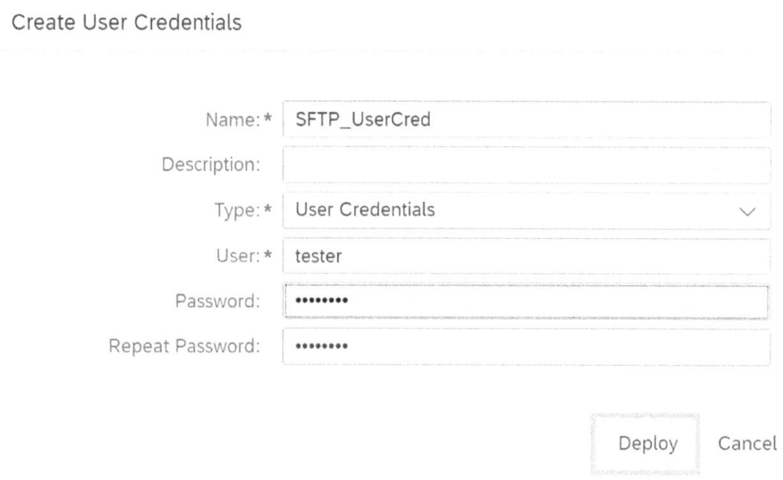

Create User Credentials

Name:*	SFTP_UserCred
Description:	
Type:*	User Credentials ⌄
User:*	tester
Password:	••••••••
Repeat Password:	••••••••

Deploy   Cancel

Figure 7-51. *Create User Credentials*

6.   Return to the tile for your Connection Test. Choose Authentication as User
     Credentials now rather than None. Host Key Verification should be set to None.
     Choose Next. You receive the test result of success, as shown in Figure 7-52.

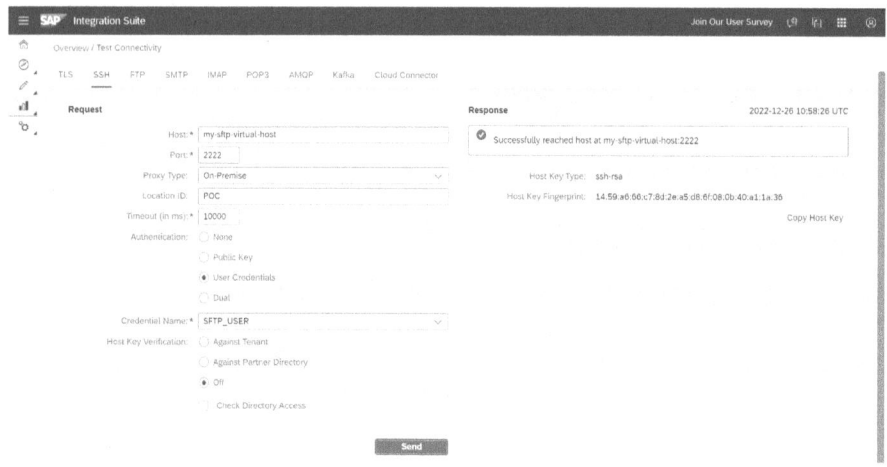

Figure 7-52. *Successful test result*

7.1.7.1.5 Create Integration Flow

The following explains how to create the integration flow.

1.   Open **Cloud Integration** and create the integration package. Create the
     integration flow from the Artifacts tab and open the integration flow in
     edit mode.

2. Connect the Sender with the Start message using SFTP adapter, as shown in Figure 7-53.

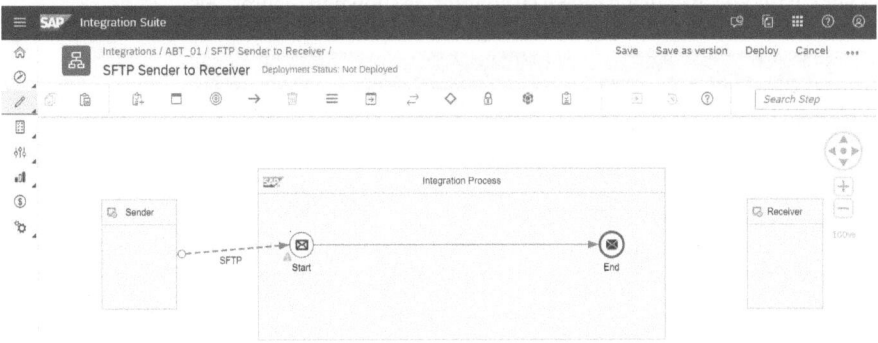

Figure 7-53. *Integration flow*

3. Specify the following details in the SFTP adapter's Source tab.

- Directory: Specify the directory of the source file in the local drive (/source)

- File Name: Specify the file name which you want to be named after transferring the file

- Address: Virtual host: VirtualPort

- Proxy Type: Since the SFTP server and the client is on-premise use the proxy type as the on-premise

- Location ID: Specify the location ID (POC)

- Authentication: User Name/Password

- Credential Name: The name of the user credential created in the previous section

The SFTP adapter Source tab is shown in Figure 7-54.

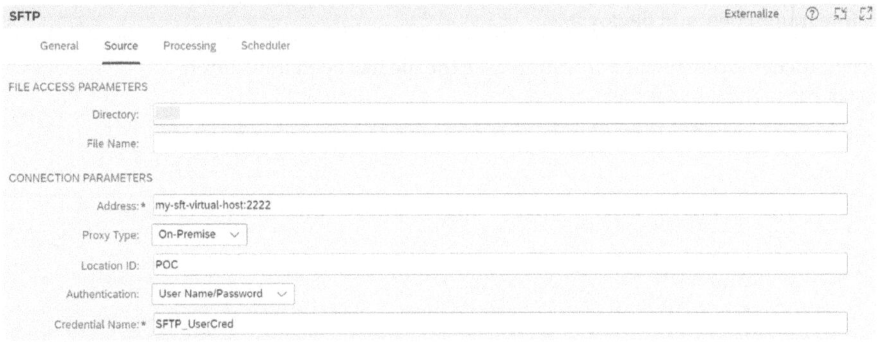

Figure 7-54. *SFTP Source tab*

4. From the external call, select Send and place it between the Send message and End.

5. Connect Send with Receiver using the SFTP adapter, as shown in Figure 7-55.

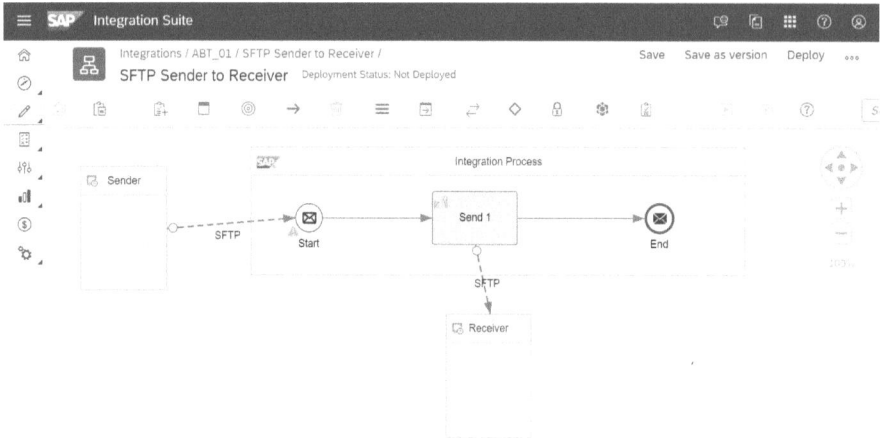

Figure 7-55. *Integration flow*

6. In the target tab of the SFTP Server, Configure the following details.

   - Directory: The Target folder in the local drive (/target) where file needs to be transferred

   - File name: Name of the file

   - Address: Virtual host: Virtual port

   - Proxy Type: On-Premise

   - Location ID: Specify the location ID

   - Authentication: User/password

   - Credential Name: Name of the user credential created in previous section

7. Save the integration flow and deploy it.

8. After successful deployed integration you can see the file has been transferred from the source to the target folder.

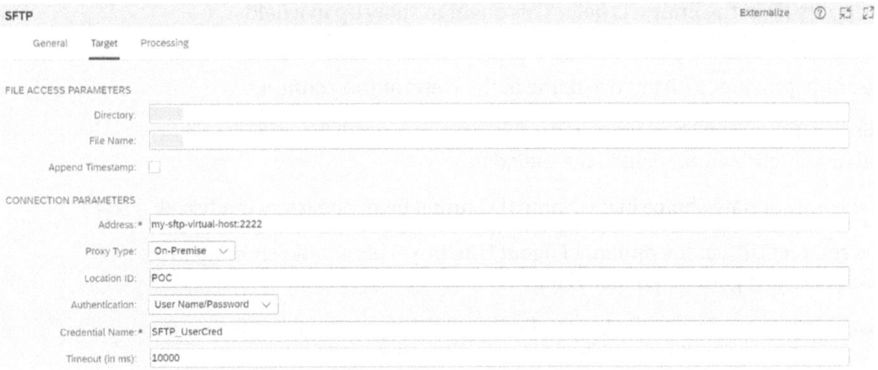

SFTP   Externalize ⑦ ⌞⌝ ⌞⌝

General   Target   Processing

FILE ACCESS PARAMETERS

Directory: [   ]

File Name: [   ]

Append Timestamp: ☐

CONNECTION PARAMETERS

Address: * [my-sftp-virtual-host:2222   ]

Proxy Type: [On-Premise ∨]

Location ID: [POC   ]

Authentication: [User Name/Password ∨]

Credential Name: * [SFTP_UserCred   ]

Timeout (in ms): [10000   ]

Figure 7-56. *SFTP Target tab*

7.1.8 Security

A cloud-based connection platform called SAP Open Connectors enables users to link their company systems with other software and data sources. In order to secure sensitive customer data from unauthorized access or harmful assaults, security in SAP Open Connectors is a crucial component of the platform.

7.1.8.1 Configure Single Sign-On (SSO)

The following explains how to configure single sign-on.

1. Identification may be found by selecting the Security option from the toolbar on the left.

2. On the Add Trusted Identity Provider page, click Application Identity Provider after selecting that option, as shown in Figure 7-57.

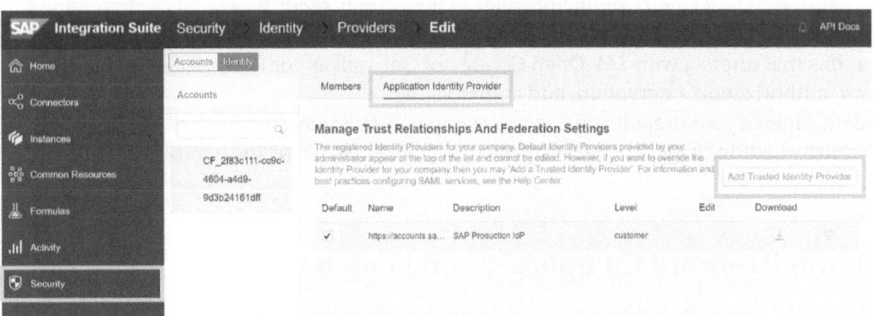

Figure 7-57. *Add Trust Identity Provider*

3. Choose the SAP Open Connectors level at which you want this trusted SSO to be present in the Level box. The choices are account and organization. The rest of the setup data is based on the identity provider; this is the only information that is truly special to SAP Open Connectors.

4. Type the entity ID in the Entity ID field. This ought to show up in a field associated with the Entity Descriptor in the metadata.

5. Give the entity provider a distinctive name in the Description column.

6. The redirect URL should be entered in the SSO URL column. This ought to be included in a single sign-on field in the metadata.

7. Choose the appropriate choice in the Name ID Format field; persistent is advised.

8. Enter the redirect URL in the optional Logout URL box. This should show up in a field connected to logout in the metadata.

9. You must also give a value for the Logout Redirect URL if you enter a value for the Logout URL. When a user logs out, they is sent to this URL.

10. Enter the contents of the certificate in the Signing Certificate area. This should exist in a field in the metadata with a name that sounds like "x509 certificate."

11. Choose the algorithm to be used for signing from the Signature Algorithm menu.

12. Click Save.

13. To create and download the provider XML file, click the download icon for your provider.

14. To establish the connection, upload the provider's XML file to the SSO provider's end.

15. Members who plan on using SSO should be added. To do this, use the Identity tab's Members page.

16. The user ought should be able to sign in using SSO after finishing these procedures.

7.1.8.2 Manage Security Settings

The Manage Security settings in SAP Open Connectors refer to the various security-related configurations and controls that are available in the platform. These settings allow you to manage access and permissions for users and applications that interact with SAP Open Connectors, as well as configure security features such as authentication, authorization, encryption, and audit logging.

To manage the data, change your organization secret, set up two-factor authentication, and modify passwords, use the Security Settings page. The Security page is only accessible by the organization administrator.

7.1.8.2.1 Minimum Password Complexity and Length

While not controlled from the Security Settings page, it is necessary that user passwords contain at least ten characters, at least three of which must be one of the following four types.

- a lowercase letter
- an uppercase letter
- a number
- a special character

7.1.8.2.2 Manage Account Lockouts

User accounts are briefly locked out for 30 minutes after five failed login attempts, and login information cannot be changed via the "Lost Password?" link on the login screen. After thirty minutes, users can choose Forgot Password? to reset their password if no more unsuccessful login attempts are made.

The lockout must be lifted by their org admin if three additional failed login attempts are made on the restricted account.

To unlock a user in your organization as the org admin, take the following actions.

1. Choose the Accounts tab on the Security page.

2. Locate the locked user using the Lock column in the Accounts tab's Users section.

3. To unlock the user account, click the Edit icon followed by the key icon.

7.1.8.2.3 Use Two Factors to Authenticate

SMS and Google Authenticator are two methods of two-factor authentication that SAP Open Connectors enable. After entering their username and password correctly using any method, users are required to enter authentication codes. Although users are permitted to retry the code, SAP Open Connectors lock them out if they input it incorrectly three times.

The following explains how to set up two-factor authentication.

1. The Security page is reachable.

2. Choose a two-factor authentication technique from Two-Factor Authentication.

3. Click Update.

7.1.8.2.4 Reset Organization Token by UI

In SAP Open Connectors, the organization token is a security token that is used to authenticate requests made by applications that interact with the platform. If for some reason, you need to reset the organization token, you can do so using the UI in SAP Open Connectors.

Every time you access one of our platform APIs, such as /instances, /organizations, or /formulas, SAP Open Connectors needs to know the organization token, also known as the organization secret, as well as each individual user secret. Your organization token can be reset at any moment, generating a fresh random string. Remember to use the new organization token in all your API request headers.

The following explains how to use the UI to reset an organization token.

1. Go to the Security page.

2. Reset the organization token by clicking it in the Profile section and then confirming.

7.1.8.2.5 Reset Organization Token by API

Use the POST /authentication/user-secret-reset and POST /authentication/organization-secret-reset endpoints, respectively, to reset an organization token or user token through API. These secrets can be reset at any time, and the previous ones are rendered useless.

7.1.9 Summary

This chapter explained SAP Open Connectors, which is a platform that enables integration between SAP and non-SAP applications. You learned about the terms and resources, user profiles, authentication, base URLs, and error codes.

The chapter covered the working of connectors, the Connectors Catalog, and how to view Connector API docs. You learned about creating custom connectors, which involves configuring properties, authentication, configuration and parameters, hooks, events, and resources. You also learned about non-SAP connectivity and SAP on-premises connectivity.

The next chapter focuses on a more recent SAP Integration Suite capability: Migration Assessment.

■ ■ ■

CHAPTER 8

Migration Assessment

With the ever-evolving landscape of technology, there comes a time when an organization must consider migrating to a newer version of the SAP Integration Suite.

This chapter focuses on Migration Assessment, which is a critical step in any SAP Integration Suite migration project. The chapter explored the key features of the Migration Assessment process and its importance in ensuring a smooth and successful migration. It also delves into the process, discussing the steps involved in conducting a thorough Migration Assessment, including adding an SAP Process Orchestration system, establishing a data extraction request, and requesting a scenario evaluation.

This chapter also discusses the migration tooling available for the SAP Integration Suite and its features, looking at the supported components, templates, and limitations of the migration tooling and Integrated Configuration Objects Migration. Additionally, it touches on the security considerations that organizations must consider when planning a migration.

8.1 Overview of Migration Assessment

The SAP Integration Suite's Cloud Integration feature now includes a new capability called Migration Assessment. This capability assesses how different integration situations can be moved and assists you in estimating the technical efforts required for the conversion process. Its release is an effort by SAP to consider moving to the next generation of integration capabilities by migrating from the on-premises SAP Process Orchestration system to the SAP BTP–based Integration Suite.

The Migration Assessment analyses potential migration paths for different integration situations and aids in estimating the technical work required for the migration process, as shown in Figure 8-1.

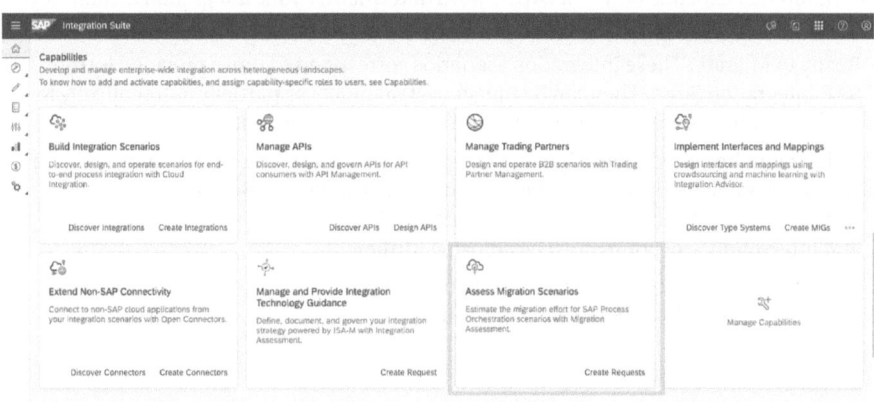

Figure 8-1. *Migration Assessment in SAP Integration Suite*

J. Bagga, *Introduction to Integration Suite Capabilities*, https://doi.org/10.1007/978-1-4842-9630-1_8

8.1.1 Features of Migration Assessment

The following are SAP Migration Assessment features:

- Takes information out of your current SAP Process Orchestration 7.5 system.

- Analyzes the gathered data.

- Determines how much work it could take to transfer the integrated configurable items from the SAP Process Orchestration 7.5 systems to the SAP Integration Suite.

- Integration scenarios are automatically extracted and evaluated using SAP Process Orchestration.

- Reduces the time to conduct an assessment from days or weeks to just a few hours.

- Actively lowers the risk associated with a migration project and assists in making decisions about integration scenarios gradually.

- Upfront disclosure of anticipated relocation costs.

8.1.2 Deep-Dive into Migration Assessment

Table 8-1 lists and describes Migration Assessment terms that you need to know before moving further in the chapter.

Table 8-1. *Concept of Migration Assessment*

Concept	Description
Data extraction	A procedure in which the application gets data from a connected on-premises system and prepares it for evaluation, such as integration scenarios.
Migration template	A pattern with a unique ID that specifies how an integrated configurable object need to be migrated, for instance, because of their linked sending and receiving channels, flow steps, and properties
	Every integrated configurable object is given a template ID during the scenario evaluation. Several integrated configuration objects can use the same template ID.
Assessment categories	A classification that explains how to move forward and whether your integration instances are prepared to be moved to SAP Integration Suite.
	• Ready to Migrate: These integration scenarios correspond to those provided by the SAP Integration Suite. They can be transferred manually or partially automatically to SAP Integration Suite. More configuration steps may be needed.
	• Adjustment Required: These integration scenarios must be adjusted to align with the scenarios provided by SAP Integration Suite. They can be transferred manually or partially automatically to SAP Integration Suite. Based on best practices, the end-to-end integration procedure must be adjusted further.
	• Evaluation required: Some items need to be evaluated further before moving these integration scenarios to SAP Integration Suite.

(continued)

Table 8-1. (*continued*)

Concept	Description
Rule	A group of characteristics that the application uses to determine how much work is involved and whether an integration instance may be moved.
	Many parameters make up a rule, each given a particular weight. Some parameters and rules have a stronger impact on the final computation than others because the program bases the estimated effort determination on such weights.
Scenario evaluation	A procedure in which the application applies predefined rules to assess the information gathered during a previous data extraction to determine whether the extracted integration instances can be migrated, how difficult the migration is expected to be, and which migration templates can be used.

8.1.2.1 Add an SAP Process Orchestration System

To leverage the information from your system in later steps, establish a connection between Migration Assessment and your SAP Process Orchestration system.

To connect your SAP Process Orchestration system and the Migration Assessment application, complete the following steps in your subaccount in the SAP BTP cockpit and the Migration Assessment application.

1. Choose **Settings** from the Migration Assessment application's menu.

2. Choose **Add** in the Process Orchestration Systems table.

3. Select **Create** after entering a System Name and Description.

4. To configure the destinations, log in to the SAP BTP cockpit and go to your subaccount. Your connectivity configuration is stored on the BTP side by the Destination service.

5. Choose **New Destination** in Connectivity Destinations from the navigation menu.

6. Create a new destination and configure it using the information in Table 8-2 according to your system.

Table 8-2. *Destination Configuration*

Field	Content
Name	PO_<systemname>_DIR
Type	HTTP
Description	Provide a pertinent description
URL	Include the system URL
Proxy Type	OnPremise
Authentication	Basic Authentication
Location ID	Provide your location ID
User	Provide your SAP Process Orchestration system's technical user ID
Password	Incorporate the user's password
systemname	Provide the name of the system you chose in the Migration Assessment (Optional: If you use a central ESR and need to construct an ESR destination, include the destination's name here.)
ESRdestination	Provide the system's name that you made in Migration Assessment (Optional: Add the name of the destination here if you utilize a central ESR and have to construct an ESR destination (see step 7).)

7. You only need the Integration Directory destination if you don't utilize a central ESR.

8. Choose **Save**.

8.1.2.2 Establish a Data Extraction Request

Use a data extraction request to obtain data from your preferred system, as shown in Figure 8-2.

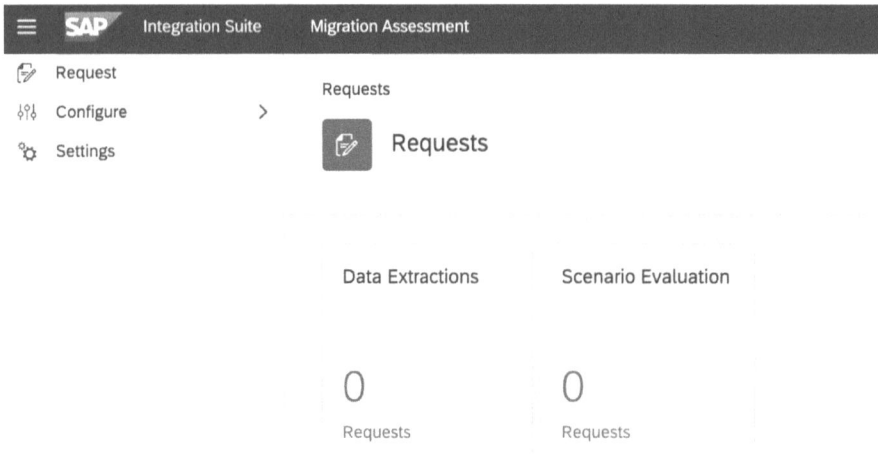

Figure 8-2. *Request in SAP Migration Assessment*

The following steps establish a data extraction request.

1. Go to Request Data Extraction in the Migration Assessment application.

2. Choose **Create**.

3. Choose the system you want to connect to and enter a Request Name.

4. Choose **Create**.

5. Extraction of data begins. When the extraction is complete, the new request shows up with the Finished with in the collection of data extraction requests.

6. Choose Log to view the data extraction log, which contains details on the data extraction.

8.1.2.3 Make a Request for a Scenario Evaluation

Using the data from data extraction requests, evaluate your integration scenarios.

1. Choose Request Scenario Evaluation from the Migration Assessment application's menu.

2. Select **Create**.

3. Choose a Data Extraction Request you previously carried out and enter a Request Name.

4. Name the evaluation run and describe the scenario evaluation. For instance, you can begin a fresh evaluation run anytime to compare fresh and historical data.

5. Click **Create**. A scenario evaluation is conducted after the new request appears on the list of requests for scenario evaluations.

6. For a request for a scenario examination, the following actions may be taken.

   • Visit the dashboard to see and download a summary of your model evaluation runs that include information relevant to your integration scenarios, such as adapters, potential migration templates, and a summary of the evaluation rules. You can toggle between the results of every run completed so far for the situation evaluation request.

   • Do a trigger analysis. Based on recent data, plan a fresh evaluation run.

   • Download the following information regarding the most recent evaluation run in one of two formats.

   • An XLSX file contains a list of all the integration scenarios that have been requested, along with information about the effort and progress of the migration, the rules that applied to it, and the templates that may be used during the migration.

   • A PDF file also offers a textual overview of the adapter and the assessment, with charts and figures as visual aids. This document would work well as a summary report.

8.1.3 Migration Tooling

You can migrate integration cases from SAP Process Orchestration to SAP Integration Suite using the Migration Tooling feature in the Cloud Integration capability. Prior to proceeding with the migration of your integration scenario, it is essential to evaluate its success through the SAP Integration Suite's Migration Assessment feature. This ensures a smooth and hassle-free migration experience.

Manually moving integration scenarios can be a labor-intensive and error-prone from SAP Process Orchestration to SAP Integration Suite. A migration solution to help simplify this process was necessary.

The migration tool automatically generates interfaces for SAP Cloud Integration based on design-time artifacts from SAP Process Orchestration and supported current scenarios. When the artifacts are compatible with the available scenarios, most migrating may be automated, while some manual changes may still be necessary. The migration tool's ultimate objective is to give an automated migration that takes about 60–70% less time.

8.1.3.1 Features of Migration Tooling

The following are some of the key features of migration tooling.

- Transform your SAP Process Orchestration system's integration artifacts into SAP Integration Suite's flows. Integrated Configuration Object (ICO) migration is supported.

- Only SAP Process Orchestration versions 7.5 SP06 and higher are supported for migration.

- ICOs may be in one of the following migration states: Evaluation Required, Adjustment Required, or Ready to Migrate. Each status's specifics are described in Concepts. Only ICOs in the Ready to Migrate and Adjustment Required statuses are supported in the current scope for migration.

8.1.3.1.1 Supported Components

The following ICO components fall inside the current scope of migration tooling support.

- HTTP, REST, SOAP, IDOC, FTP, SFTP, and XI adapter are examples of communication channels.

- Events–Timer

- Message mapping, XML to JSON conversion, JSON to XML conversion, and router are steps in the flow.

8.1.3.1.2 Supported Template

The migration tools convert the sending and receiving channels from the original ICO to the correspondingly compatible ones inside the SAP Integration Suite. The equivalent sender and receiver channels are filled in in the new integration flow. The same is true for the events and flow steps. The migration tooling contains templates that may be customized to fit any possible combination of supporting channels of communication, events, and flow phases.

8.1.3.1.3 Limitations

Be aware of the migration tooling's restrictions.
Table 8-3 describes the constraints of using migration tooling.

Table 8-3. *Limitations*

Components	Limitations
Message Mapping	It is not possible to migrate a message mapping containing User-defined Functions (UDFs) that have function libraries, imported archives, or parameters.
	Importing, viewing, and editing a local Java UDF is possible.
XSLT Mapping	The XSLT mapping does not support extensions. To prevent mistakes, it is advised that you eliminate any such references from your resources, the xmlns:ext reference, for instance.

8.1.3.2 Integrated Configuration Objects Migration

Upgrade your SAP Process Orchestration system's supported integration artifacts to the SAP Integration Suite as integration flows.

1. Select **Edit Migrate** from the integration package you generated. It launches the migration wizard.

2. Choose the SAP Process Orchestration system's name from the Process Orchestration System tab.

3. To check the system's connectivity, select **Connect**. After the test connection is successful, you may view the location of the ES Repository and Integration Directory for the selected system based on your settings. Click on Move.

4. Choose the name of the ICO that you wish to move in the Process Orchestration Artifacts tab. To narrow your search, choose Display Filters.

5. Click **Next**.

6. A template connected to your initial coin offering (ICO) is immediately preselected in the Template tab. Choose your preferred template if there are numerous templates associated.

7. Click **Next**.

8. Give to the integration flow in Integration Suite a Name and an ID in the Integration Flow tab.

9. Select **Review**.

10. Verify each of your entries in the Review tab. Use the associated tab's Edit button to adjust if necessary.

11. Select **Migrate**. An integration flow similar to ICO is created. The properties from the ICO are also moved, along with the sender and recipient channels and additional flow steps like mappings. All sender and receiver ICO properties are externalized for the adapters in the recently developed integration flow. This design enables you to customize the Integration Suite's parameters without changing the integration flow.

12.   On the Migration Success page, keep an eye out for helpful information. You may view the sender and receiver channel mappings in the Channel Mappings section. The comparable adapter types the migration tools generate in the integration flow for the receiver and sender adapter categories used in the source ICO.

8.1.3.3 Migration Tooling

Configure the following fields if you provide input for migration tooling in the sender system (see Figure 8-3).

- Communication Party
- Communication Component
- Interface
- Namespace

Configure the following fields in the Inbound Processing tab.

- Communication Channel
- Adapter Type
- Adapter Engine
- Software Component Version of Sender Interface
- Virus Scan
- Schema Validation

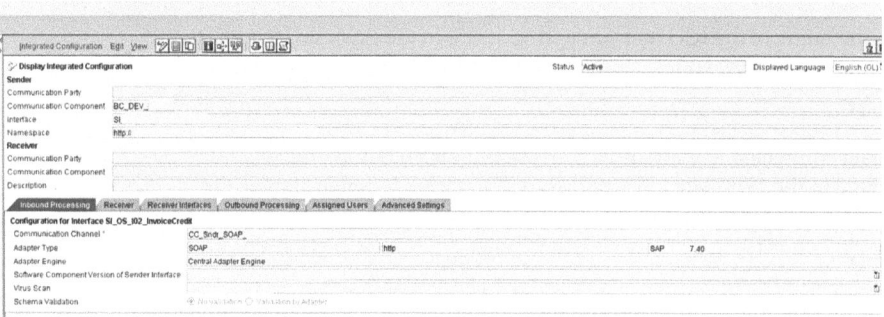

Figure 8-3. *Integrated Configuration*

The migration tooling creates an iFlow in SAP Cloud Integration with SOAP Sender and some Mapping with SFTP receiver, as shown in Figure 8-4.

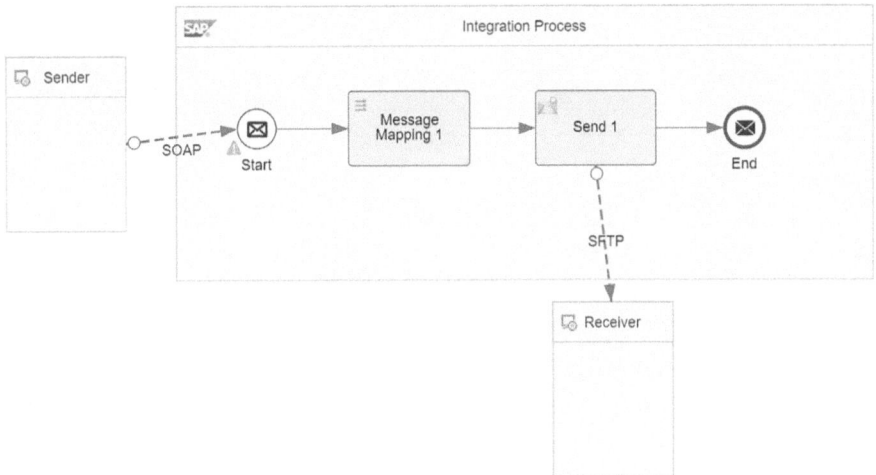

Figure 8-4. *iFlow*

8.1.4 Security

For Migration Assessment, there are two role sets that encompass several personas and responsibilities.

- Administrative duties (e.g., configuring system settings) are carried out by PIMAS_Administrator.

- Executive duties (e.g., reading and executing evaluations and downloading results) are carried out by PIMAS_IntegrationAnalyst.

8.1.5 Summary

In this chapter, you learned how to migrate an SAP system to the cloud and Migration Assessment features. You also learned how to add an SAP Process Orchestration system, establish a data extraction request, and make a request for scenario evaluation.

You gained insight into migration tooling and its features, supported components, supported templates, and limitations. Additionally, the chapter covered ICO migration and security considerations. Overall, the chapter provides a comprehensive guide for readers on migrating an SAP system to the cloud.

As *A Practical Guide to SAP Integration Suite* wraps up, I hope it has provided you with a comprehensive understanding of APIs and how to leverage the SAP Integration Suite to integrate your business systems. From setting up the SAP Integration Suite to using SAP API Management, Open Connectors, and other capabilities, I have covered many topics to help you streamline your integration processes.

This final chapter went through Migration Assessment, an essential step in any integration project if you want to migrate from SAP PO 7.5 system. By understanding the challenges and risks involved in migration, you can ensure a smooth transition from your existing systems to the SAP Integration Suite.

This concludes your SAP Integration Suite journey. I want to thank you for taking the time to read this book, and I hope that you have found it informative and helpful. If you have any questions or feedback, please feel free to contact me.

Index

R

S